NITRIC OXIDE SYNTHASE - SIMPLE ENZYME-COMPLEX ROLES

Edited by **Seyed Soheil Saeedi Saravi**

Nitric Oxide Synthase - Simple Enzyme-Complex Roles
http://dx.doi.org/10.5772/63170
Edited by Seyed Soheil Saeedi Saravi

Contributors

Xiangming Mao, Qingfeng Yu, Tieqiu Li, Jingping Li, Liren Zhong, Emine Atakisi, Oguz Merhan, Masoumeh Kourosh Arami, Seyed Akbar Moosavi, Behnam Jameie, Melih Dagdeviren, Elizabeth Ferreira, Ricardo Serafim, Saadat Huseynova, Nushaba Panakhova, Safikhan Hasanov, Mehman Guliyev, Jean-François Laurent Bodart, Michal Jeseta, Markéta Sedmikova, Hun-Kuk Park, Gi-Ja Lee, Carmen Matas, Florentin-Daniel Staicu, Surekha Rani Hanumanth, Mohini Aiyengar Tupurani, Chiranjeevi Padala, Malgorzata Burek, Kinga Blecharz, Hugo Castro Faria Neto, Patricia Reis, Tatiana Maron-Gutierrez, Cassiano Felippe Gonçalves-De-Albuquerque, Adriana Silva

Published by InTech
Janeza Trdine 9, 51000 Rijeka, Croatia

Publishing Process Manager Iva Simcic
Technical Editor SPi Global
Cover InTech Design team

First published May, 2017
Printed in Croatia
Legal deposit, Croatia: National and University Library in Zagreb

Additional hard copies can be obtained from orders@intechopen.com

Nitric Oxide Synthase - Simple Enzyme-Complex Roles, Edited by Seyed Soheil Saeedi Saravi
p. cm.
Print ISBN 978-953-51-3163-2
Online ISBN 978-953-51-3164-9

For EU product safety concerns:
IN TECH d.o.o., Prolaz Marije Krucifikse Kozulić 3, 51000 Rijeka, Croatia,
info@intechopen.com or visit our website at intechopen.com.

Meet the editor

Seyed Soheil Saeedi Saravi (PharmD, PhD) graduated from the School of Medicine, Tehran University of Medical Sciences, Iran, and has served as an assistant professor of Pharmacology. He has been awarded a postdoctoral fellowship of cardiovascular medicine at Harvard University. Saeedi Saravi has authored and coauthored over 110 publications in the fields of neuro- and cardiovascular pharmacology and written or edited 8 books and chapters for international publishers. He has served as a member of the editorial board and reviewer of over 15 professional journals and of over 10 national scientific associations. He has been first ranked in the National Exam of PhD in Pharmacology and is a recipient of the National Medical Student, National Young Thesis Editor, National Young Researcher, and National Elites Foundation of Iran Awards.

Contents

Preface

The book Nitric Oxide Synthase - Simple Enzyme—Complex Roles is a compilation of 12 chapters focused on the role of nitric oxide synthase and its product, nitric oxide, in neuronal function and disorders; cardiovascular diseases like hypertension; male and female reproduction; and cancer and infection.

The first book section (Chapters 1–4) discusses about the isoforms of nitric oxide synthase and addresses the involvement of nitric oxide synthase and nitric oxide pathway in neurodevelopment and prenatal brain injuries; normal brain functions including memory, learning, cognition, and hearing/vision senses; as well as pathophysiology of neuropsychiatric and neurodegenerative diseases and brain tumors. Moreover, the role of nitric oxide signaling system in the function of the central nervous system in healthy and infectious conditions, along with the pharmacological interactions of this system with other molecular signaling pathways, such as ionotropic and metabotropic glutamate receptors and kinase-linked receptors, is discussed.

The second book section (Chapters 5 and 6) is dedicated to the nitric oxide synthase impacts on cardiovascular system. The role of endothelial nitric oxide synthase in the pathophysiology of hypertension under increased glucocorticoid level is experimentally studied. In addition, an experimental method for monitoring nitric oxide in myocardial tissue is established and fully discussed.

The third book section (Chapters 7–9) comments on the presence and involvement of nitric oxide in normal male sexual and reproductive function, as well as urological disorders such as erectile dysfunction, impotence, infertility, benign prostatic hyperplasia, and prostate carcinoma. Furthermore, this section's attempt is to discuss the key features of nitric oxide synthase in spermatogenesis and gametogenesis.

The fourth book section (Chapters 10 and 11) tries to notice the miscellaneous roles of nitric oxide synthase in the pathophysiology of breast cancer, as well as mechanisms of drug toxicity.

The fifth book section (Chapter 12) presents a picture of therapeutic indication of nitric oxide synthase inhibitors. Although the medical use of these agents has not yet been established, this chapter concludes experimental effects of nitric oxide synthase inhibitors in animal studies.

Since discovery of nitric oxide, as a significant gaseous transmitter, and the isoforms of nitric oxide synthase, many preclinical and clinical investigations have been performed to find the role of the molecule in physiological function of different biological systems and organs, as well as in the pathophysiology of various diseases. The addressed issues in this book associated with the involvement of nitric oxide synthase in neurological, cardiovascular, and reproductive disorders, along with cancer and drug toxicity, can summarize the important role of nitric oxide in these events. The aim of this book is to present a comprehensive overview of found functions of nitric oxide synthase to scientists, researchers, and students of fields of cell and molecular biology, physiology, pharmacology, toxicology, neuroscience, cardiology, urology, and endocrinology.

The present book was made due to the valuable efforts and expertise of the contributing authors, who are gratefully acknowledged.

Dr. Seyed Soheil Saeedi Saravi
School of Medicine, Tehran University of Medical Sciences,
Teheran, Iran

Section 1

Nitric Oxide Synthase in Central Nervouss System (CNS)

Chapter 1

Neuronal Nitric Oxide Synthase

Kourosh Masoumeh Arami, Behnam Jameie and
Seyed Akbar Moosavi

Additional information is available at the end of the chapter

http://dx.doi.org/10.5772/67494

Abstract

Nitric oxide synthase (NOS), a flavo-hemoprotein, regulates nitric oxide (NO) synthesis that has dual biological activities: as an important signalling molecule in vasodilatation and neurotransmission at low concentrations and at higher concentrations as a defensive cytotoxin. In central and peripheral nervous system, neuronal NOS (nNOS) produces NO that has been implicated in modulating physiological functions such as synaptic plasticity, learning, memory and neurogenesis as well as some pathological conditions in which overproduction of NO may lead to the generation of highly reactive species, such as peroxynitrite and stable nitrosothiols, which may cause irreversible cell damage in excitotoxicity, ischaemia, Parkinson, Alzheimer's disease (AD) and depression. NOS-derived NO also involves in regulation of blood pressure, smooth muscle relaxation and gut peristalsis via peripheral nitrergic nerves.

Keywords: neuronal nitric oxide synthase, nitric oxide

1. Introduction

After discovery of nitric oxide as a biological mediator many researchers have focused on the importance of nitric oxide in the physiology of the nervous system [1–3].

NO, as the smallest signalling molecule, is produced by three types of NO synthase arising from three different genes referred to as neuronal nitric oxide synthase (also known as nNOS, Type I, NOS-I and NOS-1) that is found in neuronal tissues, inducible nitric oxide synthase (also known as iNOS, Type II, NOS-II and NOS-2) that is synthesized after formation of pro-inflammatory cytokines or endotoxin and endothelial nitric oxide synthase (also known as eNOS, Type III, NOS-III and NOS-3) that is found in endothelial cells [4]. The nNOS and eNOS are constitutively expressed and considered to be calcium-dependent, but when the activity of iNOS is fully activated at basal intracellular calcium concentration, it would be calcium-independent.

The main difference between three NOS isoform regarding the reactions achieved lies in the rate of the nicotinamide-adenine-dinucleotide phosphate (NADPH) oxidation, termed the uncoupled reaction. Moreover, nNOS carry on transferring electrons to the haem and, hence, oxidase NADPH at a high rate, while in eNOS and iNOS this reaction can happen at a much slower rate [5].

The nNOS constitutes the principal source of NO in distinct populations of neurons and synaptic spines in the brain and the peripheral nervous system while eNOS can occur in some neurons and iNOS may exist in microglia and astrocytes but usually not in neurons [6].

Interneurons expressing nNOS are involved in physiological procedures like neurovascular coupling to regulate neocortical blood flow [7], the homeostatic control of sleep [8], synaptic integration of adult neurons and balance of excitatory and inhibitory signalling in brain [9].

2. nNOS enzymology

The nNOS monomer with a molecular weight of 160.8 kDa and 1434 amino acids is inactive and can be activated after dimerization by tetrahydrobiopterin (BH_4), haem and L-arginine (L-Arg) binding [10]. Each nNOS monomer has two domains including a reductase domain (C-terminal) and an oxygenase domain (N-terminal) which can be separated by a calmodulin-binding motif [11]. The reductase domain which attaches the substrate NADPH comprises a binding site for flavin adenine dinucleotide (FAD) and flavin mononucleotide (FMN) [12, 13]. An autoinhibitory loop within the FMN-binding domain regulates nNOS activity [10]. The oxygenase domain which binds the substrate L-arginine contains BH_4 binding site, cytochrome P-450-type haem active site and Zn binding site for nNOS dimerization facilitation (**Figure 1**).

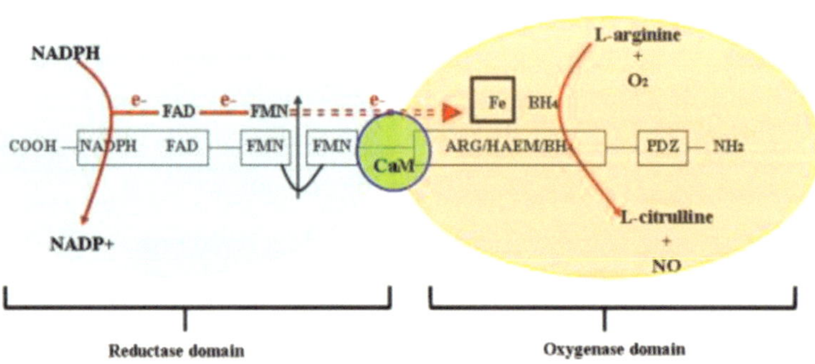

Figure 1. Schematic demonstration of nNOS structure and the metabolic formation of nitric oxide by nNOS include an oxygenase domain (N-terminal) and a reductase domain (C-terminal) which can be separated by a calmodulin-binding motif. The reductase domain which binds NADPH includes a binding site for FMN, FAD and the oxygenase domain which binds L-arginine contains a tetrahydrobiopterin (BH_4) binding site and a cytochrome P-450-type haem active site. NADPH electrons (e^-) via FMN and FAD transfer from the reductase domain to the oxygenase domain. nNOS catalyzes the oxidation of L-arginine to form L-citrulline and NO (Reproduced with permission from Dong-Ya Zhu).

http://dx.doi.org/10.5772/67494

All NOS proteins comprise a zinc–thiolate cluster formed by a zinc ion that is tetrahedrally coordinated to two Cys motifs (one donated by each monomer) at the NOS dimer interface. Zinc in NOS has a catalytic activity [10].

3. NOS-catalyzed reaction

All forms of NOS use L-arginine and molecular oxygen as the substrate and reduced nicotinamide-adenine-dinucleotide phosphate (NADPH) as co-substrates to produce citrulline NO. FAD and FMN with BH_4 are cofactors of all isozymes. All NOS proteins are homodimers. A functional NOS transfers electrons from NADPH, via FAD and FMN in the carboxy-terminal reductase domain to the haem in the amino-terminal oxygenase domain. This flowing of electron (NADPH → FAD → FMN → haem) can be facilitated by Ca^{2+}/CaM binding. The oxygenase domain also binds the cofactor BH_4, molecular oxygen and L-arginine. At the haem site, the electrons are used to activate O_2 to oxidize L-arginine to L-citrulline and NO. Sequences near the cysteine ligand of the haem are also implicated in L-arginine and BH_4 binding. The NOS enzyme is monooxygenases, generating NO and citrulline from L-arginine (L-Arg), NADPH and O_2.

The NOS catalysis has two-step as follows: In first step, the substrate, L-Arg, is converted to N-hydroxy-L-arginine (NOHA), that in turn is converted to NO and citrulline in the second step. The nitrogen atom of NO is derived from the guanidino group of the l-Arg substrate, and the oxygen atom is derived from dioxygen [10, 14]. NO can activate some transduction pathways, such as, activation of guanylyl cyclase and conversion of Guanosine-5′-triphosphate (GTP) into Cyclic guanosine monophosphate (cGMP) and

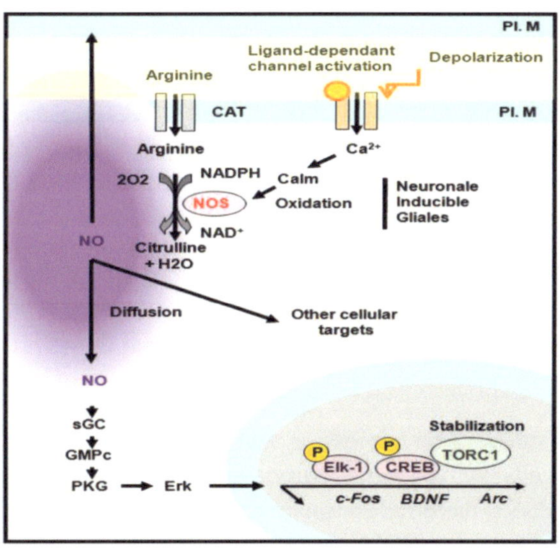

Figure 2. Signal transduction of nNOS. nNOS is activated by a calcium-dependent calmodulin. nNOS produces NO from oxidation of arginine into citrulline. NO diffuses and act on pre-synaptic or post-synaptic targets. NO activates guanylyl cyclase (GC) that triggers a protein kinase G (PKG) resulting in Erk activation and the stabilization of TORC1 a CREB co-activator. PL. M, plasma membrane; CAT, cation and anion transporter. (Adapted from Gallo and Iadecola).

consequent activation of protein kinase G (PKG). PKG activity leads to Erk activation of early genes such as c-fos, Arc and Brain-derived neurotrophic factor (BDNF) (**Figure 2**). NO/cGMP pathway is implicated in various neurophysiological processes including neuronal synaptic modulation, development, learning and memory. Some effects of NO are cGMP-independent. For instance, several pre-synaptic targets such as SNAP25, synthaxin Ia, n-Sec 1, neurogranin as well as the post-synaptic targets ADP ribosyltransferase and NMDA receptors have been identified for NO [15].

4. Histological distribution

The nNOS has been found in neurons, astrocytes, the adventitia of brain blood vessels and cardiac myocytes. Besides brain tissue, nNOS has been distinguished by immunohistochemistry in the spinal cord, sympathetic ganglia and adrenal glands, peripheral nitrergic nerves, skeletal and cardiac myocytes, epithelial cells of different organs, kidney macula densa cells, pancreatic islet cells, parasympathetic ganglia, nonadrenergic noncholinergic peripheral autonomic nerve fibres and the vascular smooth muscle and endothelial cells [16]. In mammalians, the largest origin of nNOS regarding tissue mass is the skeletal muscle [4, 16, 17].

Since NO cannot be stored in the cells, new synthesis is necessary to have its activities. Thus nNOS should be bonded to the plasma membrane directly or through adapter proteins. Fractionation studies have shown that brain nNOS occurs in particulate and soluble forms in cytosol far from membranes in a patch-like form. Furthermore, through the early six days in the cultured cerebral cortical astrocytes of rats, nNOS immunoreactivity mostly appeared in the cytoplasm. Nevertheless, at day 7, nNOS immunoreactivity was predominantly expressed in the nucleus, and this nuclear localization continued about 10 h. Then, nNOS immunoreactivity was mainly expressed in the cytoplasm again. Recently, some researchers showed nNOS nuclear localization without cytoplasmic staining of nNOS in some parts of neural and glial cells. Therefore, diverse functions of nNOS in the cell may arise from differential subcellular localization [10].

Adapter proteins are involved in transfer of nNOS to distinct sites. For instance, nNOS is anchored to membranes by binding to syntrophin, PSD95/SAP90 or PSD93. CAPON, another adapter protein for nNOS, comprises a C-terminal domain that binds to the PDZ domain of nNOS.

In rats and mice, five interneuron expressing nNOS have been found as follows: (1) neurogliaform cells, (2) Ivy cells (IvC), (3) interneurons expressing the vasoactive intestinal peptide (VIP) and calretinin (CR), (4) interneurons expressing PV and (5) projection cells close to the subiculum. PV cells expressing nNOS mainly exist in the dentate gyrus (DG) of hippocampus. Though species differences between rat and mouse have been noted that coexpression of nNOS and PV in rat DG is much lower than that in mouse. Somatostatin-expressing interneurons that express nNOS are resided in hippocampus.

5. Cofactors and prosthetic groups that impact on nNOS activity

5.1. Phosphorylation

Kinases increase nNOS activity by phosphorylation, whereas, phosphatases decrease nNOS activity by dephosphorylation [18]. nNOS phosphorylation is important for the enzyme activity that is regulated by some kinases and phosphatases, for example, calmodulin-dependent kinases, protein kinase C (PKC), protein kinase A (PKA) and phosphatase 1, which are in fact modulated by extracellular and intracellular factors [19]. Nonetheless, phosphorylation at different sites of nNOS affects its activity differently. The protein kinase CaMKII can phosphorylate nNOS at Ser847 that diminishes nNOS activity by Ca^{2+}–CaM binding inhibition. Ser847-PO_4 is found in the autoinhibitory loop which inhibits the movement of the loop even in the occurrence of high concentrations of Ca^{2+}–CaM, thus decreasing nNOS activity [20]. In contrast, the protein phosphatase 1 decreases phosphorylation level of nNOS at Ser847, leading to an increased nNOS activity. Another phosphorylation site of nNOS is at Ser1412 in endothelial nitric oxide synthase [21]. In addition, CaM-KI deregulates nNOS activity by phosphorylation at Ser741 in transfected cells. Since the expression of CaM-KI declines with the brain development, it still needs to be demonstrated whether nNOS is phosphorylated at Ser741 by CaM-KI *in vivo* [10].

5.2. Dimerization

Despite the ability of the reductase and oxygenase domains that function independently in some circumstances, NOS activity is accomplished by the homodimer [5].

Active nNOS is a dimer with two high-affinity binding sites for BH_4 and L-arginine. Two cysteine residues make a disulphide bridge or ligate a zinc ion to connect the two monomers. Furthermore, an 'N-terminal hook' domain sustains the dimer. BH_4 as well as haem and L-arginine make nNOS a stable dimer.

High-affinity binding sites for BH_4 and L-arginine and facilitating electron flow that occurs in dimerization raise nNOS activity. The electron seems to transfer from one monomer to another, which may be the main reason why the nNOS monomer is inactive. PIN (inhibitor of nNOS) destabilizes the nNOS dimer, thus inhibiting nNOS activity. Dimer stabilization preserves nNOS from proteolysis. Destabilization of dimeric nNOS makes it more vulnerable to be phosphorylated by protein kinase C and hydrolyzed by trypsin [5, 10]. nNOS monomers are able to catalyze the cytochrome c reduction. This shows that the electrons transfer within the reductase domain from NADPH by two flavins is independent of dimeric structure [5]. The haem plays a crucial role in dimerization. In absence of haem, NOS is monomer which is principally normal with respect to secondary structure [5].

5.3. Calcium and calmodulin

Dependence on Ca^{2+} is the main characteristic between the constitutive and inducible isoforms. eNOS and nNOS are both triggered by an elevation in intracellular Ca^{2+}, followed by the consequent binding of Ca^{2+}/CaM [5]. Calmodulin acts as an allosteric activator of all forms of NOS that facilitates electron flow transferring from NADPH to the reductase domain flavins and from the reductase domain to the haem center. Calmodulin binding is brought about by an increase in intracellular Ca^{2+} [22, 23].

When intracellular Ca^{2+} concentrations decline to basal levels, calmodulin detaches from nNOS, and it becomes inactive again. Hence, nNOS activity is primarily controlled by intracellular Ca^{2+} concentrations and so calmodulin-binding effect on nNOS activity.

5.4. Proteins binding to nNOS PDZ domain

PDZ (Post Synaptic Density proteins, discs-large, ZO-1) domain of the NH_2 terminal involves in dimerization, activation and interaction of nNOS with many other proteins in specific areas of the cell [24]. These interactions determine the sub-cellular distribution and the function of the enzyme [4, 10].

nNOS PDZ domain contains two separate binding sites, one site binds to PDZ domains of other proteins and another site binds to COOH-terminal peptide ligands. Proteins containing PDZ domains are important in connecting constituents of signal transduction pathways in multiple complexes [25]. NO signalling is modified by nNOS attachment to membrane or cytosolic protein by direct PDZ–PDZ domain or C-terminal-PDZ interactions. PSD95 (post-synaptic density protein-95), a multivalent synaptic protein and main component of the post-synaptic density, can connect nNOS to *N*-methyl-D-aspartate receptor (NMDAR), and elucidate the NMDAR stimulation effect on nNOS activation. Post-synaptic targeting of nNOS is directed by binding to PSD95 (**Figure 3**). nNOS–PSD95 is a typical PDZ–PDZ binding, which needs the intact tertiary structure of both domains and a 30-amino acid extension.

CAPON, another adapter protein, comprises a C-terminal PDZ domain that binds to the N-terminal of nNOS PDZ domain and an N-terminal phosphotyrosine binding (PTB) domain. CAPON interacts with a component of the Ras family of small G proteins, Dexras1, which is induced by dexamethasone. Interaction of CAPON with nNOS delivers NO to Dexras1, leading to S-nitrosylation of Dexras1 on cysteine-11. Dexras1 binds to the peripheral benzodiazepine receptor-associated protein (PAP7), and PAP7 binds to the divalent metal transporter (DMT1), an iron channel mediating iron uptake in neurons [26]. Abnormally high cellular iron levels may lead to disordered neuronal function [27].

5.5. nNOS inhibitors

It has been shown that the N-terminus of nNOS could bind to a protein termed PIN which can inhibit nNOS activity. It is found that PIN destabilizes nNOS dimers and inhibits nNOS activity. Recently, it has been demonstrated that PIN inhibits production of NO and O_2, not nNOS dimerization [28].

Another protein that inhibits NO productions is nitric oxide synthase interacting protein (NOSIP) [29]. NOSIP and nNOS co-localize in different areas of the central and peripheral nervous systems. NOSIP negatively affects nNOS activity in a neuroepithelioma cell line stably expressing nNOS. In addition, over-expression of NOSIP in cultured primary neurons

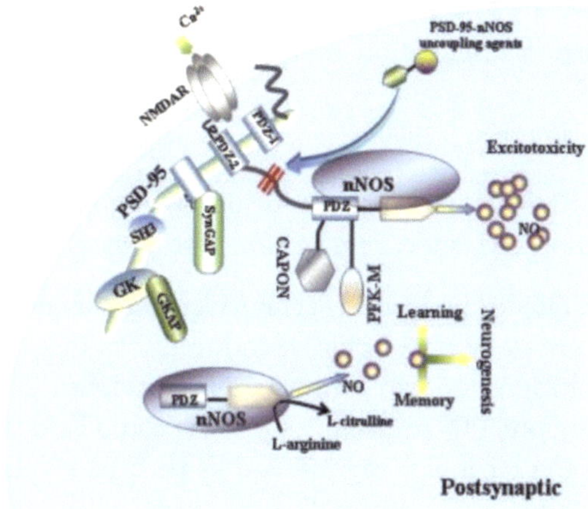

Figure 3. PSD95–nNOS uncoupling agents dissociates the PSD95–nNOS interaction. Glutamate produces NMDAR activation, result in NMDAR/PSD95/nNOS complex formation, and thereby recruiting nNOS to the calcium pore of the NMDA receptor, which makes a principal component of excitotoxicity. PSD95–nNOS uncoupling agents may dissociate the PSD95–nNOS complex, thus, having a neuroprotective effect. More, particulary, PSD95–nNOS as uncoupling agent does not influence other pathway of PSD95 or nNOS, such as PSD95–GKAP, PSD95–SynGAP, nNOS–CAPON and nNOS–PFK-M, therefore, physiological functions of nNOS, for example, learning, memory and neurogenesis, were not affected (Reproduced with permission from Dong-Ya Zhu).

limits nNOS trafficking to terminal dendrites and direct nNOS to the soma. These findings suggest that NOSIP regulates NO production in the nervous system by regulating the activity and localization of nNOS. NOSIP upregulation by neuronal activity may prevent NO production in neurons [10].

6. Physiological and pathophysiological functions of nNOS

6.1. Physiological functions of nNOS

Even though nNOS-derived NO has many functions in neuronal signalling, it varies from a physiological neuromodulator to a neurotoxic factor when extra amount of NO is generated. Thus nNOS may play an important role in many physiological and pathological conditions [10]. Neuronal functions of nNOS are modulating physiological functions such as learning, memory and neurogenesis. In the central nervous system (CNS), nNOS-derived NO causes long-term regulation of synaptic transmission (long-term potentiation and long-term inhibition) [30], while in acute neurotransmission there is no involvement. NOS inhibitors diminish learning and produce amnesia in animal models, so it involves in memory formation. Also in the CNS, nNOS-derived NO is implicated in the central regulation of blood pressure. Blockade of nNOS activity in the medulla and hypothalamus makes systemic hypertension [31].

The nNOS-derived NO as an important neurotransmitter is participated in neuronal plasticity (especially in memory formation), peripheral and central transmission of pain signals, regulation of central nervous system blood flow, neurotransmitter release from cholinergic nerve fibres and the functional regulation of organs with nitrergic innervation [32].

6.1.1. Pancreatic nNOS and insulin regulation

Pancreatic β-cells express nNOS that controls insulin secretion through two catalytic activities: nitric oxide production and cytochrome c reductase activity [33].

PIN is primarily incorporated with insulin secretory granules and co-located with nNOS. In addition, PIN overproduction increases glucose-induced insulin secretion, which is reversed by NO donor, sodium nitroprusside. In contrast, nNOS inhibitor increased insulin secretion induced by glucose. Therefore, PIN insulinotropic effect could be related to its co-localization with the actin-based molecular motor myosin and as such be involved in the physiological regulation of insulin secretion at the exocytotic machinery [33].

6.1.2. Cardiovascular regulation

Sarcoplasmic reticulum and mitochondria functions are regulated by nNOS through Ca^{2+} maintaining, which are directly related to myocardial injury. nNOS overexpression protects mouse hearts from injury and nNOS deficiency increases ventricular arrhythmia and mortality after myocardial infarction [34].

It is accepted that the local regulation of vascular tone in health is mainly regulated by eNOS-derived NO [35]. However, some studies have suggested that nNOS-derived NO may also be implicated in this process and plays an important role in the local regulation of basal microvascular tone as well as in the vasodilator response to mental stress [35]. nNOS can regulate vascular tone, independent of its effects in the central nervous system [35] and by direct effects on vascular smooth muscle. In the kidney *S*-methyl-L-thiocitrulline (SMTC) as a selective nNOS-inhibitor can decrease basal afferent and efferent arteriolar tone. These studies showed that macula densa (the main source of renal nNOS) removal eliminate the vasoconstrictor effects of SMTC.

In cerebral vessels, nNOS may regulate vascular tone reflex especially in response to hypoxia and/or hypotension. For example, Bauser-Heaton et al. have shown that selective inhibition of nNOS with *N*-(4*S*)-(4-amino-5-[aminoethyl]aminopentyl)-*N*'-nitroguanidine decreased basal cerebral arterial diameter and eliminated the vasodilator response to hypoxia.

In skeletal muscle nNOS is located at the cell membrane bound to the cytoskeletal protein dystrophin. Dystrophin absence in patients with Duchenne muscular dystrophy (DMD) makes a significant reduction in skeletal muscle nNOS expression and also in blood flow.

In coronary microvascular system, nNOS and eNOS have distinct local roles in the physiologic regulation. While nNOS may be principal factor in regulation of basal vasomotor tone and blood flow, eNOS-derived NO facilitates dynamic alterations in blood-flow distribution and has anti-atherosclerotic effects at endothelium.

Many smooth muscle tissues in the periphery are innervated by nitrergic nerves, that is, nerves that include nNOS that produce and release NO [4]. For example, relaxation of corpus cavernosum and penile erection is occurred by nitrergic nerves. Phosphodiesterase-5 inhibitors (sildenafil, vardenafil and tadalafil) need at least a residual nNOS activity for their action [4].

Some studies indicated that nNOS-derived NO may change vascular tone by perivascular sympathetic nerve inhibition. Hatanaka et al. indicated that selective nNOS inhibitor, vinyl-L-5-(1-imino-3-butenyl)-L-ornithine (L-VNIO) raises arterial vasoconstriction and local norepinephrine concentration in perivascular nerve stimulation in isolated rat mesenteric arteries without endothelium. However, L-VNIO could not alter vasoconstrictor response to exogenous norepinephrine, showing that nNOS-derived NO may affect release of neurotransmitter from perivascular sympathetic nerves.

In other studies nNOS-derived NO in central systems may change peripheral vascular tone. For instance, nitric oxidergic neurons in nucleus tractus solitaries influence on blood pressure in diabetic rats. Unilateral microinjection of sodium nitroprusside (100 mmol/60 nL) into the nucleus raised blood pressure in diabetic rats [36, 37]. In another study, it is revealed that nitric oxide in the nucleus raphe magnus modulates cutaneous blood flow in rats during hypothermia [38, 39].

6.1.3. Neurogenesis

NO as a paracrine messenger in newly produced neurons regulates the proliferation and differentiation of mouse brain neural progenitor cells. L-NAME, NO synthase inhibitor, raises cell proliferation and reduces neuronal differentiation [40].

The subventricular zone (SVZ) and the subgranular zone (SGZ) of dentate gyrus are two principal neurogenesis sites in the adult brain. In the adult mouse SVZ and olfactory bulb, NO has a negative control on the size of the undifferentiated precursor pool and enhances neuronal differentiation so acts as a physiological inhibitor of neurogenesis [41].

Mice treated with 7-NI display a rise in the number of mitotic cells in the SVZ, the olfactory bulb and the rostral migratory stream, but not in the DG. Though, a recent research established that nNOS inhibition elevated progenitor cells proliferation in the DG. Also, the anti-proliferative role of nNOS-derived NO on SVZ and DG has been shown in cerebral ischaemia. Thus, endogenous NO-derived nNOS can inhibit SVZ neurogenesis. Nevertheless, the role of nNOS in hippocampal neurogenesis is arguable. While nNOS-derived NO has anti-proliferative effect in adult animals, NO donor administration induces neurogenesis. It may be due to the different experimental protocols. Intravenous or hippocampal administration of a NO donor which can result in raised NO levels enhances cerebral blood flow, indirectly influencing neurogenesis [10].

6.1.4. Cerebral maps formation

NO has been involved in the cerebral map formation. In visual system, NO prompts synaptic refinement of immature synaptic connections at retinothalamic and retinocollicular levels. Normal organization of the somatosensory cortex and barrel field plasticity were found by daily injection of nitroarginine before the period of ocular dominance column formation. However, NO may still contribute in establishing and refining neocortical connectivity. Definitely, when NADPHd activity is reformed in the barrel field, abnormal separation of thalamocortical axons happens. In these animals thalamocortical axons show fewer branch points in layer IV and abnormally expansive thalamocortical arbors. These results propose that NO could promote thalamocortical sprouting and participates in the consolidation of synaptic strength in layer IV of the primary somatosensory cortex [40].

6.1.5. Synchronization and coordination

Also, regulation of gap junctions is mediated by NO. Rorig and colleagues have shown that sodium nitroprusside (an NO donor) reduced the number of gap-junction-coupled neurons. Nonetheless, NO can affect electrical coupling, synchronization of metabolic states and coordination of transcriptional activity between connected neurons [40].

6.1.6. Neurotransmitter release and plasticity

Release of several neurotransmitters comprising acetyl choline, catecholamines, glutamate and gamma-Aminobutyric acid (GABA) are regulated by endogenous NO [40]. Furthermore, NO involves in balancing between GABAergic and glutamatergic synaptic transmission in early post-natal development. Disruption of this balance precipitates pathological disorders such as epilepsy, autism and schizophrenia [42, 43]. Moreover, NO is involved in fine-tuning synchronous network activity in the developing hippocampus [40].

NO plays an important role in memory formation in hippocampus [44] and NOS inhibition impedes learning and/or memory [45] while some studies failed to find any effect on learning and/or memory [10]. In mature hippocampus, NO regulates LTP at the Schaffer collateral/CA1 synapses and acts as a retrograde messenger. This occurs via the activation of post-synaptic NMDA receptors, synthesis of NO by NOS expressed in pyramidal cells and then retrograde activation of guanylate cyclase located in axon terminals. In contrast, in the cerebellum NO serves as an anterograde messenger that is produced in parallel fibre terminals or cerebellar interneurons and then diffuses to the post-synaptic Purkinje cell to induce long term depression (LTD) through a cGMP-dependent mechanism [40]. Additionally, NO involves in experience-dependent plasticity in the barrel cortex by reduction of bicuculine-induced activation of Erk and increment of c-Fos, Egr-1 and Arc.

In water maze, 8-arm radial maze, passive-avoidance and elevated plus-maze, 7-NI, at a dose inhibiting nNOS but not affecting blood pressure, induced amnesic effects. Before training in avoidance conditioning in goldfish anterograde amnesia was produced. However, immediately after training retrograde amnesia was formed. Moreover, genetic inhibition of nNOS indicated spatial performance impairment in the Morris water maze [46]. The hippocampus of nNOS knockout mice showed an abnormal expression of a synaptosomal-associated protein of the exocytotic machinery, glycolytic enzymes, T-complex protein 1, the signalling structure guanine nucleotide-binding protein G and heterogeneous nuclear ribonucleoprotein H of the splicing machinery. Therefore, in nNOS knockout mice spatial memory in the Morris water maze may impair by specific hippocampal protein derangements [10].

6.1.7. Gastrointestinal tract

The most obvious phenotype of nNOS knockout mice is stomach enlargement, several times the normal size, proving a role for nNOS in smooth muscle relaxation of the pyloric sphincter. Ablation of exon 6 in another nNOS knockout mice results in severe pyloric stenosis and reproductive endocrine abnormalities.

Citrulline measurements showed that nNOS activity in exon 2-deficient mice is 0.5% of that in wild-type, compared to 3% in exon 6-deficient mice [47].

6.2. Pathophysiological functions of nNOS

Abnormal NO signalling involved in some neurodegenerative pathologies that include excitotoxicity results in stroke, Parkinson's and Alzheimer's diseases and multiple scleroses. NO can involve in excitotoxicity, probably by peroxynitrite activation of poly-ADP-ribose polymerase (PARP) and/or mitochondrial permeability transition. High levels of NO can also produce energy reduction, caused by mitochondrial respiration inhibition and glycolysis. Some disorders of smooth muscle tone in the gastrointestinal tract (e.g. gastro-oesophageal reflux disease) may also drive from an excessive NO production by nNOS in peripheral nitrergic nerves [4].

An important mode of inactivation of NO is its reaction with superoxide anion (O_2^-) to form the potent oxidant peroxynitrite ($ONOO^-$). This can make oxidative damage, nitration and S-nitrosylation of biomolecules including proteins, lipids and DNA [48]. Nitrosative stress by $ONOO^-$ has been involved in DNA single-strand breakage, followed by poly-ADP-ribose polymerase (PARP) activation [49].

6.2.1. Role of nNOS in neurodegeneration

Previous studies have shown that NO implicate in the pathogenesis of some neuroinflammatory/degenerative diseases. The constitutive NOS make the cuprizone-induced model of demyelination/remyelination [50]. Previous results demonstrate that demyelination was mainly prevented in mice lacking nNOS. In $eNOS^{-/-}$ mice, demyelination increased to the same level as in wild type, but they showed a slight delay in spontaneous remyelination [50].

6.2.1.1. Role of nNOS in the pathophysiology of Parkinson's disease

A progressive loss of dopaminergic input from the substantia nigra pars compacta leads to overactivity in Parkinson's disease (PD) which creates extrapyramidal motor dysfunction, including bradykinesia, rigidity and tremor. 1-methyl-4-phenyl-1,2,3,6-tetrahydropyridine (MPTP) yield Parkinson-like symptoms and has been used to assess the mechanism of PD pathogenesis.

Recent studies indicate that resistance to MPTP neurotoxicity in the nNOS knockout mice are more than wild-type ones. This was established by 7-NI that can protect against this neurotoxicity in experimental animals. Furthermore, a great expression of nNOS was reported in basal ganglia and the respiratory burst of circulating neutrophils of PD patients. However, NO production and protein tyrosine nitration were considerably increased [10].

L-DOPA is the most used drug in Parkinson's disease treatment. Nitric oxide is a satisfying goal for the decrease of L-DOPA-induced dyskinesia in PD [51]. Thus, nNOS in the pathogenesis of Parkinson's disease is important.

6.2.1.2. Role of nNOS in the pathophysiology of Alzheimer's disease

Disruption of neuronal nitric oxide synthase dimerization contributes to the development of Alzheimer's disease (AD). nNOS-Ser^{293}, a potential site of cyclin-dependent kinase-5 phosphorylation, may be participated in the nNOS dimerization reduction and, hence, the development of AD [52].

Extracellular accumulation of amyloid β-peptide (Aβ) causing the neuritic plaques and intracellular neurofibrillary tangles is due to the tau protein hyperphosphorylation considered to be an important feature of AD. Chronic infusion of Aβ1-40 results in ONOO-formation and subsequent tyrosine nitration of proteins. Nitrotyrosine found in AD was highly co-localized with nNOS in cortical pyramidal cells. Moreover, all three isoforms of NOS are raised up in AD; therefore, NOS inhibitors could be useful for AD treatment [10].

6.2.1.3. Role of nNOS in the pathophysiology of multiple sclerosis

Multiple sclerosis (MS) is characterized by demyelination associated with an infiltration of mononuclear white blood cells within the CNS. The demyelination leads to diminishing conduction of the action potential in neurons.

nNOS can be induced in nerve injury. It has been revealed that nitric oxide deriving from nNOS may be toxic to oligodendrocytes and induce axonal degeneration. Additionally, it is shown that nitrate as NO degradation product is increased in cerebrospinal fluid of MS patients. Also, the oxidised agent, peroxynitrite is found within active MS lesions. Additionally, NO scavengers have been revealed to reduce the severity of an MS-like disease model [53].

6.2.2. Role of nNOS in the neurodevelopment

Previous studies indicated that nitric oxide derived from nNOS have been associated in social interaction. They have shown that nNOS deletion decreased anxiety-like behaviour, augmented general locomotor activity, reduced spatial learning and memory, and diminished preference for social novelty which are characteristics of autism spectrum disorder [54].

Attention deficit/hyperactivity disorder (ADHD) is a psychiatric disorder with inattention, hyperactivity and impulsivity as main signs, and is frequently associated with learning disability, substance abuse, epilepsy and other psychiatric disorders such as anxiety and disruption of circadian rhythm [55]. In other study, they reported that NOS1 Ex1f-VNTR is involved in impulsive and empathic personality traits and associated with self-rated impulsiveness and venturesomeness [56].

NOS1 KO exhibited higher locomotor activity than wild-type in a novel environment, as measured by open-field test. NOS1 KO mimics certain ADHD-like behaviours and could potentially serve as a novel rodent model for ADHD.

Recent results propose changes in NO-signalling pathways may be associated with ADHD in humans. Neuronal nitric oxide synthase (NOS1) is a main enzyme responsible for the neuronal creation of NO. Previous studies show that 28% of adult ADHD patients were found to be homozygous for a risk allele in the NOS1 promoter region (termed ex1f-VNTR) that diminishes NOS1 expression. This allele is apparently associated with developed prefrontal cortex and ventral striatal roles, which are involved in impulsive and aggressive behaviours associated with ADHD. Moreover, past findings showed that NOS1 knockout mice (NOS1 KO) could be a candidate model for ADHD. Behaviourally, NOS1 KO is apparently hyperactive and exhibits abnormal social, aggressive and impulsive behaviours as well as deficits in learning and memory [55].

6.2.3. Role of nNOS in excitotoxicity

Nitric oxide that occurs naturally in the brain without causing overt toxicity, implicated in cell death. One account is that ischaemia creates overproduction of NO, allowing it to react with superoxide to form the potent oxidant peroxynitrite [6].

Brain ischaemia is a great pre-synaptic glutamate release or post-synaptic stimulation of its membrane receptors, reproduce neuronal damage or death. Glutamate by binding to four major types of receptors, metabotropic receptors, NMDA receptors, alpha-amino-3-hydroxy-5-methyl-4-isoxazolepropionic acid (AMPA) receptors, and kainate receptors do its action post-synaptically. Following focal ischaemia, NMDA receptor activation causes more glutamate excitotoxicity and neuronal injury than other receptors owning to their high calcium permeability. Great stimulation of NMDAR results in Ca^{2+} overload in the cell, and therefore triggering the Ca^{2+}-sensitive enzymes. nNOS, as a Ca^{2+}-sensitive enzyme, indicates a principal role in excitotoxicity. In primary cortical neuronal cultures of $nNOS^{-/-}$ mice, these neurons would be resistant to NMDA neurotoxicity and to oxygen-glucose deprivation compared with wild-type cultures. These *in vitro* studies indicate that nNOS-derived NO is the principal source of neurotoxicity in neurons [10]. $nNOS^{-/-}$ mice have a reduced infarct size, under focal ischaemia. nNOS knockout mice are resistant to focal and global cerebral ischaemia, consistent with a role for nNOS-derived NO in cellular injury following ischaemia.

In addition, 7-nitroindazole and ARL 17477, as selective nNOS inhibitors, can also decrease the infarct size focal ischaemia. Recent studies demonstrate that nNOS expression and enzymatic activity in the hippocampus of mice was decreased under focal cerebral ischaemia. The nNOS reduction following ischaemia stimulated cell proliferation in the DG. Therefore, nNOS inhibition can improve ischemic injury [10].

Even though NMDA activates NOS, the NOS-containing neurons resist toxic effects of NMDA and form NO that is released to kill adjacent non-NOS neurons. The unique resistance of NOS neurons to NMDA toxicity seems to be associated with their very high content of manganese superoxide dismutase, which blocks interactions of NO with superoxide to form the peroxynitrite [10].

Since NO measuring and localizing is difficult, citrulline as a marker of NO synthase activity localized completely to nNOS-containing neurons and is eliminated following NOS inhibitors treatment. Additionally, no other enzyme capable of synthesizing citrulline has been found in the brain. Thus, citrulline staining supplies a useful approach to evaluate NO turnover [6].

Ischaemia triggers a pronounced enhancement in citrulline immunoreactivity but more so in a large population of the neuronal isoform of NO synthase (nNOS) in the peri-infarct than the infarcted tissue. In contrast, 3-nitrotyrosine (a marker for peroxynitrite formation) is confined to the infarcted tissue and is not present in the peri-infarct tissue.

6.2.4. Mood disorders

NO may play an important role in mediating the effect social interactions have on anxiety. Inhibition of nNOS diminishes anxiety-like responses to pair housing [57]. Also, anxiety-like behaviours in aged mice are recovered by modifying nNOS expression levels in the hippocampus or cerebellum [58].

Major depressive disorder is a mental disorder characterized by at least two weeks of low mood that is present across most situations. Paroxetine, a typical anti-depressant inhibited NOS in humans and animals. Moreover, imipramine (IMI, anti-depressant) significantly diminished NOS activity. IMI withdrawal significantly amplified NOS activity. Furthermore, plasma nitrate concentrations (indicator of NO production) were highly greater in depressed patients. This finding showed likely participation of NO in depression, in line with the observation with those 7-NI made anti-depressant–like effects in the Forced Swimming Test. Interestingly, immobilization-produced stress elevated nNOS mRNA and protein expression in hypothalamic-pituitary-adrenal axis in rats. Meanwhile, a recent research showed that chronic mild stress (CMS) augmented nNOS expression in the hippocampus. nNOS inhibition prevented CMS-produced depression. In addition, mice with targeted deletion of the genes encoding nNOS were resistant to the CMS-induced depression [10].

6.2.5. Pain modulation

It is revealed that central anti-nociceptive effect of tapentadolis augmented by 7NI. Neuronal NOS impact on the anti-nociceptive action of tapentadol at the spinal and supraspinal level [59].

Acknowledgements

The authors thank Neuroscience Research Center of Iran University of Medical Sciences for the support of this study.

Author details

Kourosh Masoumeh Arami[1,2*], Behnam Jameie[1,2] and Seyed Akbar Moosavi[3]

*Address all correspondence to: kourosharami.m@iums.ac.ir

1 Neuroscience Research Center, Iran University of Medical Sciences, Tehran, Iran

2 Department of Basic Sciences, Faculty of Allied Medicine, Iran University of Medical Sciences, Tehran, Iran

3 Department of Medical laboratory Sciences, Faculty of Allied Medicine, Iran University of Medical Sciences, Tehran, Iran

References

[1] Garthwaite J., *Concepts of neural nitric oxide-mediated transmission.* European Journal of Neuroscience, 2008. **27**(11): pp. 2783–2802.

[2] Kang Y.C., et al., *Regulation of programmed cell death in neuronal cells by nitric oxide.* In Vivo, 2004. **18**(3): pp. 367–376.

[3] Moncada S. and Bolaños J.P., *Nitric oxide, cell bioenergetics and neurodegeneration.* Journal of Neurochemistry, 2006. **97**(6): pp. 1676–1689.

[4] Förstermann U. and Sessa W.C., *Nitric oxide synthases: regulation and function.* European Heart Journal, 2012. **33**(7): pp. 829–837.

[5] Andrew P.J. and Mayer B., *Enzymatic function of nitric oxide synthases.* Cardiovascular Research, 1999. **43**(3): pp. 521–531.

[6] Eliasson M.J., et al., *Neuronal nitric oxide synthase activation and peroxynitrite formation in ischemic stroke linked to neural damage.* The Journal of Neuroscience, 1999. **19**(14): pp. 5910–5918.

[7] Perrenoud Q., et al., *Activation of cortical 5-HT (3) receptor-expressing interneurons induces NO mediated vasodilatations and NPY mediated vasoconstrictions.* 2012; 6: 82.

[8] Kilduff T.S., Cauli B. and Gerashchenko D., *Activation of cortical interneurons during sleep: an anatomical link to homeostatic sleep regulation?.* Trends in Neurosciences, 2011. **34**(1): pp. 10–19.

[9] Somogyi, P. and Klausberger T., *Defined types of cortical interneurone structure space and spike timing in the hippocampus.* The Journal of Physiology, 2005. **562**(1): pp. 9–26.

[10] Zhou, L. and Zhu D.Y., *Neuronal nitric oxide synthase: structure, subcellular localization, regulation, and clinical implications*. Nitric Oxide, 2009. **20**(4): pp. 223–230.

[11] Cho, H.J., et al., *Calmodulin is a subunit of nitric oxide synthase from macrophages.* The Journal of Experimental Medicine, 1992. **176**(2): pp. 599–604.

[12] Crane, B.R., et al., *Structure of nitric oxide synthase oxygenase dimer with pterin and substrate.* Science, 1998. **279**(5359): pp. 2121–2126.

[13] Sagami, I., Daff S. and Shimizu T., *Intra-subunit and inter-subunit electron transfer in neuronal nitric-oxide synthase effect of calmodulin on heterodimer catalysis*. Journal of Biological Chemistry, 2001. **276**(32): pp. 30036–30042.

[14] Feng, C., *Mechanism of nitric oxide synthase regulation: electron transfer and interdomain interactions*. Coordination Chemistry Reviews, 2012. **256**(3): pp. 393–411.

[15] Gallo, E.F. and Iadecola C., *Neuronal nitric oxide contributes to neuroplasticity-associated protein expression through cGMP, protein kinase G, and extracellular signal-regulated kinase.* The Journal of Neuroscience, 2011. **31**(19): pp. 6947–6955.

[16] Förstermann, U., et al., *Nitric oxide synthase isozymes. Characterization, purification, molecular cloning, and functions*. Hypertension, 1994. **23**(6 Pt 2): pp. 1121–1131.

[17] Nakane, M., et al., *Cloned human brain nitric oxide synthase is highly expressed in skeletal muscle*. FEBS Letters, 1993. **316**(2): pp. 175–180.

[18] Rameau, G.A., et al., *Biphasic coupling of neuronal nitric oxide synthase phosphorylation to the NMDA receptor regulates AMPA receptor trafficking and neuronal cell death*. The Journal of Neuroscience, 2007. **27**(13): pp. 3445–3455.

[19] El-Mlili, N., et al., *Chronic hyperammonemia reduces the activity of neuronal nitric oxide synthase in cerebellum by altering its localization and increasing its phosphorylation by calcium-calmodulin kinase II*. Journal of Neurochemistry, 2008. **106**(3): pp. 1440–1449.

[20] Rameau, G.A., Chiu L.Y. and Ziff E.B., *Bidirectional regulation of neuronal nitric-oxide synthase phosphorylation at serine 847 by the N-methyl-D-aspartate receptor*. Journal of Biological Chemistry, 2004. **279**(14): pp. 14307–14314.

[21] Adak, S., et al., *Neuronal nitric-oxide synthase mutant (Ser-1412→ Asp) demonstrates surprising connections between heme reduction, NO complex formation, and catalysis*. Journal of Biological Chemistry, 2001. **276**(2): pp. 1244–1252.

[22] Li, H. and Poulos T.L., *Structure–function studies on nitric oxide synthases*. Journal of Inorganic Biochemistry, 2005. **99**(1): pp. 293–305.

[23] Yamamoto, Y., et al., *Ca^{2+}, calmodulin and phospholipids regulate nitricoxide synthase activity in the rabbit submandibular gland*. Journal of Comparative Physiology B, 2004. **174**(8): pp. 593–599.

[24] Cui, H., et al., *PDZ protein interactions underlying NMDA receptor-mediated excitotoxicity and neuroprotection by PSD-95 inhibitors*. The Journal of Neuroscience, 2007. **27**(37): pp. 9901–9915.

[25] Chanrion, B., et al., *Physical interaction between the serotonin transporter and neuronal nitric oxide synthase underlies reciprocal modulation of their activity*. Proceedings of the National Academy of Sciences, 2007. **104**(19): pp. 8119–8124.

[26] Cheah, J.H., et al., *NMDA receptor-nitric oxide transmission mediates neuronal iron homeostasis via the GTPase Dexras1*. Neuron, 2006. **51**(4): pp. 431–440.

[27] Moos, T. and Morgan E.H., *The metabolism of neuronal iron and its pathogenic role in neurological disease: review*. Annals of the New York Academy of Sciences, 2004. **1012**(1): pp. 14–26.

[28] Xia, Y., Berlowitz C.O. and Zweier J.L., *PIN inhibits nitric oxide and superoxide production from purified neuronal nitric oxide synthase*. Biochimica et Biophysica Acta (BBA)-General Subjects, 2006. **1760**(9): pp. 1445–1449.

[29] Dreyer, J., et al., *Nitric oxide synthase (NOS)-interacting protein interacts with neuronal NOS and regulates its distribution and activity*. The Journal of Neuroscience, 2004. **24**(46): pp. 10454–10465.

[30] Izumi, Y. and Zorumski C., *Nitric oxide and long-term synaptic depression in the rat hippocampus*. Neuroreport, 1993. **4**(9): pp. 1131–1134.

[31] Toda, N., Ayajiki K. and Okamura T., *Control of systemic and pulmonary blood pressure by nitric oxide formed through neuronal nitric oxide synthase*. Journal of Hypertension, 2009. **27**(10): pp. 1929–1940.

[32] Böhme, G.A., et al., *Altered synaptic plasticity and memory formation in nitric oxide synthase inhibitor-treated rats*. Proceedings of the National Academy of Sciences, 1993. **90**(19): pp. 9191–9194.

[33] Lajoix, A.D., et al., *Protein inhibitor of neuronal nitric oxide synthase (PIN) is a new regulator of glucose-induced insulin secretion*. Diabetes, 2006. **55**(12): pp. 3279–3288.

[34] Hu, L., et al., *Ischemic postconditioning protects the heart against ischemia–reperfusion injury via neuronal nitric oxide synthase in the sarcoplasmic reticulum and mitochondria*. Cell Death & Disease, 2016. **7**(5): p. e2222.

[35] Melikian, N., et al., *Neuronal nitric oxide synthase and human vascular regulation*. Trends in Cardiovascular Medicine, 2009. **19**(8): pp. 256–262.

[36] Arami, M.K., et al., *The effect of hyperglycemia on nitric oxidergic neurons in nucleus tractus solitarius and blood pressure regulation in rats with induced diabetes*. Journal of Diabetes and Metabolic Disorders, 2005. **4**: p. 106.

[37] Kourosh Arami, M., et al., *The effect of nucleus tractus solitarius nitric oxidergic neurons on blood pressure in diabetic rats*. Iranian Biomedical Journal, 2006. **10**(1): pp. 15–19.

[38] Arami, M.K., *Nitric oxide in the nucleus raphe magnus modulates cutaneous blood flow in rats during hypothermia*. Iranian Journal of Basic Medical Sciences, 2015. **18**(10): p. 989.

[39] Malakouti, S.M., et al., *Reversible inactivation and excitation of nucleus raphe magnus can modulate tail blood flow of male wistar rats in response to hypothermia*. Iranian Biomedical Journal, 2008. **12**(4): pp. 237–240.

[40] Tricoire, L. and Vitalis T., Neuronal nitric oxide synthase expressing neurons: a journey from birth to neuronal circuits. 2012; 6: 82

[41] Moreno-López, B., et al., *Nitric oxide is a physiological inhibitor of neurogenesis in the adult mouse subventricular zone and olfactory bulb*. The Journal of Neuroscience, 2004. **24**(1): pp. 85–95.

[42] Lewis, D.A., et al., *Perisomatic inhibition and cortical circuit dysfunction in schizophrenia*. Current Opinion in Neurobiology, 2011. **21**(6): pp. 866–872.

[43] Marín, O., *Interneuron dysfunction in psychiatric disorders*. Nature Reviews Neuroscience, 2012. **13**(2): pp. 107–120.

[44] O'dell, T.J., et al., *Tests of the roles of two diffusible substances in long-term potentiation: evidence for nitric oxide as a possible early retrograde messenger*. Proceedings of the National Academy of Sciences, 1991. **88**(24): pp. 11285–11289.

[45] Koylu, E.O., et al., *Effects of nitric oxide synthase inhibition on spatial discrimination learning and central DA2 and mACh receptors*. Pharmacology Biochemistry and Behavior, 2005. **81**(1): pp. 32–40.

[46] Akar, F.Y., et al., *7-Nitroindazole, a neuronal nitric oxide synthase inhibitor, impairs passive-avoidance and elevated plus-maze memory performance in rats*. Pharmacology Biochemistry and Behavior, 2007. **87**(4): pp. 434–443.

[47] Liu, V.W. and Huang P.L., *Cardiovascular roles of nitric oxide: a review of insights from nitric oxide synthase gene disrupted mice*. Cardiovascular Research, 2008. **77**(1): pp. 19–29.

[48] Mikkelsen, R.B. and Wardman P., *Biological chemistry of reactive oxygen and nitrogen and radiation-induced signal transduction mechanisms*. Oncogene, 2003. **22**(37): pp. 5734–5754.

[49] Ridnour, L.A., et al., *The chemistry of nitrosative stress induced by nitric oxide and reactive nitrogen oxide species. Putting perspective on stressful biological situations*. Biological Chemistry, 2004. **385**(1): pp. 1–10.

[50] Liñares, D., et al., *Neuronal nitric oxide synthase plays a key role in CNS demyelination*. The Journal of Neuroscience, 2006. **26**(49): pp. 12672–12681.

[51] Aquilano, K., et al., *Role of nitric oxide synthases in Parkinson's disease: a review on the antioxidant and anti-inflammatory activity of polyphenols*. Neurochemical Research, 2008. **33**(12): pp. 2416–2426.

[52] Kwon, K.J., et al., *Disruption of neuronal nitric oxide synthase dimerization contributes to the development of Alzheimer's disease: Involvement of cyclin-dependent kinase 5-mediated phosphorylation of neuronal nitric oxide synthase at Ser293*. Neurochemistry International, 2016. **99**.

[53] Tajouri, L., et al., *Investigation of a neuronal nitric oxide synthase gene (NOS1) polymorphism in a multiple sclerosis population*. Journal of the Neurological Sciences, 2004. **218**(1): pp. 25–28.

[54] Walton, J.C., et al., *Neuronal nitric oxide synthase and NADPH oxidase interact to affect cognitive, affective, and social behaviors in mice*. Behavioural Brain Research, 2013. **256**: pp. 320–327.

[55] Gao, Y. and Heldt S.A., *Lack of neuronal nitric oxide synthase results in attention deficit hyperactivity disorder-like behaviors in mice*. Behavioral Neuroscience, 2015. **129**(1): p. 50.

[56] Retz, W., et al., *Association of a functional variant of neuronal nitric oxide synthase gene with self-reported impulsiveness, venturesomeness and empathy in male offenders*. Journal of Neural Transmission, 2010. **117**(3): pp. 321–324.

[57] Workman, J.L., et al., *Inhibition of neuronal nitric oxide reduces anxiety-like responses to pair housing*. Behavioural Brain Research, 2008. **187**(1): pp. 109–115.

[58] Tomiga, Y., et al., *Effects of environmental enrichment in aged mice on anxiety-like behaviors and neuronal nitric oxide synthase expression in the brain*. Biochemical and Biophysical Research Communications, 2016. **476**.

[59] Bujalska-Zadrozny, M., et al., *Central antinociceptive effect of tapentadol is increased by nitric oxide synthase inhibitors*. Behavioural Pharmacology, 2016. **27**.

Chapter 2

Endothelial Nitric Oxide Synthase and Neurodevelopmental Disorders

Saadat Huseynova, Nushaba Panakhova,
Safikhan Hasanov and Mehman Guliyev

Additional information is available at the end of the chapter

http://dx.doi.org/10.5772/67814

Abstract

Endothelial activity reflects the balance of endogenous factors regulating vasoconstriction and vasodilation. Among these factors, nitric oxide (NO) is the most important contributor to the acute regulation of vascular tone. Altered nitric oxide synthesis by the vascular endothelium plays several important roles in the pathogenesis of neonatal disease through its effects on vascular homeostasis. However, the role of NO in the pathogenesis of perinatal brain injury has not been fully investigated. The present chapter explores how NO synthesis is regulated under physiological and pathological conditions, the impact of acute and chronic hypoxia on NO synthase activity in the vascular endothelium, and the role of perinatal endothelial dysfunction in the pathogenesis of neurodevelopmental disorders later in life.

Keywords: endothelial dysfunction, eNOS, perinatal hypoxia, neuronal injury, neurodevelopmental disorders

1. Introduction

Endothelial function and the associated production of nitric oxide (NO) play a key role in the pathogenesis of diseases involving the disturbance of vascular homeostasis [1]. Enzymatic generation of NO in mammalian systems is accomplished by the oxidation of L-arginine to L-citrulline with the participation of NADPH as a cofactor. Thus, NO is produced by NO synthase isoforms including endothelial NO synthase (eNOS) and neuronal NO synthase (nNOS), with eNOS being the dominant isoform in the vasculature under physiological conditions [2]. eNOS, also known as nitric oxide synthase 3 (NOS3), participates in the regulation of vascular tone and has a wide range of actions that control cerebral blood flow and metabolism. In contrast, the main role of nNOS is the production of NO for retrograde signaling across neuronal synapses.

Altered NO production by the vascular endothelium contributes to the pathogenesis of neonatal disease and may influence developmental growth [3]. The actions of eNOS in endothelial dysfunction lead to vascular and metabolic disorders and are also implicated in hypoxic-ischemic brain injury [4]. Studies have shown that hypoxic brain injury is characterized by changes in vascular growth and endothelial dysfunction [5–8]. Despite the widespread confirmation a significant role for NO in physiological and pathological vascular homeostasis, the role of NO in the pathogenesis of perinatal brain injury has not been fully investigated. Specifically, the impact of chronic hypoxia on NO synthase isoenzymes in the neonatal brain is unknown. Therefore, the goal of this chapter is to present the results of recent investigations of the pathological role of eNOS in endothelial dysfunction in preterm infants with hypoxic-ischemic encephalopathy (HIE) and in early-age neurodevelopmental disorders.

2. Specification and function of vascular endothelium in fetoplacental circulation

The formation of the mammalian vasculature involves many interdependent processes, including the maturation of multiple cell types within tissue compartments, pulsatile blood flow, blood pressure, the activity of smooth muscle cells in vessel walls, and the transmigration of immune cells. Scientific investigations of endothelial remodeling have confirmed its relevance to vascular barrier function, inflammation, and vascular disease [9].

During embryonic development, the first endothelial cells are derived from the extraembryonic mesoderm and appear around embryonic day 7. The placental barrier to the maternal blood is gradually breached between 8 and 12 weeks of gestation owing to invasion of the uteroplacental spiral arteries of the placental bed by the extravillous trophoblast. Accordingly, placental oxygen tension rises and leads to a phase of branching angiogenesis that lasts 24 weeks [10]. The fetoplacental endothelium is continuous with the fetal circulation, such that its function and potential dysfunction have a profound impact on fetal development [11]. To this end, successful pregnancies are highly dependent on effective vasculogenesis.

The regulatory sites and mechanisms responsible for endothelial function in the uteroplacental and fetal circulation remain unclear; however, it is obvious that endothelial activity is regulated through the balanced production and action of local endogenous constricting and dilating factors. Among vasodilatory factors, NO appears to be a chief regulator of acute vascular tone. NO generated by NO synthase expressed in the uterine artery endothelium is a diffusible gas molecule that produces smooth muscle relaxation and therefore vasodilation in a cGMP-dependent manner [12]. In general, NO is essential for the formation of endothelial function. In pregnancy, NO promotes endovascular invasion by the cytotrophoblast; interstitial trophoblasts produce NO as they invade the maternal spiral arteries in the uterine wall in order to maintain a low-resistance and high-caliber uteroplacental unit. If this process fails, endothelial dysfunction associated with increased vascular resistance and reduced fetoplacental blood flow results in placental ischemia, pregnancy complications, and restrictive effects on fetal growth [13, 14]. Moreover, placental hypoperfusion and ischemia lead to the release of antiangiogenic factors that cause oxidative stress and inflammation, further contributing to endothelial dysfunction.

eNOS is well established as a primary physiological source of NO. eNOS affects vascular tone, reduces uteroplacental resistance, regulates uterine and fetoplacental blood flow, and is involved in uterine quiescence prior to parturition in normal pregnancy. Several studies have confirmed that eNOS activity is increased in the uterine artery during pregnancy in several species. Yet, investigations of the role of NO modulators in normal and abnormal pregnancies have shown conflicting results. The concentration of NO in the fetoplacental system depends on many factors including L-arginine availability, the activity levels of NO synthase isoforms, the presence of endogenous NO synthase inhibitors, and species-dependent variation. While some studies have reported lower eNOS expression in preeclamptic syncytiotrophoblasts than in normal syncytiotrophoblasts [15, 16], a series of clinical studies revealed that increased NO concentration primarily caused altered fetoplacental circulation, endothelial dysfunction, and reduced flow-mediated vasodilatation in different pregnancy pathologies [17–19]. Moreover, these studies identified increased endothelial permeability and decreased eNOS expression in the peripheral vasculature under pathological conditions [17]. Norris et al. reported increased NO production in the uteroplacental and fetoplacental circulation during preeclampsia compared to normotensive pregnancies and reasoned that this increase was a compensatory mechanism to offset the pathological effects of preeclampsia [18]. In support of this hypothesis, uterine arteries of pregnant rats exposed to plasma from women with preeclampsia were found to have increased eNOS expression and decreased inducible nitric oxide synthase (iNOS) expression [19]. In contrast, Leiva et al. purported that the bioavailability of NO in the fetoplacental system is decreased in pregnancy pathologies such as preeclampsia, gestational diabetes mellitus, and maternal supraphysiological hypercholesterolemia [20]; the authors hypothesized that altered NO synthesis and bioavailability in these cases are owing to the transcriptional and posttranslational modulation of NO synthases during hypoxia and oxidative stress.

One controversial question regards the putative effect of eNOS depression on fetoplacental blood flow in acute and chronic pathological processes. This topic was investigated in a series of experimental animal models; systemic NO synthase inhibitor administration was found to decrease uteroplacental blood flow and increase peripheral vascular resistance in several species [21]. Rosenfeld and Roy argued that the uteroplacental vasculature is less sensitive to prolonged systemic NO synthase inhibition than the peripheral circulation, which might be explained by the activation of compensatory mechanisms such as those reported for NO synthase in ovine uterine artery smooth muscle [22].

Several studies have investigated eNOS gene polymorphisms and their effects in different pregnancy pathologies. Whereas chronic hypoxia selectively augments the pregnancy-associated upregulation of eNOS gene expression and endothelium-dependent relaxation of the uterine artery [23], women with eNOS gene mutations were found to be at risk for developing preeclampsia in a study of Egyptian families [24]. However, other studies do not support a major role for eNOS gene variants in preeclampsia [25, 26]. Comparing the results of studies conducted worldwide, Ma et al. concluded that an eNOS gene polymorphism was related to pregnancy-induced hypertension risk in Asian populations but not in European and American populations [27].

In summary, altered functionality of the fetal endothelium likely contributes to the formation of extrauterine pathologies from the neonatal period onward. However, the mechanisms underlying fetoplacental vascular development and pathologies thereof remain incompletely defined, such that further studies are necessary to understand the exact role of eNOS in pregnancy pathologies and fetal growth problems.

3. The role of endothelial nitric oxide synthase activity in the pathophysiology of perinatal brain injury

Perinatal hypoxic-ischemic brain injury is a major cause of neonatal death and long-term disability. Approximately 15–25% of newborns with hypoxic-ischemic encephalopathy (HIE) die during the postnatal period, and surviving infants are at risk for the development of severe and permanent neuropsychological sequelae such as cerebral palsy, seizures, visual impairment, mental retardation, and learning and cognitive impairments [28–30]. Decreased cerebral perfusion, hypoxia, hypoglycemia, and severe anemia can cause critical energy shortages in newborn infants, and accordingly severe hypoxia/ischemia can also affect other tissues of the body [31]. Disorders affecting the peripheral organs are often caused by hemodynamic disturbances resulting from the centralization of the bloodstream and/or poor circulation to the internal organs [32].

In the early days of extrauterine life, the vascular endothelium is exposed to high concentrations of inflammatory stimuli and can become dysfunctional if exposed to a hypoxic environment [33]. Cerebral ischemia induces an inflammatory response in the brain parenchyma and systemic circulation [34, 35], resulting in the augmented secretion of proinflammatory cytokines and chemotactic molecules by the vascular endothelium in newborn infants with hypoxic-ischemic injury. Hence, cytokines are important upstream effector of brain injury after ischemia [36]. Vasoregulatory mechanisms play essential roles in brain injury and tissue reperfusion in critically ill children; endothelial dysfunction results in an imbalance between vasoconstriction and vasodilatation, which causes tissue reperfusion, cytotoxic edema, and brain injury [37]. A previous study determined that hypoxic inflammation was regulated via bioactive mediators synthesized by endothelium, whereas NO and the sources of its synthesis play a special role in the pathophysiology of leukocyte-endothelial interactions [38]. To this end, studies show that growth-retarded fetuses and infants with severe and long-lasting neuronal injuries exhibit decreased vascular growth and endothelial dysfunction [39].

Of note, brain injury after hypoxic-ischemic injury progresses over many days even after reperfusion has been achieved. For example, oxygenated blood flow is restored to ischemic brain areas after severe perinatal asphyxia; however, while reperfusion temporarily corrects energy failure, excitotoxicity, and the generation of reactive oxygen species during the ischemic period are together responsible for a significant degree of brain damage. Brain damage after hypoxia-ischemia includes the primary insult and secondary damage such as delayed neuronal death related to cerebral edema [40]. Primary perinatal insults resulting from hypoxia-ischemia are associated with the failure of ATPase-dependent ions channels, which can disrupt synaptic function and lead to the accumulation of extracellular glutamate [41]. The increased availability of reactive oxygen metabolites after reperfusion is also directly involved in augmented glutamate release after injury. Increase in extracellular glutamate concentrations and the activation of glutamate receptors lead to excitotoxicity [42], which involves increased intracellular flux of calcium through open NMDA receptor channels and the release of calcium from intracellular stores. Elevations in intracellular calcium activate lipases, proteases, and endonucleases that lead to cellular damage and death [43]. Moreover, the post-hypoxic reperfusion process results in oxidative stress; energy failure activates nNOS and increases NO production, increasing the likelihood of its reaction with superoxide anion to form the powerful oxidant peroxynitrite [44]. Together, cellular energy failure, acidosis, glutamate excitotoxicity, and oxidative stress lead to cytotoxic edema and neuronal death after hypoxia-ischemia injury [45]. Additionally, there is a continuum of necrosis and apoptosis after such injury: often, early (primary) cell death appears necrotic, whereas later (secondary) cell death appears apoptotic. Therefore, while severe insult results in cell necrosis, more moderate asphyxia can cause delayed neuronal death through apoptosis [46]. Secondary apoptosis involves multiple pathophysiological processes such as excitatory neurotransmission, altered growth factor production, and changes in protein synthesis [47].

NO serves diverse functions in the perinatal brain, including neuronal differentiation and survival and synaptic formation and plasticity [48]. NO also affects these processes in pathological contexts by (in part) mediating neuronal death and neurodegeneration [49]. Previous studies in growth-restricted infants demonstrated that elevated NO production was associated with decreased endogenous antioxidant activity, increased lipid peroxidation, and impaired neuronal function [50]. NO supplementation was also found to increase uteroplacental circulation and decrease biomarkers of neuronal injury in the cord blood of infants diagnosed with intrauterine growth retardation [51, 52]. Therefore, it is difficult to interpret the role of NO in the pathogenesis of perinatal neuronal injury: is the concentration of NO increased as a defensive mechanism or does it point to a more profound impairment? Clinical and experimental investigations describing the roles of different NO sources in the pathogenesis of brain injury have provided insight on this problem. In the prospective clinical trial conducted by the Azerbaijan Medical University Neonatology group (ACTRN12612000342819), NO, eNOS, and endotelin-1 were quantified in 240 preterm infants with high risk for perinatal HIE; the results indicated that while eNOS expression was reduced, NO concentrations were increased in accordance with the severity of HIE (**Figure 1**).

This result provided foundation evidence for nonendothelial sources of NO synthesis in tissue hypoperfusion and hypoxia. Thereafter, the balance of NO/eNOS and its effect on neuronal injury in preterm infants was investigated. An important finding was that infants with severe HIE had higher NO/eNOS ratios compared with mild/moderate HIE and control infants, suggesting a relationship between nonendothelial NO production and neuronal injury (**Figure 2**).

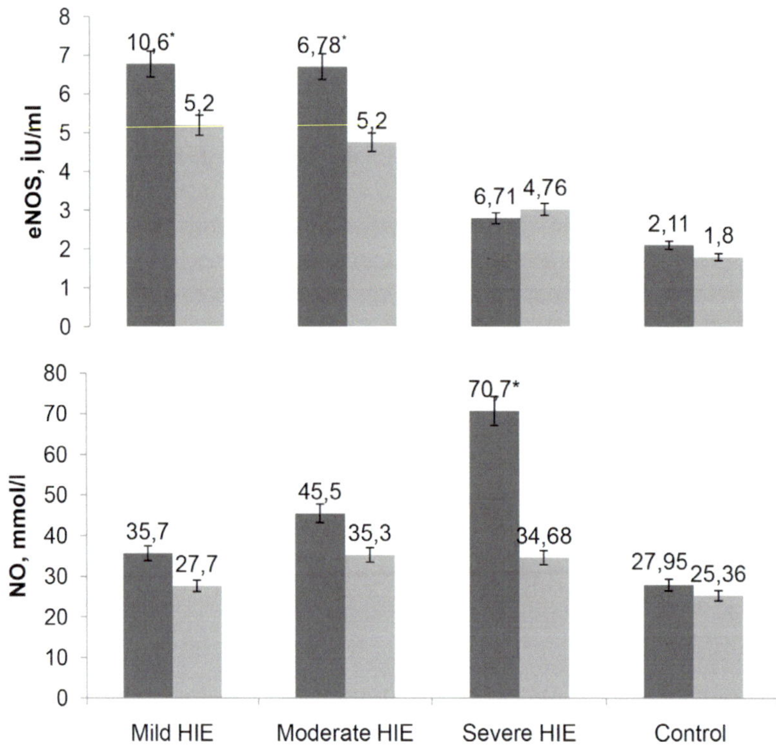

Figure 1. Mean total eNOS and NO values in preterm infants with HIE. Error bars indicate the standard error of the mean. Black bars show results from days 1 to 3 and grey bars show results from days 5 to 7. *$p < 0.05$ compared with the control group.

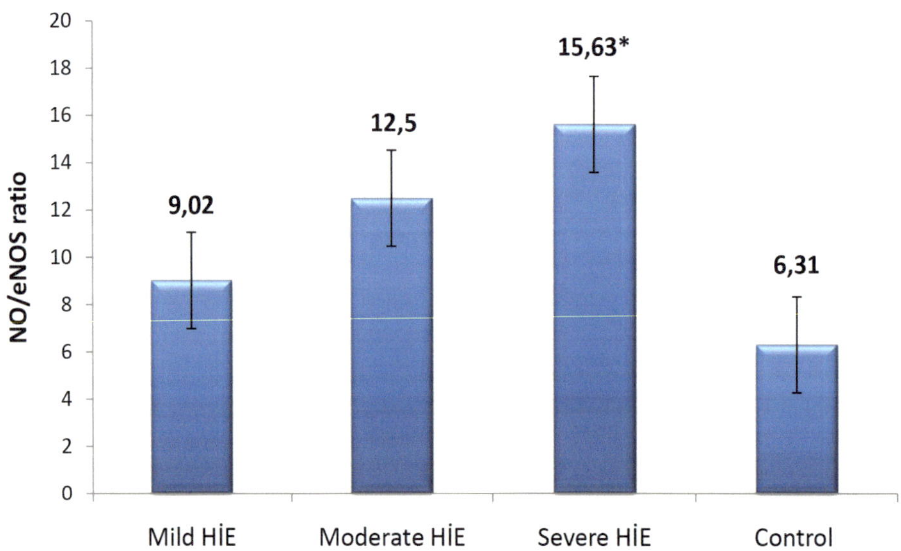

Figure 2. NO/eNOS ratios in preterm infants with HIE. *$p < 0.05$ compared with the control group.

http://dx.doi.org/10.5772/67814

Significantly higher NO/eNOS ratios in preterm infants with severe HIE suggest that the activation of neuronal and inducible NO synthases is related to long-term and severe intrauterine and birth distress in infants. Moreover, increased NO in tandem with eNOS activation in infants with low risk for perinatal HIE might represent a compensatory or defensive strategy in the preterm brain. It should be noted that increased NO generation is not necessarily solely derived from areas of neuronal injury in HIE. Under hypoxic conditions, NO is also produced by the activated endothelium in all injured vasculature. Therefore, it might be difficult to accept the idea that NO/eNOS balance is a good predictor of neuronal injury. Yet, consistent with our previous investigation [53], we observed a statistically significant positive correlation between neuron-specific enolase (NSE) and NO/eNOS ratio, which suggests that decreased synthesis of NO by endothelial sources is related to more severe hypoxic changes and neuronal injury (**Figure 3**).

It was also found that growth-restricted infants are subject to significant endothelial dysfunction and eNOS depression, implicating NO in the pathogenesis of intrauterine hypoxic injury [53]. Together, these results provide a strong support for NO/eNOS balance as a marker of endothelial inflammation under hypoxic conditions. Previous experimental and clinical investigations have demonstrated that eNOS is responsible for preserving the functional integrity of the neurovascular unit [54, 55] and may have antiinflammatory effects in aging and other pathological contexts [56, 57].

The follow-up of newborn infants in the aforementioned study identified significant relationships between peripheral endothelial vasoregulatory markers in the perinatal period and the onset of neurodevelopmental disorders at an early age. It was found that, in the presence of high concentrations of NO, early eNOS activity was insufficient in infants diagnosed with cerebral palsy later in life compared to neonates who did not show neurodevelopmental delays associated with HIE (**Figure 4**). These findings suggest that depressed eNOS activity and increased non-endothelial NO synthesis play important roles in the formation of developmental impairments.

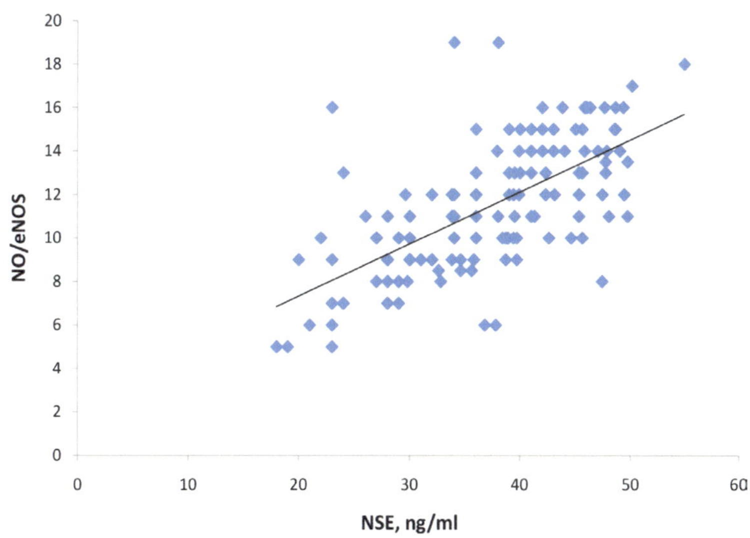

Figure 3. Spearman rank-order correlation between NO/eNOS ratio and NSE in preterm infants with HIE (r = 0.67; p = 0.001).

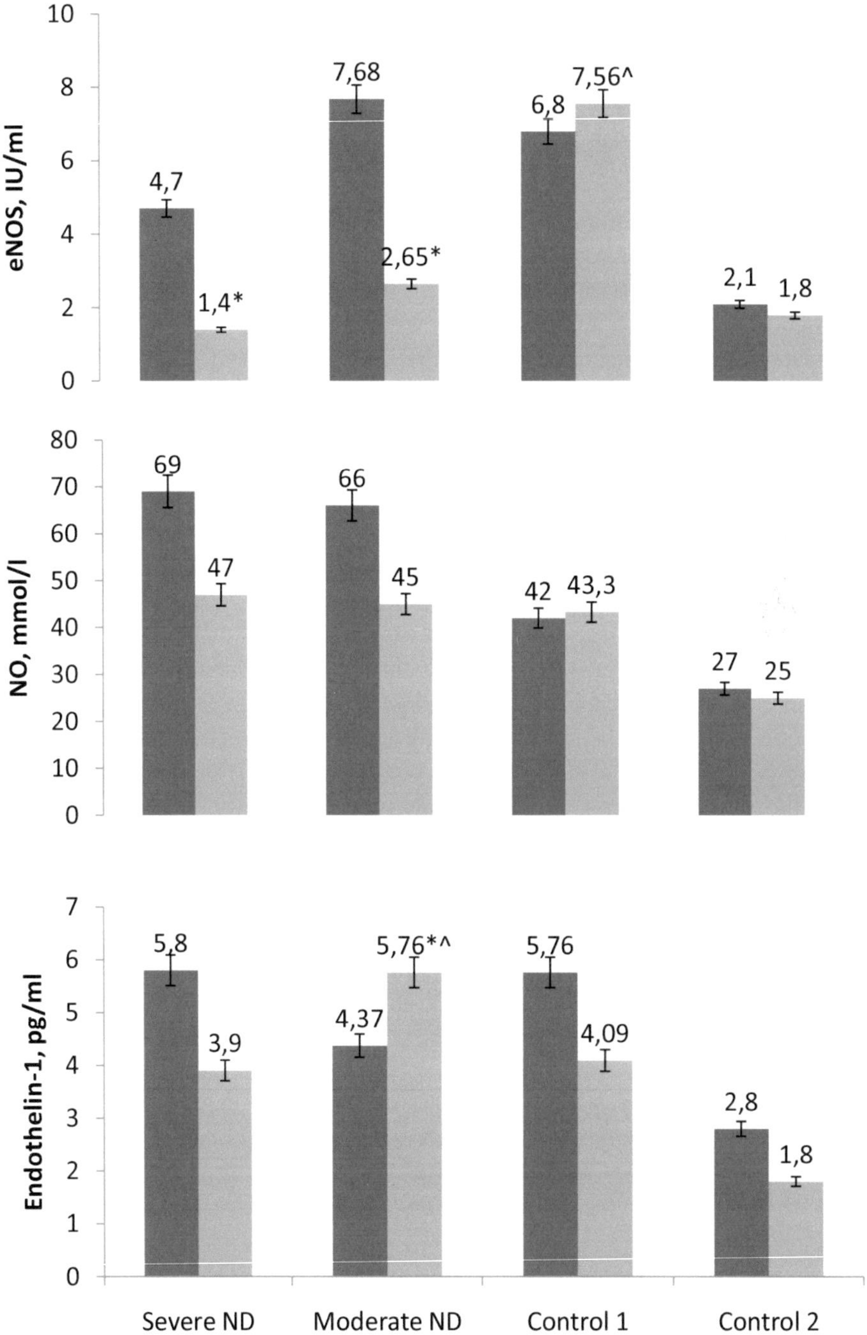

Figure 4. Peripheral vasoregulatory markers in the early neonatal period in preterm infants with neurodevelopmental disorders (NDs). Error bars indicate the standard error of the mean. Black bars show results from days 1 to 3 and grey bars show results from days 5 to 7. Control 1: data from infants with HIE who did not develop a ND; Control 2: data from healthy infants. *$p < 0.05$ compared with Control 1; ^$p < 0.05$ compared with Control 2.

As shown in **Figure 4**, cases with substantial eNOS and endothelin-1 depression during the perinatal period exhibited more profound neurodevelopmental delay and cerebral palsy. In contrast, when eNOS was depressed but vasoconstriction was maintained (i.e., increased endothelin-1 expression), functional impairments were more moderate, including mild motor and cognitive deviations and minimal brain dysfunction later in life. Therefore, insufficient eNOS activation in combination with the absence of a compensatory mechanism (e.g., peripheral vasospasm and/or the centralization of circulation in vital organs during the early stages of pathology) might ultimately drive the more serious and irreversible injury of brain tissue.

The abovementioned findings are consistent with those of several studies of different NO synthases in the pathogenesis of brain injury. In one study, chronic hypoxia decreased eNOS expression in the hippocampus and increased nNOS expression in neuronal and glial cells of the thalamus [5]. Moreover, in addition to elevated glutamate synthesis, long-term and severe hypoxemic processes have been reported to alter NO synthase enzyme activity in a manner related to DNA structure, resulting in iNOS and nNOS activation [58, 59]. Wei et al. determined that endothelial NO production by eNOS can decrease ischemic injury by inducing vasodilation, while neuronal NO production can exacerbate neuronal injury [6]. Therefore, several researchers have suggested the potential neuroprotective utility of nNOS inhibitors after brain injury [7, 8].

Many specific biochemical markers of neuronal injury are being investigated as indicators of brain damage in neonates [60]. Some neuron-specific proteins and cytokines show promise for identifying infants who are at risk for perinatal encephalopathy, although the exact value of these markers for predicting severe brain damage and neurodevelopmental disorders remains controversial [61, 62]. The early assessment of acute cerebral lesions in preterm infants may provide useful information regarding appropriate therapeutic intervention strategies and allow the prevention of future neurological complications. One possibility is that the severity of brain injury in newborns can be assessed by measuring the activity of NO synthases. Future studies are required to validate this hypothesis and better elucidate the clinical significance of NO synthesis in perinatal injury.

4. Conclusion

Autoregulation in the neonatal brain is tightly coupled with neuronal and endothelial regulatory mechanisms. Clinical and experimental investigations confirm that neuronal injury is in part mediated by the activation of endothelial and nonendothelial sources of NO synthesis. eNOS activity plays a fundamental role in the autoregulation of vascular tone in the perinatal period and is additionally involved in the formation of hypoxic brain damage during this period; however, the various roles of NO in neuroprotection and metabolism in the brain complicate our exact understanding of the relationship between NO and brain injury. Recent investigations suggest that eNOS plays a protective role in perinatal brain injury whereas other endogenous sources of NO (e.g., iNOS and nNOS) may participate in the pathogenesis of perinatal pathologies and neurodevelopmental disorders. Future studies should further delineate the molecular pathways responsible for the roles of NO synthase isoforms in brain injury and neuroprotection.

Finally, the ratio of NO/eNOS expression may indicate the severity of neuronal injury and have clinical utility for predicting long-term outcomes in infants after perinatal brain injury.

Author details

Saadat Huseynova[1*], Nushaba Panakhova[1], Safikhan Hasanov[2] and Mehman Guliyev[3]

*Address all correspondence to: sadi_0105@mail.ru

1 Department of 3rd Pediatrics, Azerbaijan Medical University, Baku, Azerbaijan

2 Department of 2nd Pediatrics, Azerbaijan Medical University, Baku, Azerbaijan

3 Department of Biochemistry, Azerbaijan Medical University, Baku, Azerbaijan

References

[1] Toda N, Ayajiki K, Okamura T. Cerebral blood flow regulation by nitric oxide. Recent advances. Pharmacol Rev. 2009;**61**:62–97. DOI: 10.1124/pr.108.000547

[2] Mayer B, Hemmens B. Biosynthesis and action of nitric oxide in mammalian cells. Trends Biochem Sci. 1997;**22**:477–481.

[3] Vannuci SJ, Hagberg H. Hypoxia–ischemia in the immature brain. J Exp Biol. 2004;**207**: 3149–3154. DOI: 10.1242/jeb.01064

[4] Buonocore G, Perrone S, Turrisi G, Kramer BW, Balduini W. New pharmacological approaches in infants with hypoxic-ischemic encephalopathy. Curr Pharm Des. 2012;**18**:3086–3100.

[5] Dong Y, Yu Z, Sun Y, Zhou H, Stites J, Newell K, Weiner CO. Chronic fetal hypoxia produces selective brain injury associated with altered nitric oxide synthases. Am J Obstet Gynecol. 2011;**3**:16–28. DOI: 10.1016/j.ajog.2010.11.032

[6] Wei G, Dawson VL, Zweier JL. Role of neuronal and endothelial nitric oxide synthase in nitric oxide generation in the brain following cerebral ischemia. Biochim Biophys Acta. 1999;**20**:23–34. DOI: 10.1016/S0925-4439(99)00051-4

[7] Pham H, Duy AP, Pansiot J, Bollen B, Gallego J, Charriaut-Marlangue C, Baud O. Impact of inhaled nitric oxide on white matter damage in growth-restricted neonatal rats. Pediatr Res. 2015;**4**:563–569. DOI: 10.1038/pr.2015.4

[8] Liu PK, Robertson CS, Valadka A. The association between neuronal nitric oxide synthase and neuronal sensitivity in the brain after brain injury. Ann NY Acad Sci. 2002;**962**:226–241.

[9] Dorland YL, Huveneers S. Cell-cell junctional mechanotransduction in endothelial remodeling. Cell Mol Life Sci. 2017;**74**:279–292. DOI: 10.1007/s00018-016-2325-8

[10] Kingdom JCP, Kaufmann P. Oxygen and placental vascular development. In: Roach RC, Wagner PD, Hackett PH, editors. Hypoxia. Advances in Experimental Medicine and Biology. New York: Springer; 1999. p 259–275.

[11] Wadsack C, Desoye G, Hiden U. The feto-placental endothelium in pregnancy pathologies. Wien Med Wochenschr. 2012;**162**:220–224. DOI: 10.1007/s10354-012-0075-2

[12] Kublickiene KR, Cockell AP, Nisell H, Poston L. Role of nitric oxide in the regulation of vascular tone in pressurized and perfused resistance myometrial arteries from term pregnant women. Am J Obstet Gynecol. 1997;**177**:1263–1269.

[13] Jones S, Bischof H, Lang I, Desoye G, Greenwood SL, Johnstone ED, Wareing M, Sibley CP, Brownbill P. Dysregulated flow-mediated vasodilatation in the human placenta in fetal growth restriction. J Physiol. 2015;**15**:593:3077–3092. DOI: 10.1113/JP270495

[14] Khalil A, Hardman L, O Brien P. The role of arginine, homoarginine and nitric oxide in pregnancy. Amino Acids. 2015;**47**:1715–1727. DOI: 10.1007/s00726-015-2014-1

[15] Kim YJ, Park HS, Lee HY, Ha EH, Suh SH, Oh SK, Yoo HS. Reduced L-arginine level and decreased placental eNOS activity in preeclampsia. Placenta. 2006;**27**:438–444. DOI:10.1016/j.placenta.2005.04.011

[16] Bartlett SR, Bennett PR, Campa JS, Dennes WJ, Slater DM, Mann GE, Poston L, Poston R. Expression of nitric oxide synthase isoforms in pregnant human myometrium. J Physiol. 1999;**521**:705–716.

[17] Wang Y, Gu Y, Zhang Y, Lewis DF. Evidence of endothelial dysfunction in preeclampsia: decreased endothelial nitric oxide synthase expression is associated with increased cell permeability in endothelial cells from preeclampsia. Am J Obstet Gynecol. 2004;**190**:817–824. DOI: 10.1016/j.ajog.2003.09.049

[18] Norris LA, Higgins JR, Darling MR, Walshe JJ, Bonnar J. Nitric oxide in the uteroplacental, fetoplacental, and peripheral circulations in preeclampsia. Obstet Gynecol. 1999;**93**:958–963.

[19] Kao CK, Morton JS, Quon AL, Reyes LM, Lopez-Jaramillo P, Davidge ST. Mechanism of vascular dysfunction due to circulating factors in women with pre-eclampsia. Clin Sci (Lond). 2016;**130**:539–549. DOI: 10.1042/CS20150678

[20] Leiva A, Fuenzalida B, Barros E, Sobrevia B, Salsoso R, Sáez T, Villalobos R, Silva L, Chiarello I, Toledo F, Gutiérrez J, Sanhueza C, Pardo F, Sobrevia L. Nitric oxide is a central common metabolite in vascular dysfunction associated with diseases of human pregnancy. Curr Vasc Pharmacol. 2016;**14**:237–259. DOI: 10.2174/1570161114666160222115158

[21] Kulandavelu S, Whiteley KJ, Bainbridge SA, Qu D, Adamson SL. Endothelial NO bsynthase augments fetoplacental Blood flow, placental vascularization, and fetal growth in mice. Hypertension. 2013;**61**:259–266. DOI: 10.1161/HYPERTENSIONAHA.112.201996

[22] Rosenfeld CR, Roy T. Prolonged uterine artery nitric oxide synthase inhibition modestly alters basal uteroplacental vasodilation in the last third of ovine pregnancy. Am J Physiol Heart Circ Physiol. 2014;**307**:1196–1203. DOI: 10.1152/ajpheart.00996.2013

[23] Xiao D, Bird IM, Magness RR, Longo LD, Zhang L. Upregulation of eNOS in pregnant ovine uterine arteries by chronic hypoxia. Am J Physiol Heart Circ Physiol. 2001;**280**:812–820.

[24] El-Sherbiny WS, Nasr AS, Soliman A. Endothelial nitric oxide synthase (eNOS) (Glu298Asp) and urotensin II (UTS2 S89N) gene polymorphisms in preeclampsia: prediction and correlation with severity in Egyptian females. Hypertens Pregnancy. 2013;**32**:292–303. DOI: 10.3109/10641955.2013.806539

[25] Zdoukopoulos N, Doxani C, Messinis IE, Stefanidis I, Zintzaras E. Polymorphisms of the endothelial nitric oxide synthase (NOS3) gene in preeclampsia: a candidate-gene association study. BMC Pregnancy Childbirth. 2011;**3**:11:89. DOI: 10.1186/1471-2393-11-89

[26] Lade JA, Moses EK, Guo G, Wilton AN, Grehan M, Cooper DW, Brennecke SP. The eNOS gene: a candidate for the preeclampsia susceptibility locus? Hypertens Pregnancy. 1999;**18**:81–93.

[27] Ma Q, Lv J, Huang K, Guo H, Yang W, Luo W, Qiu J, Yang L. Endothelial nitric oxide synthase gene G894T polymorphism and risk assessment for pregnancy-induced hypertension: evidence from 11 700 subjects. Hypertens Res. 2016;**39**:899–906. DOI: 10.1038/hr.2016.95

[28] Liu L, Oza S, Hogan D, Perin J, Rudan I, Lawn JE, et al. Global, regional, and national causes of child mortality in 2000-13, with projections to inform post-2015 priorities: an updated systematic analysis. Lancet. 2015;**385:**430–440. DOI: 10.1016/S0140-6736(14)61698-6.

[29] Glass HC, FerrieroDM. Treatment of hypoxic-ischemic encephalopathy in newborns. Curr Treat Options Neurol. 2007;**9**:414–423. DOI:10.1007/s11940-007-0043-0

[30] Teixeira L, Soares H, Flor-de-Lima F, Vilan A, Guimarães H. Perinatal hypoxic-ischemic encephalopathy: severity determinants and outcomes. J Pediatr Neonatal Individualized Med. 2014;**2**:e030204. DOI: 10.7363/030204

[31] Cowan F, Rutherford M, Groenendaal F, Eken P, Mercuri E, Bydder GM, Meiners LC, Dubowitz LM, de Vries LS. Origin and timing of brain lesion in term infants with neonatal encephalopathy. Lancet. 2003;**361**:736–742. DOI:10.1016/S0140-6736(03)12658-X

[32] Volpe JJ. Neonatal encephalopathy: an inadequate term for hypoxic–ischemic encephalopathy. Ann Neurol. 2012;**72**:156–166. DOI: 10.1002/ana.23647

[33] Sunshine P. Epidemiology, pathophysiology and pathogenesis of fetal and neonatal brain injury. Neonatal encephalopathy: epidemiology and overview. In: David KS, William EB, Philip S, Susan RH, Maurice LD, editors. Fetal and Neonatal Brain Injury. 3rd ed. Cambridge: Cambridge University Press; 2009. p. 1–13. DOI: 10.1017/CBO9780511581281

[34] Mirabelli-Badenier M, Braunersreuther V, Lenglet S, Galan K, Veneselli E, Viviani GL, Mach F, Montecucco F. Pathophysiological role of inflammatory molecules in paediatric ischaemic brain injury. Eur J Clin Invest. 2012;**42**:784–794. DOI: 10.1111/j.1365-2362.2012.02640

[35] Huseynova S, panakhova N, Orujova P, Hasanov S, Guliyev M, Orujov A. Elevated levels of serum sICAM-1 in asphyxiated low birth weight newborns. Sci Rep. 2014;**4**:6850. DOI: 10.1038/srep06850

[36] Albertsson A, Bi D, Duan L, Zhang X, Leavenworth JW, Qiao L, Zhu C, Cardell S, Cantor H, Hagberg H, Mallard C, Wang X. The immune response after hypoxia-ischemia in a mouse model of preterm brain injury. J Neuroinflammation. 2014;**11**:153. DOI: 10.1186/s12974-014-0153-z

[37] Bonestroo HJ, Heijnen CJ, Groenendaal F, van Bel F, Nijboer CH. Development of cerebral gray and white matter injury and cerebral inflammation over time after inflammatory perinatal asphyxia. Dev Neurosci. 2015;**37**:78–94. DOI: 10.1159/000368770

[38] Yu L, Derrick M, Ji H, Silverman RB, Whitsett J, Vásquez-Vivar J, Tan S. Neuronal nitric oxide synthase inhibition prevents cerebral palsy following hypoxia-ischemia in fetal rabbits: comparison between JI-8 and 7-nitroindazole. Dev Neurosci. 2011;**33**:312–319. DOI: 10.1159/000327244

[39] Rozance PJ, Seedorf GJ, Brown A, Roe G, O'Meara MC, Gien J, Tang JR, Abman SH. Intrauterine growth restriction decreases pulmonary alveolar and vessel growth and causes pulmonary artery endothelial cell dysfunction in vitro in fetal sheep. AJP Lung Physiol. 2011;**6**:860–871. DOI: 10.1152/ajplung.00197.2011

[40] Lai MC, Yang SN. Perinatal hypoxic-ischemic encephalopathy. J Biomed Biotechnol. 2011;**2011**:609813. DOI: 10.1155/2011/609813

[41] Ferriero DM. Neonatal brain injury. N Engl J Med. 2004;**351**:1985–1995. DOI: 10.1056/NEJMra041996

[42] Burd I, Welling J, Kannan G, Johnston MV. Excitotoxicity as a common mechanism for fetal neuronal injury with hypoxia and intrauterine inflammation. Adv Pharmacol. 2016;**76**:85–101. DOI: 10.1016/bs.apha.2016.02.003

[43] Jin C, Londono I, Mallard C, Lodygensky GA. New means to assess neonatal inflammatory brain injury. J Neuroinflammation. 2015;**12**:180. DOI: 10.1186/s12974-015-0397-2

[44] Dawson VL. Nitric oxide: role in neurotoxicity. Clin Exp Pharmacol Physiol. 1995;**22**:305–308. DOI: 10.1111/j.1440-1681.1995.tb02005.x

[45] MacDonald JF, Xiong ZG, Jackson MF. Paradox of Ca2+ signaling, cell death and stroke. Trends Neurosci. 2006;**29**:75–81. DOI 10.1016/j.tins.2005.12.001

[46] Northington FJ, Chavez-Valdez R, Martin LJ. Neuronal cell death in neonatal hypoxia-ischemia. Ann Neurol. 2011;**69**:743–758. doi: 10.1002/ana.22419

[47] Hagberg H, Mallard C, Ferriero DM, Vannucci SJ, Levison SW, Vexler ZS, Gressens P. Nat Rev Neurol. 2015;**11**:192–208. DOI: 10.1038/nrneurol.2015.13.

[48] Lev-Ram V, Jiang T, Wood J, Lawrence DS, Tsien RY. Synergies and coincidence requirements between NO, cGMP, and Ca2+ in the induction of cerebellar long-term depression. Neuron. 1997;**18**:1025–1038.

[49] Dawson VL, Dawson TM. Nitric oxide in neurodegeneration. Prog Brain Res. 1998;**118**: 215–229.

[50] Hracsko Z, Hermesz E, Ferencz A, Orvos H, Novak Z, Pal A, Varga IS. Endothelial nitric oxide synthase is up-regulated in the umbilical cord in pregnancies complicated with intrauterine growth retardation. In Vivo. 2009;**23**:727–732.

[51] Lyall F, Greer IA, Young A, Myatt L. Nitric oxide concentrations are increased in the feto-placental circulation in intrauterine growth restriction. Placenta. 1996;**17**:165–168.

[52] Gazzolo D, Bruschettini M, Di Lorio R, Marinoni E, Lituania M, Marras M, Sarli R, Bruschettini PL, Michetti F. Maternal nitric oxide supplementation decreases cord blood S100B in intrauterine growth-retarded fetuses. Clin Chem. 2002;**4**:647–650.

[53] Huseynova S, Panakhova N, Orujova P, Hasanov S, Guliyev M, Yagubova V. Altered endothelial nitric oxide synthesis in preterm and small for gestational age infants. Pediatr Int. 2015;**57**:269–275. DOI: 10.1111/ped.12520

[54] Faraci FM. Protecting against vascular disease in brain. Am J Physiol Heart Circ Physiol. 2011;**300**:1566–1582. DOI: 10.1152/ajpheart.01310.2010

[55] Katusic ZS, Austin SA. Neurovascular protective function of endothelial nitric oxide – recent advances. Circ J. 2016;**80**:1499–1503. DOI: 10.1253/circj.CJ-16-0423

[56] Dimmeler S, Zeiher AM. Nitric oxide-an endothelial cell survival factor. Cell Death Differ. 1999;**6**:964–968. DOI: 10.1038/sj.cdd.4400581

[57] Austin SA, Santhanam AV, Hinton DJ, Choi DS, Katusic ZS. Endothelial nitric oxide deficiency promotes Alzheimer's disease pathology. J Neurochem. 2013;**127**:691–700. DOI: 10.1111/jnc.12334

[58] Mifsud G, Zammit C, Muscat R, Di Giovanni G, Valentino M. Oligodendrocyte pathophysiology and treatment strategies in cerebral ischemia. CNS Neurosci Ther. 2014; **20**:603–612. DOI:10.1111/cns.12263

[59] Degos V, Loron G, Mantz J, Gressens P. Neuroprotective strategies for the neonatal brain. Anesth Analg. 2008;**106**:1670–1680. DOI:10.1213/ane.0b013e3181733f6f

[60] Naithani M, Simalti AK. Biochemical markers in perinatal asphyxia. J Nepal Pediatr Soc. 2011;**31**:151–156. DOI: 10.3126/jnps.v31i2.4155

[61] Nagdyman N, Grimmer I, Scholz T, Muller C, Obladen M. Predictive value of brain-specific proteins in serum for neurodevelopmental outcome after birth asphyxia. Pediatr Res. 2003;**54**:270–275. DOI: 10.1203/01.PDR.0000072518.98189.A0

[62] Wijnberger LD, Nikkels PG, van Dongen AJ, Noorlander CW, Mulder EJ, Schrama LH, Visser GH. Expression in the placenta of neuronal markers for perinatal brain damage. Pediatr Res. 2002;**4**:492–496. DOI: 10.1203/00006450-200204000-00015

Chapter 3

Role of Nitric Oxide Synthase in Normal Brain Function and Pathophysiology of Neural Diseases

Melih Dagdeviren

Additional information is available at the end of the chapter

http://dx.doi.org/10.5772/67267

Abstract

Nitric oxide synthase has three isoforms; according to their roles and tissues or cells they are involved. Neuronal NOS (nNOS) takes place in neuronal signalling, endothelial NOS (eNOS) takes place in vasodilation and inducible NOS (iNOS) takes place in immune responses. nNOS and eNOS are dominant but all isoforms have various roles in the central nervous system. nNOS and eNOS separately or together works in healthy brain during cognitive processes and in unhealthy brain during the pathology of related diseases. These roles were shown by inhibitor applied or by transgenic animal model studies and also by investigating the diseases at the molecular level. Besides, it is possible to say that iNOS has roles in some neurological pathologies creating immune responses. Three different isoforms mainly serve in different systems so there are lots of researchers from various disciplines working collaterally not knowing the others related works about NOSs. Because of this, a comprehensive chapter will be valuable for neuroscientists working with either healthy or unhealthy brains. The purpose of this chapter is to gather an overview of NOSs duties during the normal processes of the brain like learning and memory formation and abnormal processes such as depression, schizophrenia and brain cancers.

Keywords: NOS, learning, depression, brain cancers

1. Introduction

Nitric oxide (NO) also known as nitrogen oxide or nitrogen monoxide is a small molecule, which is a gaseous secondary messenger in mammalian cells [1]. Since the early 1990s, the importance of that molecule for biological systems has been investigated by various branches of related fields. Robert F. Furchgott, Louis J. Ignarro and Farid Murad earned a Nobel Prize in physiology or medicine in the year 1998 about the findings of the signalling properties of NO in cardiovascular systems [2]. Thus, the importance of that molecule for biological systems was emphasized.

The pathways that create NO in an organism can differ from system to system and tissue to tissue. Inside the mammalian cells NO is produced as a co-product of a biochemical activity catalysed by the nitric oxide synthase (NOS) enzymes. NOS enzymes are flavoenzymes that contain iron-heme, and these enzymes need nicotinamide-adenine-dinucleotide phosphate (NADPH), flavin adenine dinucleotide (FAD), flavin mononucleotide (FMN) and (6R-)5, 6, 7, 8-tetrahydro-L-biopterin (BH_4) as cofactors to convert substrate L-arginine to L-citrulline. During that reaction NO is formed. As a water-soluble gas, NO can easily diffuse to neighbouring cells; however, the diffusion is limited because of the short half-life of NO [3].

NOS enzyme has three isoforms, two of them are constitutive isoforms: endothelial NOS (eNOS/NOS3) essentially found in the vascular system and neuronal NOS (nNOS/NOS1) essentially found in the nervous system. The third isoform is inducible isoform: inducible NOS (iNOS/NOS2) principally found in immune system cells [4]. NOSs are homodimeric enzymes and for NO producing reactions NOSs transfer electrons from NADPH to heme via FAD and FMN in the amino terminal. The oxygenase domain of the enzyme also binds the BH_4, O_2 and L-arginine. L-arginine is oxidized to L-citrullin and NO. All three isoforms bind calmodulin, which work as a molecular switch for NOSs [5].

eNOS is one of the constitutive isoforms, which is generally present in the vascular endothelial cells. eNOS is dominantly expressed by endothelial cells and expressed in little amounts by some other cardiovascular system cells such as cardiomyocytes, erythrocytes, leucocytes, platelets and microparticles in the blood. Ca^{2+} concentration is an important factor for eNOS activation, also haemodynamic forces, hypoxia, catecholamines, exercise, G-protein activation and post-translational modifications activate eNOS. To response these various stimulants, eNOSs sometimes gather in special sections inside cells, these special sections have caveolin-binding activity. These various stimulant response adaptations make eNOS differ from the other isoforms. Because the other two isoforms are regulated by a smaller number of factors [6], NO is very important to keep the vascular homeostasis at a balance. NO synthesized freshly, travels to the neighbouring cells. Then in the vessel walls, inside the cells, NO binds to guanylate cyclase (GC), thus cyclic guanosine monophosphate (cGMP) concentration increase, Ca^{2+} channels start to open and when calcium inflows the smooth muscle relaxation is triggered [7].

The three isoforms have small differences between each other. nNOS was first identified in brain and have the biggest molecular weight than the other isoforms with 160 kDa, and eNOS and iNOS having 133 and 131 kDa, respectively. The difference is caused by PDZ domain of nNOS, and caveolin-binding site of eNOS [8]. nNOS is mainly expressed in nervous system cells and is very important for neural functions. The primary nNOS expressing cells are neurons in the brain and also neurons of hypothalamus, pineal gland, spinal cord and nerves innervating other organs. Besides nNOS is also expressed in myocytes, epithelial cells, macula densa cells, testis-urethra cells, mast cells and neutrophils of various mammals [9]. nNOS is also Ca^{2+}-calmodulin dependent, and the activation of nNOS is also regulated by the phosphorylation and neurotransmitter activity. N-methyl-D-aspartate (NMDA) receptors have the key role for nNOS activation in neurons. Sometimes nNOS activation is triggered by presynaptic neurons; glutamate released from the presynaptic neuron increases the Ca^{2+} in the post-synaptic neuron through NMDA receptor, which increased Ca^{2+} activating the nNOS in the post-synaptic neuron. Then nNOS produce NO, which diffuses back to the presynaptic neuron and triggers soluble GC [4, 10].

There are several answers to the question why NO is important for the brain. That freely diffusible gaseous secondary messenger can be a lifesaver or villain for neurological processes [11]. Thus, nNOS became as important as NO in the brain. nNOS plays an important role in neuronal function, memory formation, sexually different behaviours, neurological regulations and for therapeutics.

nNOS in the whole brain and additionally eNOS in the hippocampus and somatosensory cortex are involved in NO and cGMP formation. These secondary messengers activate synaptic plasticity in the hippocampus, cerebral cortex, amygdala, striatum and cerebellum [12]. In several murine, nNOS inhibition studies and an nNOS knockout mice study revealed that it is crucial for memory formation and recognition of memory [13].

There are lots of studies that try to explain the link between nNOS and behaviour at the molecular level. Sex difference is a very important variable for behavioural studies. There is a significant difference between male and female mice according to their aggressive behaviour if their nNOS gene were knocked-out. Besides in a NOS inhibition study, it was shown that male rats tended to show female-like perceptual style behaviour [14, 15]. It is possible to say that NO and nNOS are important molecules for sex difference dependent behavioural studies.

There are lots of inhibitors for NOSs; arginine-like chemicals bind NOSs instead of L-arginine to inhibit the reaction catalysed by NOSs. Most widely used inhibitors are nitro-L-arginine (L-NA), nitro-L-arginine methyl ester (L-NAME) and N-monomethyl L-arginine (L-NMMA). These inhibitors are not selective for a specific isoform, but there are also selective inhibitors available. The non-selective inhibitors are important as selective ones. These non-selective inhibitors helped to suppress both eNOS and nNOS, because both isoforms take place in regulating vascular activities in the brain. With that inhibitors it was shown that NO/NOS pathway is very important for regulating the cerebral blood flow within the healthy brain, and also very important for the ischaemic brain. The studies done with the ischaemic brain have revealed that NO is also a double-edged sword for that pathology. Scientists trying to find a treatment for the ischaemic brain over NO pathway should consider the risk of manipulating that pathway [16].

There have been some other studies on NO/nNOS pathway for therapeutic concerns in the ischaemic brain. Inhibiting or activating nNOS is a point of view where scientists are dealing with the internal sources of the organism. However, there is another option: to supply NO from outside of the organism. There are some studies that supply NO to the organism by inhalation to treat injured brains. According to the results of these studies, the endogenous sources are generally a better target to manipulate. But for the ischaemic brain, the external NO supply is useful at the ischaemia phase, not in the reperfusion phase. It is possible to interpret that it is not easy to separate the ischaemic and reperfusion phases if it is not a controlled operation. Therefore, external NO inhalation seems not efficient for ischaemic conditions [17].

There are lots of therapeutic strategies to overcome different pathologies involving NOSs. But before developing a technique or a new technology for treatment, the molecular mechanisms behind the normal conditions and/or pathological conditions must be exhibited thoroughly to avoid the devastating side effects of the potential therapy.

The only inducible isoform iNOS is not readily expressed inside the cells. This isoform is mostly expressed in macrophages when there is a pathogenic condition for host defence. There are some differences between constitutive and inducible forms. iNOSs are not Ca^{2+} dependent, also when induced they produce NO continuously, they are affected by glucocorticoids and less labile. The action of iNOS enzyme starts with binding of cytokines and/or lipopolysaccharides to cell surface receptors instead of calcium influx. Then the similar reaction occurs to produce NO. For stopping the activity, glucocorticoids bind the secondary messengers of cytokine-triggered cascade [18]. The stimuli activating the mice iNOS are not activating the human iNOS [19]. This situation should mislead the researchers while developing a therapeutic strategy. It is more serious if the strategy will be depending on only the animal experiment results.

2. Roles of nitric oxide synthase enzymes in healthy brain

It is important to put forth the molecular mechanism of a molecule for normal conditions. Under this heading, it was aimed to introduce and discuss the roles of NOS enzymes in the healthy brain functions. NOS enzymes have several roles in the healthy brain; especially NO is crucial for learning and memory formation. Besides, NOS enzymes have roles in retinal function, hearing and molecular mechanisms in the cochlea.

2.1. Roles of NOSs in learning and memory formation

One of the main duties of NOS enzymes in the healthy brain is learning and memory formation. Learning is a complex behaviour but the molecular mechanisms of memory formation are mostly enlightened. During the second half of the 1990s, scientists proposed that NO has a role in learning and memory and in neuronal plasticity. Experiments were designed to look for the role of NO/NOSs in learning. NO has a part in both long-term potentiations (LTP) in the hippocampus and long-term depression (LTD) in cerebellum, which are basic mechanisms for memory formation [20].

Neuron-neuron interaction isles called synapses are very plastic; LTP is a form of that plasticity. LTP occur due to the repetitive stimulation of the presynaptic neuron. A consequence of that repetitive stimulation is the influx of calcium to post-synaptic neuron via NMDA receptor, which in turn increases the Ca^{2+} concentration in the cell. For completing the circuit of LTP, there should be a retrograde messenger. Studies done with knockout mice and inhibitors, revealed that NO is that retrograde messenger for LTP in the hippocampus. But both nNOS and eNOS take place to from LTP [21]. Selective inhibitors for nNOS also showed that there is a deficiency in memory formation in various vertebrates [22]. Not only NO/NOS pathway studies showed the initiation of NO in LTP, but also studies looking for the sGC activity in rat hippocampus support the findings of NO/NOS and LTP correlation [23]. LTD occurs in cerebellar purkinje cells. There are two cells docking each other, climbing fibres and purkinje cells. Also, interneurons are neighbouring that synapse. NO released from interneurons and diffuse to the purkinje cells, then sGC initiates in LTD formation with similar action of LTP but in opposite direction [24, 25].

For memory formation by NOS there is also other signalling molecules in the cascade, like extracellular signal-regulated kinase (ERK). Thus, memory formation processes become more complex [26]. In the stress-exposed rats' hippocampi nNOS activity was diminished. This stress causes a deficit in learning and memory processes in these stressed animals. Same learning and memory problems have seen in hypoxic/ischaemic hippocampi of rats [27].

Within memory-forming circuits between neurons, NO can act as a volume transmitter. Thus, that small molecule can affect the remote parts of the brain at the same time. Also for conceptual learning studies done with invertebrates, it is revealed that indirectly NO/NOS mechanisms take place in learning [28]. It was shown that NOS inhibitors hinder motor learning in adult animals, and the formation of olfactory memories. The motor learning malfunction may arise from the non-selective inhibitors [29, 30]. These are some results explaining the roles of NO/NOS pathway in memory formation.

2.2. Roles of NOSs in seeing and retinal function

Retinal function is very important for developed organisms, and retinal function has various molecular pathways. Also, NO/NOS mechanism is one of the regulators of action of seeing and functioning of the retina. Seeing is a complex process that takes place in visual cortex but the retinal function is essential to carry the visual signals to that processor area of the brain.

In the retina, NOS is found in retinal neurons, pigment epithelium, amacrine and ganglion cells, nerve fibres in the outer and inner plexiform layers and in photoreceptor ellipsoids. But in normal rats' optic nerve, there is not any NOS enzyme present [31]. In photoreceptor cells, ion channels activate the reaction. In the inner segment, Ca^{2+} concentration increases, then activate the nNOS and thus NO produced. That NO activates the sGC in the same cell not inside the other cells. Then GC pathway will be triggered in on-bipolar cells. Besides in amacrine cell-ganglion cell-bipolar cell synapse; NO produced in amacrine cell, diffuse to the ganglion cell and activate the GC and this cause depolarization [31, 32]. Depolarization in the retina is an essential step for seeing.

There are some studies ranging from developmental biology to animal behaviour that explored the roles of NO/NOS pathway in the retinal function. In the developing human foetal eye nNOS and eNOS are expressed at the same time but in different compartments. Besides, nuclear nNOS was found in progenitor cells, endothelial cells and pericytes. Because of this situation, nNOS may have a transcription regulatory role for some cells during ocular vasculogenesis and angiogenesis [33]. In a study an inhibitor was used which selectively triggered photoreceptor cell death to determine the role of NO in retinal degeneration.

Scientists found that there is a correlation between the increasing NO levels and nNOS activity and mouse retinal cell death [34]. Chick retinas removed during a study revealed the role of NO/NOS mechanism in the brain. The group looked for the expression of NOSs in visual structures of the brain. The NOSs expressions increased after retinal removal. This shows that the NO/NOS mechanism has a role in plasticity processes in visual parts of the chick brain [35].

To unveil the role of NO in optic nerve head blood flow, scientists conducted a NOS inhibitor study in healthy humans. Subjects received L-NMMA and performed isometric exercise during the study. They proposed that NO has an important role in basal optic nerve head blood flow but not in autoregulatory response induced by exercise [36].

In cultured retinal neurons, NO inhibit apoptosis via activating various kinases. In cultured chick embryonic retinal neurons, both endogenous and exogenous NO promoted AKT signalling pathway and probably survival mechanisms [37]. In goldfish optic nerve, scientists showed that NO signalling pathway through nNOS activation has a crucial role in nerve regeneration [38].

In a knockout mice study, roles of all three isoforms were investigated comparatively. It was shown that not having one of the three isoforms did not alter the intraocular pressure or number of neurons in the eye. But eNOS is crucial for endothelium-dependent dilation of murine eye arteries. In conclusion, NOSs are involved in the regulation of ocular vascular tone and blood flow [39].

2.3. Roles of NOSs in hearing and cochlea

Hearing starts at the outer ear and stimuli travels through tympanum and bones and finally arrives at the cochlea, where hair cells and nerve fibres take action. NO and NOSs have roles in hearing function and cochlear activities. And hearing process happens in the auditory cortex.

NO act as a neurotransmitter and/or neuromodulator in the cochlea [40, 41]. In recent years, it was revealed that NO has also a potassium channel modulator role in inner hair cells. Therefore, NO-potassium modulation may be responsible for high-frequency hearing impairment [42].

NO/cGMP pathway was triggered by nerve fibres innervating outer hair cells, NO was produced in these cells and released. But NO affects Deiters' cells and Heusen's cells and not the outer hair cells. Also, nNOS takes place in acoustic overstimulation condition. Inner hair cells release excess glutamate during continuous stimulation. This glutamate increase calcium influx to afferent dendrites where nNOS produce NO. Then overproduction of NO due to acoustic overstimulation kills afferent dendrites because of excitotoxicity [43]. Auditory nerve, lateral wall, vestibule and cochlear neuroepithelium are the areas where NOS activity is the highest in the auditory system. nNOS is the predominant isoform in the cochlea [44].

In a gerbil study, researchers examined the role of NO in cochlelar excitotoxicity. Cochlear compound action potentials thresholds were recorded with NOS inhibitor and glutamate exposed conditions. Overstimulation with glutamate caused NO-mediated excitotoxicity in the cochlea. NOS inhibition should be neuroprotective for cochlea [45].

http://dx.doi.org/10.5772/67267

To trigger iNOS expression in cochlea, bacterial lipopolysaccharides and tumour necrosis factor α was injected to guinea pigs. The iNOS expression was higher than eNOS and nNOS during that experiment. iNOS were localized in the cochlea's blood vessel walls of the spiral ligament and the modiolus, in the organ of Corti, in the limbus, in nerve fibres and in spiral ganglion [46]. This dispersed iNOS is caused by the bacterial lipopolysaccharide-triggered immune response. In another study with immonostaining data, the distribution of NOSs was determined in the auditory system. nNOS was dispersed in the inner and outer hair cells, spiral ganglion cells, cells of the stria vascularis, spiral ligament cells and vessel cells near the modiolus. eNOS was dispersed in vascular endothelial cells, and in spiral ganglion cells. If there were not immune stimulus there would be no iNOS in cochlea [47].

3. Roles of nitric oxide synthase in unhealthy brain

NOS enzymes have important duties during pathophysiology of unhealthy brains. Unlike healthy brains, the roles not only depend on signal transduction, but also on anabolic/catabolic mechanisms. Under that heading various diseases, pathologies, malfunctions and disabilities of brains will be evaluated from the point of view of NOSs.

3.1. Roles of NOSs in neuropsychiatric diseases

It is very hard to diagnose the neuropsychiatric diseases properly; however, there are globally accepted parameters. Although the criteria for diagnosis is generally evaluated at certain times by prestigious committees and there is still hardships for diagnosis. One of the main goals of the scientists working on neuropsychiatric disorders is to pair a marker molecule with a disease to facilitate the diagnosis. So far there is not any coupling for any NOSs for any neuropsychiatric disorder as a marker but NOSs have various roles for these diseases. Anxiety, depression/major depression and tendency for suicide are important and common neuropsychiatric disorders. NOS has roles for these abnormalities.

In patients with depression it was shown that NO expression altered via eNOS. Besides NO modulates neuropeptides, such as vasopressin, oxytocin and corticotrophin-releasing factor. In patients with depression, these neuropeptides' expression levels were altered. According to post-mortem studies on depression patients, NO signalling was impaired in their hypothalami [48].

Major depression disorder (MDD) will be one of the dominant causes of disability by the year 2020. Antidepressants generally decrease NO levels and/or inhibit NOS activity indirectly. Also NO levels and NOS expressions are higher in MDD patients. Besides, with mice studies, it was shown that there is a correlation between NOS mechanism and depression. NOS inhibitors could be researched as a new target for antidepressant strategies [4, 22].

In a population study done in Taiwan with MDD patients in which the potential genetic variations with healthy and MDD subjects according to their nNOS polymorphisms was researched. There is no difference between subjects, the frequencies are similar for controls and MDD patients [49]. In an autopsy and tissue bank-based study from Holland, there is decrease in nNOS expression in the anterior cingulate cortex of MDD-diagnosed people [50].

In mice along with stress-induced depression, nNOS expression increases in the hippocampi. Due to excitotoxicity neurogenesis in hippocampi is suppressed. To inhibit nNOS signalling may be a novel approach for depression treatment [51]. Also, iNOS is involved in stress-triggered depression. NO derived from iNOS and mRNA levels of iNOS increased in cortices of depression model applied mice [52].

Also in a population study there is no correlation between genetic polymorphisms and MDD. In Japanese population MDD patients were investigated for polymorphisms in their nNOS genes, but found no variation between controls and MDD [53]. From a Czech Republic population genetic study, there is also no correlation between eNOS and MDD [54]. A population study from United Kingdom found a correlation between single nucleotide polymorphisms in nNOS gene and psychosocial stress-triggered depression. The individuals carrying those polymorphisms have a tendency to develop depression if they face financial hardship [55].

There is a link between vascular problems, depressive behaviours and NO metabolism. When vascular dysfunction emerges after depressive symptoms, the characteristics behind it show lack of bioavailable NO. However, H_2O_2 covers up that deficit [56].

There are studies on ethnopharmacological level to find out if there is a potential drug candidate in botanical material. To investigate NO metabolism is one of the target pathways to detect for antidepressant-like and neuroprotective potential. *Aloysia grattissima* has the potential to treat depressive disorders depending on the NO metabolism manipulating properties of its extracts [57].

nNOS genes' functional promoter repeat length variant contains sites for transcription factors that has strong relation with hyperactive and aggressive behaviour. Thus, nNOS depending on population genetics studies combined with behaviour is a potential research area [58].

For anxiety-like and depression-like behaviours, there is a strong evidential pathway, hypothalamic-pituitary-adrenal axis (HPA). Also on the ecotoxicological aspect, the nutrients for newborns and expectant mothers are very important because they fall in the risk group. The bisphenol-A supplied pregnant female rats' male littermates revealed anxiety-like and depression-like behaviours according to their HPA experiment results. Those littermates' hippocampal nNOS activity was higher than the control animals [59]. nNOS knockout mice show abnormal social behaviour, hyperactivity and impaired remote spatial memory [60].

It is important to demonstrate how nitric oxide synthases are affected in the brain by psychotropic drugs. Orally treated rats with several psychotropics were sacrificed and iNOS gene expressions in the brain were detected. Psychotropics including antidepressants and anxiolytics modulate the gene expression of iNOS in rat brain [61].

It is a complex and controversial psychological situation: suicide. This behaviour has a strong genetic background. In a study it was shown that a single nucleotide polymorphism of nNOS gene has a correlation between suicides in Japanese population, especially in males [62].

Schizophrenia is a complex illness including biochemical, anatomical and genetic aspects of its pathology. In a post-mortem study, scientists showed that some polymorphic variants of nNOS have overexpression patterns in schizophrenic patients' brains [63].

3.2. Roles of NOSs in neurodegenerative diseases

Alois Alzheimer defined the illness during the early 1900s. Alzheimer's disease (AD) is the most common type of dementia. It generally affects the elderly people and is characterized by aggregating senile plaques and/or neurofibrillary tangles in the brain, which leads to progressive neuronal degeneration and death [64]. Deficits in short-term memory formation are characteristic for AD. Short-term memory formation through LTP is dependent on the NO/NOS pathway. NOSs are very important for creating the new trails between neurons. NOSs take place for activating the presynaptic neurons receptors. However, in the pathology of the AD the harmonization created by NOSs between neurons is blocked by plaques [20–22, 64].

In the brain of an AD patient, β-amyloid peptide aggregates in senile plaques and the arginine within the astrocytes accumulates. These are the classical neuropathology of the disease. Arginine-metabolizing enzymes like NOSs and their association with amyloid peptides are important. The correlation of Aβ-peptide fragments with nNOS has been searched with spectrofluorimetry. The interaction of Aβ-peptide with nNOS causes the molecular movement of two critical tryptophan residues in the structure of the enzyme [65].

Purified nNOS was incubated with Aβ-peptide fragments during 96 hours. The kinetics of the interaction was introduced; nNOS was the amyloidogenic catalyst and all Aβ-peptide fragments were inhibited nNOS [66]. Data from cell culture studies, knockout mice studies and behavioural studies showed that eNOS has a crucial role for decelerating the pathology of AD [67]. If patients with AD start to exercise, they start to increase heart rate, cerebral blood flow and then angiogenesis, which includes NO/eNOS pathway, after which neurogenesis and other self-healing mechanisms are activated [68].

James Parkinson described the disease during the 1810s. Besides the characteristic shakes and tremors, there is a huge molecular mechanism behind Parkinson's disease (PD). The disease is diagnosed generally between 50 and 70 years old people. The mechanism behind PD is still unknown. But the dopamine metabolism decreases significantly in PD; also substantia nigra is one of the potential areas of interest to study the disease.

PD is an illness affecting the dopaminergic pathway in the brain. NO inflict the injuries in dopaminergic neurons. There are several evidences from NOS inhibitor studies about this correlation. However, if a cell goes for the cell death pathway, NO accelerates the process [69]. Some neuroinflammatory responses associated with PD are arousing from NO/iNOS activity. Nitration of α-synuclein triggers the protein aggregation, which worsens the pathology. This mechanism is used to mimic the PD in cell culture [70].

Scientists are trying to create a thorough model of the disease for *in vivo* or *in vitro*, however, so far there is not a completely satisfying model. That is caused by the unknown mechanism behind the illness. To mimic PD a group of researchers castrated the male mice. They followed the NO/iNOS mechanisms to test their model. According to the iNOS results, the castration of young male mice induces PD pathologies [71].

During PD pathology, several reasons cause cell death in substantia nigra neurons and/or dopaminergic neurons. Nutrition and false diet should be a cause of the disease. As a strategy to add an antioxidant-rich nutritive to the diet may be beneficial. Rats were supplied with pomegranate juice after a PD model. The change in the diet by adding pomegranate juice enhanced the iNOS expression in the animal brain [72].

3.3. Roles of NOSs in brain cancers

There are lots of cancer types present in the head and neck area. But in this section, only cancers originating from cells inside the cranium will be discussed and not the metastatic pathologies.

There are lots of studies done with cultured cancer cells from various mammals; however, studies done with tumour samples from human are rare. Instead of discussing the cell culture studies, it was important to gather the knowledge, which was hard to reach. Most of cancer cell culture studies are done without the healthy control cell lines or experimental models. But some cancer cell culture studies will be discussed.

A biopsy study done with gliomas, and also with meningiomas, showed that all three NOS isoforms were present in the aforementioned tumours. nNOS was significantly dominant in glial cells of gliomas. However, that NOS dense status becomes sparse in the peritumoral tissues [73]. NOSs of tumour cells, opposite to the healthy cells, synthesize predominantly superoxide and peroxynitrite, which generate oxidative stress [74].

Neuroblastoma cell lines are generally used due to their ability to differentiate neuron-like cells. NOS inhibition in Neuro2a cells blocked that differentiation; it is possible to speculate that nNOS may have important roles for dissolving a neuroblastoma tumour [75].

Astrocytomas/gliomas are the origins of cancers from supporting cells of the nervous system. These types of cancers are dominant and dangerous when compared to other brain cancers. In a human biopsy study, it was shown that iNOS has a role in angiogenesis via vascular endothelial growth factor (VEGF), and there is a correlation with iNOS and VEGF for astrocytomas/gliomas but not for reactive astrogliosis samples [76]. Similar results were reported also from a similar study. iNOS expression was increased in grade I, II and III astrocytic gliomas. However, in the same study it was shown that the iNOS expression was decreased for grade IV astrocytic gliomas [77]. In a study with primary astrocytoma biopsy samples prove that eNOS and VEGF work cooperatively in tumour angiogenesis [78]. Besides, in another study, it was shown that astrocytic tumour vessels have more eNOS expression than normal vessels [79]. Also, it is known that nNOS expression increases in glioma tumours.

Craniopharyngiomas consist of the 2–5% of intracranial tumours. In a study done on rats searched for the immune responses of oily cyst content of human craniopharyngioma. Immunohistochemical studies after injection revealed that eNOS expression increased in a time course manner [80].

Medulloblastomas are highly malignant brain tumours generally affecting children and adolescents. In a study done with medulloblastoma cells revealed that NOS has important roles in medulloblastoma cell death. Scientists applied various chemotherapy agents and PDE5 inhibitors to kill cells. Then, they co-treated cells with L-NAME and found out that NOS inhibition accompanying PDE5 inhibitors suppressed cell killing. Most probably NO/NOS has a role in killing of medulloblastoma cells [81]. A study done in knockout mice exhibited that iNOS has an important role in medulloblastoma formation. *Ptch1* heterozygous and iNOS-deficient mice developed medulloblastoma two times higher than *Ptch1* heterozygous and iNOS producing mice. This situation may depend on the granule cell precursors' migration [82].

According to studies done with human ptiutary tumour biopsy samples, scientists tried to reveal the roles of NOSs with the disease. In human pituitary adenomas, eNOS expression increased, whereas iNOS and nNOS were stabile [83]. Another human biopsy study showed that highly invasive adenomas have higher upregulated iNOS, whereas noninvasive adenomas did not have upregulated iNOS. Also, eNOS had upregulation with haemorrhagic adenomas [84].

Schwannomas are benign tumours originating from the Schwann cells. It can arise in any peripheral nerve; however, the frequent version arose around the acoustic nerve. Immunohistochemical study done on human biopsy samples revealed that iNOS has a strong expression for this illness. iNOS was stained around the hyalinized vessels' infiltrating leukocytes in Antoni B areas [85].

In conclusion, it is possible to suggest that both clinical and experimental studies are important on the aspect of NOS and brain cancers. It is very hard to mimic pathologies of some brain cancers both in animals and *in vitro*. Likewise, collecting clinical samples from patients is very hard. NO/NOS pathway is important for brain cancers and more studies needed to reveal the therapeutic potential.

Author details

Melih Dagdeviren

Address all correspondence to: melih.dagdeviren@ege.edu.tr

Department of Biology, Faculty of Science, Ege University, Izmir, Turkey

References

[1] Maines MD. The heme oxygenase system: a regulator of second messenger gases. Annu Rev Paharmacol Toxicol. 1997; 37:517–54. DOI: 10.1146/annurev.pharmtox.37.1.517

[2] Zetterstrom R. The 1998 Nobel Prize-discovery of the role of nitric oxide as a signalling molecule. Acta Paediatr. 2009 Mar; 98(3):593–9. DOI: 10.1111/j.1651–2227.2008.01121.x

[3] Fedele E, Raiteri M. In vivo studies of the cerebral glutamate receptor/NO/cGMP pathway. Prog. Neurobiol. 1999 May; 58(1):89–120.

[4] Dagdeviren M, Dogan YH, Kanit L. Effects of restraint stress and nitric oxide synthase inhibition on learning and strategy preference in young adult male rats. Balkan Med J. 2012; 29: 376–80. DOI: 10.5152/balkanmedj.2012.100

[5] Forstermann U, Sessa WC. Nitric oxide synthases: regulation and function. Eur Heart J. 2012 Apr; 33(7):829–37, 837a-837d. DOI: 10.1093/eurheartj/ehr304.

[6] Andrew PJ, Mayer B. Enzymatic function of nitric oxide synthases. Cardiovasc Res. 1999 Aug 15; 43(3):521–31.

[7] Heiss C, Rodriguez-Mateos A, Kelm M. Central Role of eNOS in the maintenance of endothelial homeostasis. Antioxid Redox Signal. 2015 May 10; 22(14):1230–42. DOI: 10.1089/ars.2014.6158

[8] Steinert JR, Chernova T, Forsythe ID. Nitric oxide signalling in brain function, dysfunction, and dementia. Neuroscientist. 2010 Aug; 16(4):435–52. DOI: 10.1177/107385841036 6481

[9] Forstermann U, Boissel JP, Kleinert H. Expressional control of the constitutive isoforms of nitric oxide synthase (NOS I and NOS III). FASEB J. 1998 Jul; 12(10):773–90.

[10] Szabo C. Physiological and pathophysiological roles of nitric oxide in the central nervous system. Brain Res Bull. 1996; 41(3):131–41.

[11] Dzoljic E, Grbatinic I, Kostic V. Why is nitric oxide important for our brain? Funct Neurol. 2015 Jul-Sep; 30(3):159–63.

[12] Prast H, Philippu A. Nitric oxide as modulator of neuronal function. Prog Neurobiol. 2001 May; 64(1):51–68.

[13] Pitsikas N. The role of nitric oxide in the object recognition memory. Behav Brain Res. 2015 May 15; 285:200–7. DOI: 10.1016/j.bbr.2014.06.008.

[14] Pogun S. Sex differences in brain and behaviour: emphasis on nicotine, nitric oxide and place learning. Int J Psychophysiol. 2001 Oct; 42(2):195–208.

[15] Kanit L, Koylu EO, Yararbas G, Furedy JJ, Pogun S. The effect of nitric oxide synthase inhibition on cognitive ability and strategies employed for place learning in the water maze: sex differences. Brain Res Bull. 2003 Dec 15; 62(2):151–9.

[16] Iadecola C, Pelligrino DA, Moskowitz MA, Lassen NA. Nitric oxide synthase inhibition and cerebrovascular regulation. J Cereb Blood Flow Metab. 1994 Mar; 14(2):175–92.

[17] Charriaut-Marlangue C, Bonnin P, Pham H, Loron G, Leger PL, Gressens P, Renolleau S, Baud O. Nitric oxide signalling in the brain: a new target for inhaled nitric oxide? Ann Neurol. 2013 Apr; 73(4):442–8. DOI: 10.1002/ana.23842

[18] Garcia X, Stein F. Nitric oxide. Semin Pediatr Infect Dis. 2006 Apr; 17(2):55–7. DOI: 10.1053/j.spid.2006.04.002

[19] Bronte V, Zanovello P. Regulation of immune responses by L-arginine metabolism. Nat Rev Immunol. 2005 Aug; 5(8):641–54. DOI: 10.1038/nri1668

[20] Hawkins RD. NO honey, I don't remember. Neuron. 1996 Mar; 16(3):465–7.

[21] Huang EP. Synaptic plasticity: a role for nitric oxide in LTP. Curr Biol. 1997 Mar 1; 7(3):R141–3.

[22] Zhou L, Zhu DY. Neuronal nitric oxide synthase: structure, subcellular localization, regulation, and clinical implications. Nitric Oxide. 2009 Jun; 20(4):223–30. DOI: 10.1016/j.niox.2009.03.001

[23] Hawkins RD, Son H, Arancio O. Nitric oxide as a retrograde messenger during long-term potentiation in hippocampus. In: Mize RR, Dawson TM, Dawson VL, Friedlander MJ Eds. Progress in Brain Research. Elsevier. 1998; 118:155–72.

[24] Oh S. The generation of nitric oxide and its roles in neurotransmission and neurotoxicity. Keio J Med. 1995 Jun; 44(2):53–61.

[25] Wang DJ, Su LD, Wang YN, Yang D, Sun CL, Zhou L, Wang XX, Shen Y. Long-term potentiation at cerebellar parallel fiber–purkinje cell synapses requires presynaptic and postsynaptic signaling cascades. J Neurosci. 2014 Feb 5; 34(6):2355–64. DOI: 10.1523/JNEUROSCI.4064–13.2014

[26] Kishida KT, Klann E. Sources and targets of reactive oxygen species in synaptic plasticity and memory. Antioxid Redox Signal. 2007 Feb; 9(2):233–44. DOI: 10.1089/ars.2007.9.ft-8

[27] Paul V, Ekambaram P. Involvement of nitric oxide in learning and memory processes. Indian J Med Res. 2011 May; 133:471–8.

[28] Moroz LL, Kohn AB. Parallel evolution of nitric oxide signalling: diversity of synthesis and memory pathways. Front Biosci (Landmark Ed). 2011 Jun 1; 16:2008–51.

[29] Bredt DS. Endogenous nitric oxide synthesis: Biological functions and pathophysiology. Free Radic Res. 1999 Dec; 31(6):577–96.

[30] Contestabile A, Monti B, Contestabile A, Ciani E. Brain nitric oxide and its dual role in neurodegeneration/neuroprotection: understanding molecular mechanisms to devise drug approaches. Curr Med Chem. 2003 Oct; 10(20):2147–74.

[31] Goldstein IM, Ostwald P, Roth S. Nitric oxide: A review of its role in retinal function and disease. Vision Res. 1996 Sep; 36(18):2979–94.

[32] Blom J, Giove T, Deshpande M, Eldred WD. Characterization of nitric oxide signalling pathways in the mouse retina. J Comp Neurol. 2012 Dec 15; 520(18):4204–17. DOI: 10.1002/cne.23148

[33] McLeod DS, Baba T, Bhutto IA, Lutty G. Co-expression of endothelial and neuronal nitric oxide synthases in the developing vasculatures of the human fetal eye. Graefes Arch Clin Exp Ophthalmol. 2012 Jun; 250(6):839–48. DOI: 10.1007/s00417-012-1969-9

[34] Koriyama Y, Hisano S, Ogai K, Sugitani K, Furukawa A, Kato S. Involvement of neuronal nitric oxide synthase in N-methyl-N-nitrosurea-induced retinal degeneration in mice. J Pharmacol Sci. 2015 Mar; 127(3):394–6. DOI: 10.1016/j.jphs.2015.02.008

[35] Torrao AS, Britto LRG. Increased expression of nitric oxide synthase in visual structures of the chick brain after retinal removal. J Neurosci Res. 2004 Oct 1; 78(1):123–31. DOI: 10.1002/jnr.20238

[36] Schmidl D, Boltz A, Kaya S, Lasta M, Pemp B, Fuchsjager-Mayrl G, et al. Role of nitric oxide in optic nerve head blood flow regulation during isometric exercise in healthy humans. Invest Ophthalmol Vis Sci. 2013 Mar 15; 54(3):1964–70. DOI: 10.1167/iovs.12-11406

[37] Mejia-Garcia T, Portugal CC, Encarnacao TG, Prado MAM, Paes-de-Carvalho R. Nitric oxide regulates AKT phosphorylation and nuclear translocation in cultured retinal cells. Cell Signal. 2013 Dec; 25(12):2424–39. DOI: 10.1016/j.cellsig.2013.08.001

[38] Koriyama Y, Yasuda R, Homma K, Mawatari K, Nagashima M, Sugitani K, et al. Nitric oxide-cGMP signalling regulates axonal elongation during optic nerve regeneration in the goldfish in vitro and in vivo. J Neurochem. 2009 Aug; 110(3):890–901. DOI: 10.1111/j.1471-4159.2009.06182.x

[39] Laspas P, Goloborodko E, Sniatecki JJ, Kordasz ML, Manicam C, Wojnowski L, et al. Role of nitric oxide synthase isoforms for ophthalmic artery reactivity in mice. Exp Eye Res. 2014 Oct; 127:1–8. DOI: 10.1016/j.exer.2014.06.018

[40] Puel JL. Chemical synaptic transmission in the cochlea. Prog Neurobiol. 1995 Dec; 47 (6):449–76.

[41] Zdanski CJ, Prazma J, Petrusz P, Grossman G, Raynor E, Smith TL, et al. Nitric oxide synthase is an active enzyme in the spiral ganglion cells of the rat cochlea. Hear Res. 1994 Sep; 79(1–2):39–47.

[42] Kimitsuki T. Nitric oxide influences potassium currents in inner hair cells isolated from guinea-pig cochlea. Auris Nasus Larynx. 2015 Oct; 42(5):360–4. DOI: 10.1016/j.anl.2015.02.011

[43] Fessenden JD, Schacht J. The nitric oxide/cyclic GMP pathway: a potential major regulator of cochlear physiology. Hear Res. 1998 Apr; 118(1–2):168–76.

[44] Fessenden JD, Coling DE, Schacht J. Detection and characterization of nitric oxide synthase in the mammalian cochlea. Brain Res. 1994 Dec 30; 668(1–2):9–15.

[45] Patel MR, Stamat JC, Zdanski CJ, Ebert CS, Prazma J. Nitric oxide in glutamate-induced compound action potential threshold shifts. Hear Res. 2008 May; 239(1–2):54–9. DOI: 10.1016/j.heares.2008.01.007

[46] Hess A, Bloch W, Huverstuhl J, Su J, Stennert E, Addicks K, et al. Expression of inducible nitric oxide synthase (iNOS/NOS II) in the cochlea of guinea pigs after intratympanical endotoxin-treatment. Brain Res. 1999 May 29; 830(1):113–22.

[47] Gosepath K, Gath I, Mauer J, Pollock JS, Amedee R, Forstermann U, et al. Characterization of nitric oxide synthase isoforms expressed in different structures of the guinea pig cochlea. Brain Res. 1997 Jan 30; 747(1):26–33.

[48] Luciano M, Houlihan LM, Harris SE, Gow AJ, Hayward C, Starr JM, et al. Association of existing and new candidate genes for anxiety, depression and personality traits in older people. Behav Genet. 2010 Jul; 40(4):518–32. DOI: 10.1007/s10519-009-9326-4

[49] Yu YW, Chen TJ, Wang YC, Liou YJ, Hong CJ, Tsai SJ. Association analysis for neuronal nitric oxide synthase gene polymorphism with major depression and fluoxetine response. Neuropsychobiology. 2003; 47(3):137–40.

[50] Gao SF, Qi XR, Zhao J, Balesar R, Bao AM, Swaab DF. Decreased NOS1 expression in the anterior cingulate cortex in depression. Cereb Cortex. 2013 Dec; 23(12):2956–64. DOI: 10.1093/cercor/bhs285

[51] Zhou QG, Hu Y, Hua Y, Hu M, Luo CX, Han X, et al. Neuronal nitric oxide synthase contributes to chronic stress-induced depression by suppressing hippocampal neurogenesis. J Neurochem. 2007 Dec; 103(5):1843–54.

[52] Peng YL, Liu YN, Liu L, Wang X, Jiang CL, Wang YX. Inducible nitric oxide synthase is involved in the modulation of depressive behaviors induced by unpredictable chronic mild stress. J Neuroinflammation. 2012 Jul 6; 9:75. DOI: 10.1186/1742-2094-9-75

[53] Okumura T, Kishi T, Okochi T, Ikeda M, Kitajima T, Yamanouchi Y, et al. Genetic association analysis of functional polymorphisms in neuronal nitric oxide synthase 1 gene (NOS1) and mood disorders and fluvoxamine response in major depressive disorder in the Japanese population. Neuropsychobiology. 2010; 61(2):57–63. DOI: 10.1159/000265130

[54] Zeman M, Jachymova M, Jirak R, Vecka M, Tvrzicka E, Stankova B, et al. Polymorphims of genes for brain-derived neurotrophic factor, methylenetetrahydrofolate reductase, tyrosine hydroxylase, and endothelial nitric oxide synthase in depression and metabolic syndrome. Folia Biol (Praha). 2010; 56(1):19–26.

[55] Sarginson JE, Deakin JFW, Anderson IM, Downey D, Thomas E, Elliot R, et al. Neuronal nitric oxide synthase (NOSI) polymorphisms interact with financial hardship to affect depression risk. Neuropsychopharmacology. 2014 Nov; 39(12):2857–66. doi: 10.1038/npp.2014.137

[56] d'Audiffret AC, Frisbee SJ, Stapleton PA, Goodwill AG, Isingrini E, Frisbee JC. Depressive behaviour and vascular dysfunction: a link between clinical depression and vascular disease? J Appl Physiol (1985). 2010 May; 108(5):1041–51. DOI: 10.1152/japplphysiol.01440.2009

[57] Zeni AL, Zomkowski AD, Dal-Cim T, Marachin M, Rodrigues AL, Tasca CI. Antidepressant-like and neuroprotective effects of Aloysia gratissima: investigation of involvement of L-arginine-nitric oxide-cyclic guanosine monophosphate pathway. J Ethnopharmacol. 2011 Sep 1; 137(1):864–74. DOI: 10.1016/j.jep.2011.07.009

[58] Craig IW, Halton KE. Genetics of human aggressive behaviour. Hum Genet. 2009 Jul; 126 (1):101–13. DOI: 10.1007/s00439-009-0695-9

[59] Chen F, Zhou L, Bai Y, Zhou R, Chen L. Hypothalamic-pituitary-adrenal axis hyperactivity accounts for anxiety- and depression-like behaviors in rats perinatally exposed to bisphenol A. J Biomed Res. 2015 May; 29(3):250–8. DOI: 10.7555/JBR.29.20140058

[60] Tanda K, Nishi A, Matsuo N, Nakanishi K, Yamasaki N, Sugimoto T, et al. Abnormal social behavior, hyperactivity, impaired remote spatial memory, and increased D1-mediated dopaminergic signaling in neuronal nitric oxide synthase knockout mice. Mol Brain. 2009 Jun 18; 2:19. DOI: 10.1186/1756-6606-2-19.

[61] Suzuki E, Nakaki T, Shintani F, Kanba S, Miyaoka H. Antipsychotic, antidepressant, anxiolytic, and anticonvulsant drugs induce type II nitric oxide synthase mRNA in rat brain. Neurosci Lett. 2002 Nov 29; 333(3):217-9.

[62] Cui H, Supriyanto I, Asano M, Ueno Y, Nagasaki Y, Nishiguchi N, et al. A common polymorphism in the 3′-UTR of the NOS1 gene was associated with completed suicides in Japanese male population. Prog Neuropsychopharmacol Biol Psychiatry. 2010 Aug 16; 34(6):992–6. DOI: 10.1016/j.pnpbp.2010.04.028

[63] Silberberg G, Ben-Shachar D, Navon R. Genetic analysis of nitric oxide synthase 1 variants in schizophrenia and bipolar disorder. Am J Med Genet B Neuropsychiatr Genet. 2010 Oct 5; 153B(7):1318–28. DOI: 10.1002/ajmg.b.31112

[64] Maccioni RB, Munoz JP, Barbeito L. The molecular bases of Alzheimer's disease and other neurodegenerative disorders. Arch Med Res. 2001 Sep–Oct; 32(5):367–81.

[65] Padayachee ER, Whiteley CG. Spectrofluorimetric analysis of the interaction of amyloid peptides with neuronal nitric oxide synthase: implications in Alzheimer's disease. Biochim Biophys Acta. 2011 Dec; 1810(12):1136–40. DOI: 10.1016/j.bbagen.2011.09.002

[66] Padayachee ER, Ngqwala N, Whiteley CG. Association of β-amyloid peptide fragments with neuronal nitric oxide synthase: implications in the etiology of Alzheimer's disease. J Enzyme Inhib Med Chem. 2012 Jun; 27(3):356–64. DOI: 10.3109/14756366.2011.590805

[67] Katusic ZS, Austin SA. Endothelial nitric oxide: protector of a healthy mind. Eur Heart J. 2014 Apr; 35(14):888–94. DOI: 10.1093/eurheartj/eht544

[68] Paillard T, Rolland Y, Barreto P. Protective effects of physical exercise in Alzheimer's disease and Parkinson's disease: a narrative review. J Clin Neurol. 2015 Jul; 11(3):212–9. DOI: 10.3988/jcn.2015.11.3.212

[69] Tripathy D, Chakraborty J, Mohanakumar KP. Antagonistic pleiotropic effects of nitric oxide in the pathophysiology of Parkinson's disease. Free Radic Res. 2015; 49(9):1129–39. DOI: 10.3109/10715762.2015.1045505

[70] Stone DK, Kiyota T, Mosley RL, Gendelman HE. A model of nitric oxide induced α-synuclein misfolding in Parkinson's disease. Neurosci Lett. 2012 Aug 15; 523(2):167–73. DOI: 10.1016/j.neulet.2012.06.070

[71] Khasnavis S, Ghosh A, Roy A, Pahan K. Castration induces Parkinson disease pathologies in young male mice via inducible nitric-oxide synthase. J Biol Chem. 2013 Jul 19; 288 (29):20843–55. DOI: 10.1074/jbc.M112.443556

[72] Tapias V, Cannon JR, Greenamyre JT. Pomegranate juice exacerbates oxidative stress and nigrostriatal degeneration in Parkinson's disease. Neurobiol Aging. 2014 May; 35 (5):1162–76. DOI: 10.1016/j.neurobiolaging.2013.10.077

[73] Bakshi A, Nag TC, Wadhwa S, Mahapatra AK, Sarkar C. The expression of nitric oxide synthase in human brain tumours and peritumoral areas. J Neurol Sci. 1998 Mar 5; 155 (2):196–203.

[74] Rebender CS, Alam A, Sundaresan G, Cardnell RJ, Yakovlev VA, Mukhopadhyay ND, et al. The role of nitric oxide synthase uncoupling in tumor progression. Mol Cancer Res. 2015 Jun; 13(6):1034–43. DOI: 10.1158/1541-7786.MCR-15-0057-T

[75] Evangelopoulos ME, Wuller S, Weis J, Kruttgen A. A role of nitric oxide in neurite outgrowth of neuroblastoma cells triggered by mevastatin or serum reduction. Neurosci Lett. 2010 Jan 1; 468(1):28–33. DOI: 10.1016/j.neulet.2009.10.054

[76] Hara A, Okayasu I. Cyclooxygenase-2 and inducible nitric oxide synthase expression in human astrocytic gliomas: correlation with angiogenesis and prognostic significance. Acta Neuropathol. 2004 Jul; 108(1):43–8. DOI: 10.1007/s00401-004-0860-0

[77] Giannopoulou E, Ravazoula P, Kalofonos H, Makatsoris TH, Kardamakis D. Expression of HIF-1α and iNOS in astrocytic gliomas: a clinicopathological study. In Vivo. 2006 May–Jun; 20(3):421–5.

[78] Pan JW, Zhan RY, Tong Y, Zhou YQ, Zhang M. Expression of endothelial nitric oxide synthase and vascular endothelial growth factor in association with neovascularization in human primary astrocytoma. J Zhejiang Univ Sci B. 2005 Jul; 6(7):693–8. DOI: 10.1631/jzus.2005.B0693

[79] Iwata S, Nakagawa K, Harada H, Oka Y, Kumon Y, Skaki S. Endothelial nitric oxide synthase expression in tumor vasculature is correlated with malignancy in human supratentorial astrocytic tumors. Neurosurgery. 1999 Jul; 45(1):24–8; discussion 29.

[80] Tena-Suck ML, Hernandez-Campos ME, Ortiz-Plata, Salinas-Lara C, Colin-Gonzalez AL, Santamaria A. Intracerebral injection of oil cyst content of human craniopharyngioma (oil machinery fluid) as a toxic model in the rat brain. Acta Histochem. 2014 Apr; 116(3):448–56. DOI: 10.1016/j.acthis.2013.10.002

[81] Roberts JL, Booth L, Conley A, Cruickshanks N, Malkin M, Kukreja RC, et al. PDE5 inhibitors enhance the lethality of standart of care chemotherapy in pedatrics CNS tumor cells. Cancer Biol Ther. 2014 Jun 1; 15(6):758–67. DOI: 10.4161/cbt.28553

[82] Haag D, Zipper P, Westrich V, Karra D, Pfleger K, Toedt G, et al. Nos2 inactivation promotes the development of medulloblastoma in Ptch1± mice by deregulation of

Gap43-dependent granule cell precursor migration. PLoS Genet. 2012; 8(3):e1002572. DOI: 10.1371/journal.pgen.1002572

[83] Kruse A, Broholm H, Rubin I, Schmidt K, Lauritzen M. Nitric oxide synthase activity in human pituitary adenomas. Acta Neurol Scand. 2002 Dec; 106(6):361–6.

[84] Onishi K, Kamida T, Momii Y, Abe T, Fujiki M. The clinical and pathological significance of nitric oxide synthase in human pituitary adenomas: a comparison with MIB-1. Endocrine. 2014 May; 46(1):154–9. DOI: 10.1007/s12020-013-0046-4

[85] Yokoo H, Oishi T, Isoda K, Nakazato Y, Toykuni S. Oxidative stress is related to the formation of Antoni B patterns and eosinophilic hyaline droplets in schwannomas. Neuropathology. 2007 Jun; 27(3):237–44.

Chapter 4

Role of Nitric Oxide Synthase in the Function of the Central Nervous System under Normal and Infectious Conditions

Patricia Alves Reis,
Cassiano Felippe Gonçalves de Albuquerque,
Tatiana Maron-Gutierrez, Adriana Ribeiro Silva and
Hugo Caire de Castro Faria Neto

Additional information is available at the end of the chapter

http://dx.doi.org/10.5772/67816

Abstract

Nitric oxide (NO) was discovered as an endothelium-derived relaxing factor more than two decades ago. Since then, it has been shown to participate in many pathways. NO has been described as a key mediator of different pathways in the central nervous system (CNS) in both healthy and diseased processes. The three isoforms of nitric oxide synthase differ in their activity patterns and expression in different cells. Neuronal nitric oxide synthase (nNOS) is localized in synaptic spines, astrocytes, and the loose connective tissue surrounding blood vessels in the brain; eNOS is present in both cerebral vascular endothelial cells and motor neurons; and iNOS is induced in astrocytes and microglia under pathological conditions. During physiological processes, NO produced by eNOS/nNOS, respectively, controls blood flow activation, and act as a messenger during long-term potentiation (LTP). However, under pathological conditions, eNOS appears to be impaired, leading to a reduction in blood flow and, consequently, low oxygen/metabolites delivery, efflux of toxicological agents from the brain tissue and disturbance in the blood-brain barrier. The NO produced by iNOS in glial cells and nNOS, which triggers the NMDA-excitotoxic pathway, combines with superoxide anion and results in peroxynitrite synthesis, a potent free radical that contributes to tissue damage in the brain. Here, we intend to show the controversial role of the nitric oxide delivered by the three isoforms of the nitric oxide synthase in the CNS, assess its impact under healthy/pathological conditions and speculate on its possible sequela, particularly in long-term cognitive decline.

Keywords: nitric oxide, nitric oxide synthase, central nervous system, memory, infection

1. Introduction

Three isoforms of nitric oxide synthase (NOS) have been identified: two constitutive enzymes, neuronal NOS (nNOS) and endothelial NOS (eNOS), and one inducible enzyme (iNOS). These three isoforms of the enzyme nitric oxide synthase (NOS) that are present in the central nervous system (CNS) can produce nitric oxide (NO). eNOS is expressed in the vascular endothelium and choroid plexus; neuronal NOS is mainly expressed in neuronal cell bodies, especially in the cortex, hippocampus, hypothalamus, olfactory bulb, claustrum, amygdala, and thalamus; and inducible NOS is expressed in macrophages, glial cells, infiltrating neutrophils, and, to some extent, neurons [1]. It has also been reported that eNOS can be found in a subset of neurons and astrocytes and that nNOS can be found at low levels in astrocytes [2]. Because these enzymes may have different sites of expression and activation, they have a pivotal role in the divergent functions of NO [3].

Several studies have demonstrated that NO, a freely diffusible gaseous compound, has an important role in a variety of neurobiological processes [4]. Numerous functions of this regulatory molecule have been identified in the CNS, in the process of endothelium-dependent vasodilatation [5–8], in neurotransmission [9, 10], and in host-defense mechanisms [11, 12].

NO is produced from the oxidation of the terminal guanidine nitrogen of the amino acid arginine. This reaction is catalyzed by the NADPH-dependent enzyme, nitric oxide synthase (NOS). After its formation, NO diffuses outside the cell [13]. NO derived from eNOS maintains the CNS microcirculation [14] by inhibiting platelet aggregation and leukocyte adhesion and migration [15]. NO derived from nNOS is an important neurotransmitter related to neuronal plasticity, memory formation, regulation of CNS blood flow, and neurotransmitter release [16, 17].

2. Implications of nitric oxide synthase in the physiological central nervous system

Under physiological conditions, the concentration of NO fluctuates within the range of low values [18] and is produced mainly by nNOS and eNOS. Unlike the other two enzymes, iNOS is not expressed unless it is induced by inflammatory mediators, cytokines, and other agents, such as endotoxins [19]. Due to its calcium-independent activation, iNOS can produce a large amount (100–1000 times greater) of NO in relation to eNOS and nNOS [20]. Until the enzyme is degraded, iNOS constitutively produces NO [21].

NO binds to guanylyl-cyclase, which is a soluble NO receptor and, through cGMP-mediated signaling, acts either as a post- or a presynaptic messenger [22]. As a neurotransmitter, NO may activate the cGMP-dependent protein kinase G (PKG) pathway that phosphorylates synaptophysin, which is critical for fusion of presynaptic vesicles, thereby potentiating and facilitating neurotransmission [22] (**Figure 1**). NO also acts on inhibitory gamma-aminobutyric acid (GABA)-ergic synaptic transmission via cGMP-dependent pathways as well as on ion channels and exchangers [9].

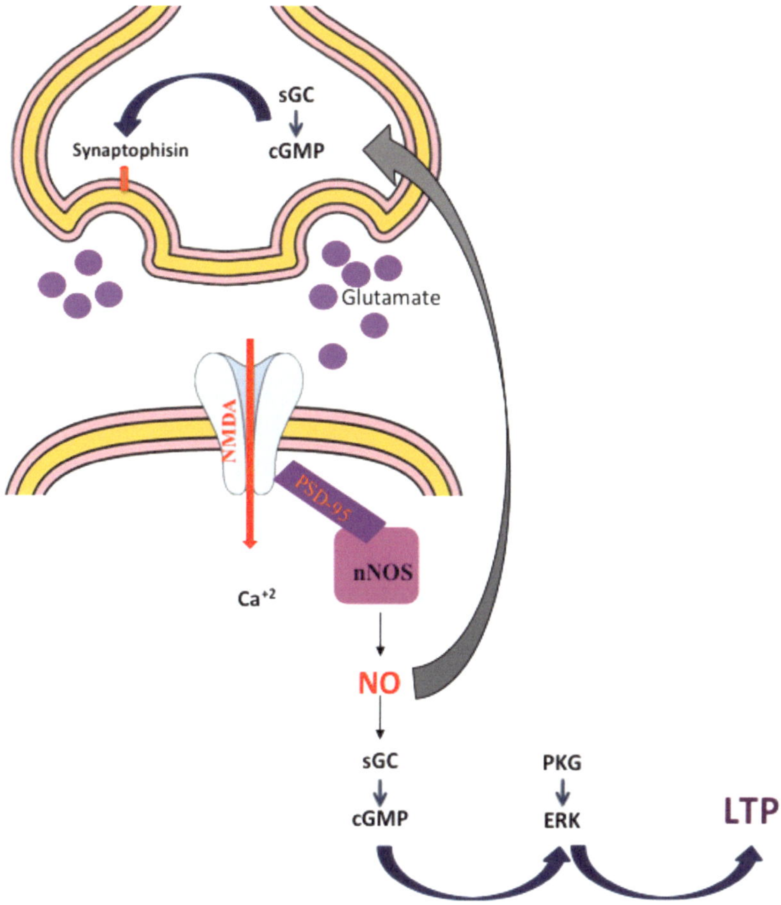

Figure 1. NO act as an unconventional neurotransmitter that is not stored in synaptic vesicles and not released upon membrane depolarization; it releases as soon it is synthesized and does not bind to any receptors, but diffuses from one neuron to another. NO stimulate soluble guanylyl-cyclase to form the second messenger molecule, cyclic guanosine monophosphate (cGMP) either as a post- or a presynaptic messenger. PKG, protein kinase G; ERK, extracellular signal-regulated kinases; LTP, long-term potentiation. Images: https://mindthegraph.com.

The brain relies on a constant and adequate supply of oxygen and glucose that is provided by blood. Cerebral blood flow is altered in response to both neural activity and humoral *stimuli* (e.g., arterial PO_2 and PCO_2). Thus, augmented neural activation results in locally increased cerebral blood flow via functional hyperemia, whereas humoral *stimuli* produce overall increase in cerebral blood flow [23]. The physiological production of endothelium-derived NO by eNOS is protective in hypoxic or ischemic brain injury.

All NOS isoforms have phosphorylation sites for different protein kinases, including protein kinase A, B, and C, and calcium-calmodulin kinase [24]. NOS enzymes are very important for the maintenance of physiological mechanisms within an organism, and the genetic ablation of NOS in mice has been informative for establishing the functional roles of NOS-generated NO in different systems [25]. For instance, nNOS-KO mice present with intense gastroparesis due to impaired vagal innervation of stomach muscle cells [26], decreased apoptosis induced by striatal N-methyl-D-aspartate (NMDA) microinjections [27], and early dysfunction of hippocampal-dependent spatial memory [28]. Additionally, disrupting the gene that encodes eNOS in mice-induced spontaneous systemic and pulmonary hypertension [29] and inhibited growth factor-mediated angiogenesis [30]. Thus, NO acts through numerous mechanisms of different physiological systems and in living cells.

Shortly after its identification, NO emerged as a possible mediator of neurovascular coupling. Neurovascular coupling is an active mechanism with vessel diameter alterations in response to increased metabolic demands from neuronal activity. Under these conditions, NO acts as a potent vasodilator that is released during enhanced neuronal activity and is well suited to mediate the coupling between neuronal activity and cerebral blood flow [31].

The importance of NO as an intermediary in cell communication in the brain is highlighted by the fact that the excitatory amino acid glutamate is an initiator of the reaction that forms NO. NO can act as a "double-edged sword". Whereas NO supports vascular homeostasis in the endothelium-dependent vasodilatation, its over- or underproduction is linked to pathological conditions [4].

3. Implications of nitric oxide synthase in the pathological central nervous system conditions

NO is also an important mediator under pathological conditions. For instance, in brain ischemia-reperfusion injury, NO formation is initially increased and has a protective function by inducing collateral perfusion as a result of its powerful stimulatory effect on vasodilatation and angiogenesis [32]. NO donors induce neuroprotective effects.

NO can exist in distinct oxidation/reduction states and present dual biological actions as either a neuroprotective or a neurotoxic molecule [4]. Under physiological conditions, nNOS produces hydrogen peroxide (H_2O_2) and superoxide ($O_2^{\bullet -}$) in addition to NO [33].

The downstream cascade in the breakdown of the BBB appears to be mediated by eNOS activity; the systemic administration of a selective eNOS inhibitor abrogates VEGF-A-induced BBB disruption and protects against neurologic damage in models of inflammatory disease [34, 35]. Nevertheless, it was suggested that eNOS-derived NO is a neuromodulator that participates in the BBB-mediated control of the cerebellum in experimental models [36].

Curiously, perivascular macrophage-derived iNOS-generated NO, that strategically localizes to leukocytes at brain penetration sites, can serve as a negative feedback regulator that prevents the unlimited influx of inflammatory cells by restoring BBB integrity [37].

In the brain, NOS regulates cerebral blood flow and neurotransmitter release, and the proper operative eNOS/NO system accounts for the microenvironment homeostasis that is essential for the normal functioning of neurons and glial cells [38]. Therefore, the NO produced by NOS results in vasodilation and controls vascular resistance, platelet adhesion and aggregation, leukocyte-to-endothelium interaction, and the maintenance of BBB integrity [36].

Several reports have indicated significant nervous system morbidity due to viral, bacterial, fungal, and parasitic infections (review [39]). During a systemic response to infections, cytokines, chemokines, and damage-associated soluble mediators of systemic inflammation can gain access to the CNS via blood flow [40]. These mediators can access the brain tissue after the disruption of the BBB. The presence of proinflammatory mediators leads to a disturbance of neuronal and glial homeostasis, with subsequent cognitive and behavioral manifestations that are common during acute infections (anorexia, malaise, depression, and decreased physical activity) and are collectively known as sickness behavior [40]. Although sickness behavior manifestations are transient and self-limited, the cognitive and behavioral changes can become permanent or long-lasting under a persistent systemic inflammatory response. For example, cognitive decline is a common consequence in sepsis and cerebral malaria survivors, as found in both clinical and experimental approaches.

Under healthy conditions, the main cell types that are present in the brain are neurons, oligodendrocytes, astrocytes, and microglia. Neurons connect to each other through long axonal processes with synapses to transmit electrical/chemical signals, thereby generating memory and emotions that are associated with learning, and to control organ and systemic functions. Oligodendrocytes support axons with myelin sheaths. Astrocytes interact with blood vessels to form the BBB and maintain neuronal synapses. Microglia form long processes to phagocyte apoptotic cells and prune inactive synapses without inducing inflammation while maintaining a type of surveillance of neurons. Systemic inflammatory conditions result in the disruption of the BBB and the efflux of proinflammatory cytokines/chemokines as well as pathogen-associated molecular patterns (PAMPs) and danger-associated molecular patterns (DAMPS), which together activate glial cells. This results in an inflammatory environment due to the release of cytokines/chemokines and reactive oxygen/nitrogen species that exert direct and indirect neuronal cytotoxic effects [40, 41]. Oligodendroglial myelin sheaths can be affected, thereby leading to axonal degeneration. Astrocytosis leads to reduced BBB and synaptic maintenance. Microgliosis results in a proinflammatory microglial phenotype with reduced tissue maintenance functions [41, 42]. Neuronal cells can develop an excitotoxicity process due to excessive glutamate in the synaptic cleft and subsequent extra-synaptic NMDA receptor activation that results in a subsequent increase of Ca^{2+} efflux in neuronal cells and the activation of proteins calpain 1 and neuronal nitric oxide. This results in mitochondrial dysfunction and oxidative damage by reactive oxygen and nitrogen species [43]. Together, these mechanisms lead to neuronal death, thereby contributing to long-term cognitive decline, which has been shown to be a consequence of several infectious diseases.

As shown in **Figure 2**, the gaseous signaling molecule NO has a variety of cellular functions, including neurotransmission, regulation of blood-vessel tone, and immunity. Under pathological conditions, free radicals may deplete NO produced by eNOS through the formation of $ONOO^-$, thus decreasing the vascular bioavailability of NO, which results in BBB dysfunction. This ultimately results in endothelial damage, edema development, and hypoxia. Furthermore, the NO produced by iNOS in glial cells or by nNOS under excitotoxic process can form with free radicals (particularly O_2^-) $ONOO^-$ and produce several deleterious effects on tissue, such as through tyrosine nitration and cysteine oxidation in various proteins. These free radicals can further decompose into highly toxic-free radicals, such as $NO_2^\bullet$ and $^\bullet OH$ (as reviewed by [44]).

Nitric oxide, as described above, is a key molecule in the regulation of physiological brain homeostasis. Nitric oxide is synthetized by neuronal vessels (eNOS), by neurons (nNOS)

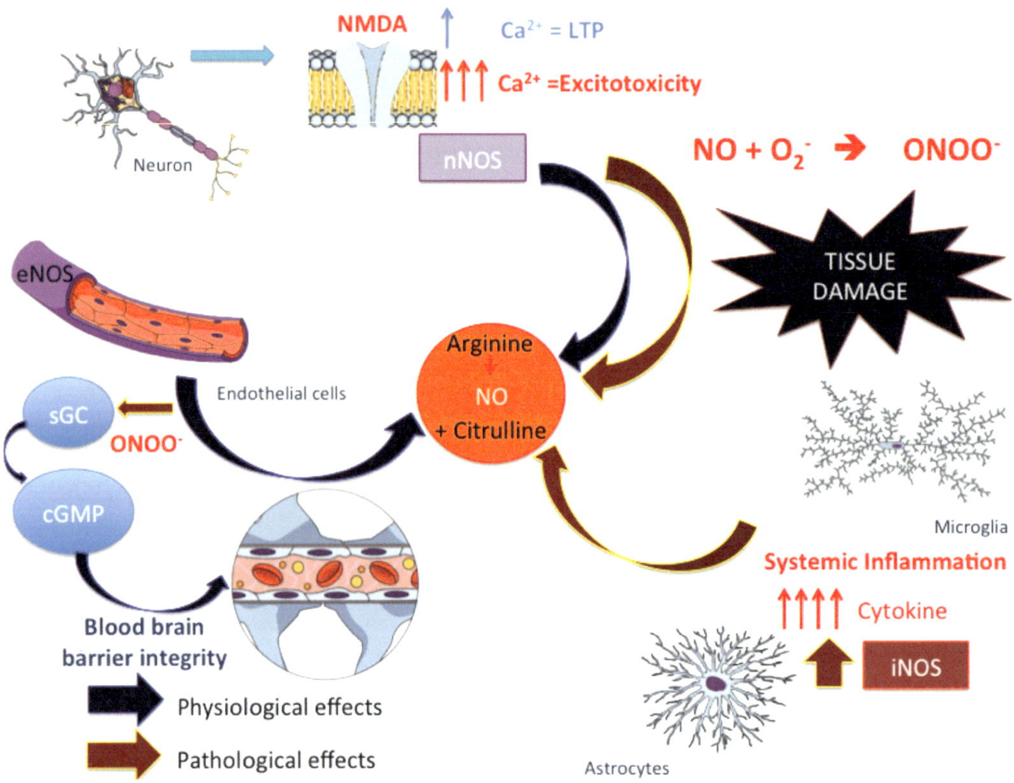

Figure 2. Different steps in the NO signaling cascade under physiological/pathological conditions in the brain. During long-term potentiation, NOS1 or neuronal NOS (nNOS) catalyze the NO synthesis after the activation of the NMDA receptor by Ca^{2+}. Under excitotoxic conditions, excessive Ca^{2+} leads to nNOS hyperactivity, whereas excessive NO production can combine with superoxide to form peroxynitrite, which is responsible for tissue damage due to several biological effects, including blockage of the eNOS pathway and BBB impairment. NO is synthesized following the transcriptional expression of a Ca^{2+}-independent NOS2 or iNOS isoform in glial cells (astrocytes and microglia) after cytokine exposure, thereby contributing to neuroinflammation and tissue damage in the brain. Intracellular Ca^{2+} activates NOS3 or eNOS to release NO from brain microvessels. This NO binds to soluble guanylyl-cyclase (sGC) receptors, which triggers a cGMP-dependent pathway and interacts with its downstream mediators of the physiological regulation of vasodilation and vascular resistance, platelet adhesion and aggregation, leukocyte-endothelial interaction, and BBB integrity maintenance. Images: https://mindthegraph.com.

under physiological/pathological *stimuli*, and by iNOS during inflammatory conditions. It is well known that high levels of nitric oxide (NO) released by the inducible NO synthase (iNOS) are critical for defense against parasites and mediate inflammatory tissue damage. However, the suppression or lack of NO production results in the impaired clearance of some types of bacteria by the host [45].

Our group and others have shown that eNOS is impaired during systemic infection, which leads to brain microcirculation dysfunction [46–48]. We showed that during different infectious diseases, there is a functional capillary rarefaction (**Figure 3**) that can be caused by the impairment of endothelial NO production by a reduction of eNOS activity or by reduced levels of the enzyme cofactor tetrahydrobiopterin—BH4 [49].

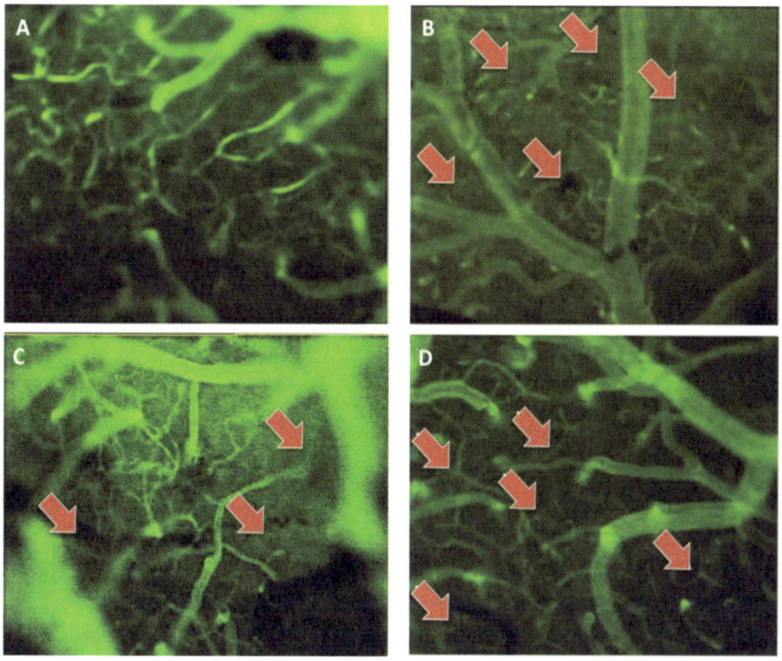

Figure 3. Photomicrography from intravital microscopy showing capillary impairment (red arrows) under healthy (A) and infectious conditions (B) malaria, (C) sepsis, and (D) Chagas' disease [47, 48].

Vascular function can be evaluated by the vasodilator response of cerebral arterioles to acetylcholine (Ach). Our group used intravital microscopy to evaluate the response to Ach in sepsis and Chagas' disease model. The vasodilation-to-Ach test is directly related to the availability of endothelium-derived NO generated from L-arginine by the action of endothelial NO synthase (eNOS). The diffusion of NO to vascular smooth muscle cells and the activation of guanylyl-cyclase resulted in cGMP-mediated vasodilation. However, the vasoconstrictive response to Ach results from a direct muscarinic smooth muscle effect [50]. This vasoconstrictive effect can lead to the slow delivery of oxygen to brain tissue, generating hypoxia and increased glycose consume by glycolytic pathway to generate ATP for neuronal functioning. Consequently, mitochondrial function is compromised by low O_2 concentration [51].

In both sepsis and Chagas' disease models, we observed a vasoconstrictive response to Ach. One mechanism of eNOS "uncoupling" from the reduction of NO production and ROS generation involves the oxidative degradation of the cofactor BH4 (tetrahydrobiopterin), especially by peroxynitrite ($ONOO^-$), which is produced in association with superoxide and nitric oxide. Several studies have shown that there is an increase in impaired eNOS-derived oxidative stress in brain tissue exposed to systemic infection, NO consumption and capillary dysfunction [46, 47, 52, 53].

Interestingly, in experimental models of malaria [48] and sepsis (Reis et al., submitted) blood flow was recovered by treatment with statins. As reviewed by Giannopoulos et al. [54], statins are shown to enhance eNOS expression and can contribute to the restoration of brain capillary function.

Resident glial cells in the CNS (i.e., astroglia and microglia) express inducible nitric oxide synthase (iNOS) and produce high levels of NO in response to a wide variety of proinflammatory and degenerative *stimuli*. Excessive NO production by glial cells evoked by inflammatory signals contributes to the pathogenesis of several neurodegenerative diseases, such as multiple sclerosis, HIV dementia, brain ischemia, trauma, Parkinson's disease, and Alzheimer's disease. NO can also be released in glial cells in response to infectious agent. The inducible form of nitric oxide synthase can be strongly associated with tissue damage, despite its activity as an antimicrobial agent. As reviewed by Saha and Pahan [55], astrocytes and microglia can express iNOS and synthetize NO. According to Radi [56], nitric oxide can combine with superoxide (produced mainly by the NADPH oxidase system and partially by mitochondria) under conditions of oxidative stress. It is common under neuroinflammatory conditions to form peroxynitrite, which binds to tyrosine to produce 3-nitrotyrosine, and a marker of reactive nitrogen species production. The reactions of peroxynitrite with biomolecules can lead to cytotoxic events and may result in apoptotic or necrotic cell death. Peroxynitrite can act via antioxidant enzyme inhibition; antioxidant depletion; protein aggregation (e.g., α-synuclein and microtubule-associated tau protein modifications in the CNS); activation of specific enzymes (e.g., matrix metalloproteinases (MMPs), cytochrome c, glutathione-S-transferase, protein kinase C-ε (PKCε), and fibrinogen); impairment of enzyme cofactors (e.g., BH4); modification of mediator pathways, receptors, and cellular signaling molecules; calcium deregulation; DNA injury; and mitochondrial dysfunction, among other processes (reviewed by [57]).

Data from our group show that the absence of the iNOS enzyme (knockout) or treatment with iNOS-specific inhibitors (e.g., aminoguanidine) has a beneficial effect on the prevention of cognitive impairment (unpublished data) in experimental cerebral malaria. This could be due to the reduced production of the peroxynitrite radical and the subsequent reduction of tissue damage. Other studies are being conducted to clarify the impact of the synthesis of nitric oxide by iNOS in infectious diseases and in the development of cognitive decline. Weberpals et al. also observed that iNOS gene deficiency prevents cognitive decline in addition to promoting a reduction in gliosis (astrocytes and microglia), proinflammatory cytokines TNF-α and IL-1β release, and reduction of synaptic dysfunction in sepsis model [58].

4. NMDA-nNOS pathway and excitotoxicity development

The NMDA receptor is a key regulator during glutamatergic long-term potentiation (LTP) response. During physiological conditions, glutamatergic ionotropic (e.g., kainate, NMDA, and AMPA) and metabotropic receptors (mGlut) are activated, delivering Ca^{2+} into neuronal cells and resulting in depolarization and the triggering of mitogen-activated protein kinase (MEK)/extracellular signal-regulated kinase (ERK) and phosphatidylinositol 3-kinase (PI3K) signaling pathways. This then activates phosphoinositide-dependent kinase 1 or 2 (PDK1/2), Akt, and mammalian target of rapamycin (mTOR) downstream signaling [59]. Additionally, calcium-dependent kinase II (CAMKII) is rapidly activated and all events together lead to the

http://dx.doi.org/10.5772/67816

activation of transcription factors (e.g., CREB) and an increase in the synthesis of neurotrophic factors (e.g., BDNF) and proteins associated with the long-term potentiation (LTP) process.

Despite the major activity of NMDA receptors in LTP, several reports have related this receptor with excitotoxicity, which is a neurodegenerative process. This process has been associated with a different class of NMDA receptor: the extra-synaptic receptor. As reviewed by Parsons and Raymond, synaptic NMDA (which expresses the subunit GlutN2A) is associated with LTP, whereas extra-synaptic receptors (which express the subunit GlutN2B) is associated with excitotoxicity and cell death [42]. NMDARs recruit the calcium-dependent enzyme nNOS via PSD95 (postsynaptic density: membrane-associated guanylate kinase (MAGUK) scaffolding protein located in neural postsynaptic densities), which is also associated with the LTP process. This is considered a key contributor to excitotoxicity lesions in both stroke and neurodegenerative diseases [60]. nNOS is activated by calcium/calmodulin signaling and is PSD95 protein-dependent [60]. During excitotoxicity, the activation of NMDA enhances intracellular calcium, leading to nNOS activation and NO production. Additionally, excessive intracellular calcium activates calpain 1, which then disrupts mitochondrial function, thereby triggering the intrinsic apoptotic pathway by releasing cytochrome C and activating apoptosomal protein complex [61]. NO can combine with superoxide, which results in peroxynitrite formation and cellular damage. Peroxynitrite can disturb cellular function by the nitration of proteins, which reduces or eliminates protein function, as described earlier, and by DNA damage via the activation of poly (ADP-ribose) polymerase 1 (PARP-1). The impact of PARP-1 on intracellular concentration of its nicotinamideadenine dinucleotide substrate (NAD), creates a bioenergetic imbalance that culminates with ATP depletion, thereby triggering necrotic neuronal death [62, 63]. In addition, NO can drive the retraction of the synaptic button via the activation of small GTPase RhoA/ROCK signaling through a paracrine/retrograde-signaling pathway [64]. Taken together, these events could contribute to neuronal dysfunction and death associated with cognitive decline.

The roles of NO in neuronal damage following insults, such as hypoxia, traumatic brain injury, and ischemia, have been well established. Recent evidence has implicated an imbalance of ROS and NO signaling in neurodegenerative disorders, such as Alzheimer's disease and Parkinson's disease, and in cognitive impairments associated with normal physiological aging [65–67]. Whereas mild oxidative/nitrosative (nitric oxide (NO)-related) stress mediates normal neuronal signaling, the accumulation of free radicals is associated with neuronal cell injury or death.

As described above, NO from eNOS modulates blood flow in the brain, and its impairment could be associated with hypoxic events in the brain. Several degenerative and infectious diseases relate hypoxia to neuronal dysfunction and cognitive decline. Using partial eNOS knockout mice model, Tan et al. [68] showed that the development of spontaneous thrombotic cerebral infarction is followed by amyloid protein deposit and cognitive decline. Cognitive decline is a major cause of disability in stroke survivors [69]. We have shown that microvascular impairment in malaria, sepsis, and Chagas' diseases may occur in response to other infectious agents [46–48]. Cerebral metabolism is dependent upon the glucose and oxygen that are delivered by blood, and it is clear that alterations in endothelial function can disrupt neuronal functions. Additionally, activation of endothelial cells by systemic cytokines, PAMP/DAMP and prooxidant molecules can contribute to eNOS dysfunction [70], blood flow disturbance and neuronal dysfunction, which subsequently results in cognitive decline.

The activation of glial cells that may be related with systemic inflammation associated to host response to pathogens can activate inducible isoforms of iNOS and subsequently generate cellular damage via the generation of peroxynitrite by the combination of NO and superoxide radicals. In this way, the induction of iNOS may result in the development of cognitive impairment [71]. Experimental models of sepsis [58] and malaria (unpublished data) have shown that the inhibition of iNOS has a beneficial effect on the central nervous system, particularly by abolishing cognitive decline.

The main enzyme target in the central nervous system seems to be nNOS. The NO generated by nNOS controls the release of neurotransmitters and is involved in synaptogenesis, synaptic plasticity, memory function, and neuroendocrine secretion. However, the overproduction of NO during NMDA-excitotoxic events can lead to neuronal death and directly impact the cognitive function.

Death pathways are activated in mouse brains during experimental model of sepsis and malaria [72, 73]. Additionally, reduced levels of neurotrophic factors, impairment of neurogenesis, and synaptic dysfunction were observed [74–78]. However, the role of excitotoxicity and nNOS delivery during infectious diseases and long-term cognitive impairment is not yet clear. Neuroinflammation can also induce cell damage/death, and the modulation of the activation of glial cells has been suggested to prevent neuronal damage and cognitive decline. Cognitive impairment is prevented by antioxidants and statin treatment [48, 53, 79], which suggests that the control of oxidative or inflammatory damage is also efficient to avoid cognitive decline.

5. Conclusion

NO is a key molecule involved in the regulation of CNS function in health and disease (**Table 1**). The impairment of enzymatic activity or the overproduction of NO under inflammatory/excitotoxic conditions can contribute to neurological sequela during systemic-infectious diseases (**Figure 3**). The NO synthase complex can be considered a target of pharmacological intervention focusing on the prevention of cognitive sequela; however, this field requires further studies.

NOS isoform	Physiological effect on CNS	Pathological effect on CNS
nNOS	Learning and memory	↑—Neuronal death by excitotoxicity
eNOS	Vasodilation and increased blood flow	↓—Hypoxia
iNOS	Sleep [33]	↑—Tissue damage

Table 1. Nitric oxide synthase enzymes functions on health/pathological conditions.

http://dx.doi.org/10.5772/67816

Author details

Patricia Alves Reis, Cassiano Felippe Gonçalves de Albuquerque, Tatiana Maron-Gutierrez, Adriana Ribeiro Silva and Hugo Caire de Castro Faria Neto*

*Address all correspondence to: hugocfneto@gmail.com

Laboratório de Imunofarmacologia, IOC, Fiocruz, Rio de Janeiro, Brazil

References

[1] Gunther M, Al Nimer F, Gahm C, Piehl F, Mathiesen T. iNOS-mediated secondary inflammatory response differs between rat strains following experimental brain contusion. Acta Neurochirurgica (Wien). 2012;**154**:689–697

[2] Brown GC. Nitric oxide and neuronal death. Nitric Oxide. 2010;**23**:153–165

[3] Garry PS, Ezra M, Rowland MJ, Westbrook J, Pattinson KT. The role of the nitric oxide pathway in brain injury and its treatment—from bench to bedside. Experimental Neurology. 2015;**263**:235–243

[4] Dzoljic E, Grbatinic I, Kostic V. Why is nitric oxide important for our brain? Functional Neurology. 2015;**30**:159–163

[5] Duncan AJ, Heales SJ. Nitric oxide and neurological disorders. Molecular Aspects of Medicine. 2005;**26**:67–96

[6] Michell AW, Lewis SJ, Foltynie T, Barker RA. Biomarkers and Parkinson's disease. Brain. 2004;**127**:1693–1705

[7] Stojanovic R, Todorovic Z, Nesic Z, Vuckovic S, Cerovac-Cosic N, et al. NG-nitro-L-arginine methyl ester-induced potentiaton of the effect of aminophylline on rat diaphragm: The role of extracellular calcium. Journal of Pharmacological Sciences. 2004;**96**:493–498

[8] Stojanovic R, Todorovic Z, Vuckovic S, Nesic Z, Prostran M. NG-nitro-L-arginine methyl ester potentiates the effect of aminophylline on the isolated rat hemidiaphragm. Journal of Pharmacological Sciences. 2003;**92**:157–162

[9] Yamamoto K, Takei H, Koyanagi Y, Koshikawa N, Kobayashi M. Presynaptic cell type-dependent regulation of GABAergic synaptic transmission by nitric oxide in rat insular cortex. Neuroscience. 2015;**284**:65–77

[10] Yassin L, Radtke-Schuller S, Asraf H, Grothe B, Hershfinkel M, et al. Nitric oxide signaling modulates synaptic inhibition in the superior paraolivary nucleus (SPN) via cGMP-dependent suppression of KCC2. Frontiers in Neural Circuits. 2014;**8**:65

[11] Akyol O, Zoroglu SS, Armutcu F, Sahin S, Gurel A. Nitric oxide as a physiopathological factor in neuropsychiatric disorders. In Vivo. 2004;**18**:377–390

[12] Boje KM. Nitric oxide neurotoxicity in neurodegenerative diseases. Frontiers in Bioscience. 2004;**9**:763–776

[13] Virarkar M, Alappat L, Bradford PG, Awad AB. L-Arginine and nitric oxide in CNS function and neurodegenerative diseases. Critical Reviews in Food Science and Nutrition. 2013;**53**:1157–1167

[14] Toda N, Ayajiki K, Okamura T. Cerebral blood flow regulation by nitric oxide in neurological disorders. Canadian Journal of Physiology and Pharmacology. 2009;**87**:581–594

[15] Broos K, Feys HB, De Meyer SF, Vanhoorelbeke K, Deckmyn H. Platelets at work in primary hemostasis. Blood Reviews. 2011;**25**:155–167

[16] Toda N, Okamura T. The pharmacology of nitric oxide in the peripheral nervous system of blood vessels. Pharmacological Reviews. 2003;**55**:271–324

[17] Schuman EM, Madison DV. Nitric oxide and synaptic function. Annual Review of Neuroscience. 1994;**17**:153–183

[18] Tieu K, Ischiropoulos H, Przedborski S. Nitric oxide and reactive oxygen species in Parkinson's disease. International Union of Biochemistry and Molecular Biology Life. 2003;**55**:329–335

[19] Zhu H, Vishwamitra D, Curry CV, Manshouri R, Diao L, et al. NPM-ALK up-regulates iNOS expression through a STAT3/microRNA-26a-dependent mechanism. Journal of Pathology. 2013;**230**:82–94

[20] Pautz A, Art J, Hahn S, Nowag S, Voss C, et al. Regulation of the expression of inducible nitric oxide synthase. Nitric Oxide. 2010;**23**:75–93

[21] MacMicking JD, Willenborg DO, Weidemann MJ, Rockett KA, Cowden WB. Elevated secretion of reactive nitrogen and oxygen intermediates by inflammatory leukocytes in hyperacute experimental autoimmune encephalomyelitis: Enhancement by the soluble products of encephalitogenic T cells. The Journal of Experimental Medicine. 1992;**176**:303–307

[22] Wang HG, Lu FM, Jin I, Udo H, Kandel ER, et al. Presynaptic and postsynaptic roles of NO, cGK, and RhoA in long-lasting potentiation and aggregation of synaptic proteins. Neuron. 2005;**45**:389–403

[23] Steinback CD, Poulin MJ. Influence of hypoxia on cerebral blood flow regulation in humans. Advances in Experimental Medicine and Biology. 2016;**903**:131–144

[24] Boehning D, Snyder SH. Novel neural modulators. Annual Review of Neuroscience. 2003;**26**:105–131

[25] Cossenza M, Socodato R, Portugal CC, Domith IC, Gladulich LF, et al. Nitric oxide in the nervous system: Biochemical, developmental, and neurobiological aspects. Vitamins and Hormones. 2014;**96**:79–125

[26] Mashimo H, Kjellin A, Goyal RK. Gastric stasis in neuronal nitric oxide synthase-deficient knockout mice. Gastroenterology. 2000;**119**:766–773

[27] Ayata C, Ayata G, Hara H, Matthews RT, Beal MF, et al. Mechanisms of reduced striatal NMDA excitotoxicity in type I nitric oxide synthase knock-out mice. Journal of Neuroscience. 1997;**17**:6908–6917

[28] Kirchner L, Weitzdoerfer R, Hoeger H, Url A, Schmidt P, et al. Impaired cognitive performance in neuronal nitric oxide synthase knockout mice is associated with hippocampal protein derangements. Nitric Oxide. 2004;**11**:316–330

[29] Huang PL, Huang Z, Mashimo H, Bloch KD, Moskowitz MA, et al. Hypertension in mice lacking the gene for endothelial nitric oxide synthase. Nature. 1995;**377**:239–242

[30] Lee PC, Salyapongse AN, Bragdon GA, Shears LL, 2nd, Watkins SC, et al. Impaired wound healing and angiogenesis in eNOS-deficient mice. The American Journal of Physiology. 1999;**277**:H1600–1608

[31] Girouard H, Iadecola C. Neurovascular coupling in the normal brain and in hypertension, stroke, and Alzheimer disease. Journal of Applied Physiology (1985). 2006;**100**:328–335

[32] Su KH, Lin SJ, Wei J, Lee KI, Zhao JF, et al. The essential role of transient receptor potential vanilloid 1 in simvastatin-induced activation of endothelial nitric oxide synthase and angiogenesis. Acta Physiologica (Oxf). 2014;**212**:191–204

[33] Costa ED, Rezende BA, Cortes SF, Lemos VS. Neuronal nitric oxide synthase in vascular physiology and diseases. Frontiers in Physiology. 2016;7:206

[34] Argaw AT, Asp L, Zhang J, Navrazhina K, Pham T, et al. Astrocyte-derived VEGF-A drives blood-brain barrier disruption in CNS inflammatory disease. The Journal of Clinical Investigation. 2012;**122**:2454–2468

[35] Park HJ, Shin JY, Kim HN, Oh SH, Song SK, et al. Mesenchymal stem cells stabilize the blood-brain barrier through regulation of astrocytes. Stem Cell Research & Therapy. 2015;**6**:187

[36] Soares ES, Mendonca MC, da Cruz-Hofling MA. eNOS uncoupling in the cerebellum after BBB disruption by exposure to phoneutria nigriventer spider venom. Toxicon. 2015;**104**:7–13

[37] Olivera GC, Ren X, Vodnala SK, Lu J, Coppo L, et al. Nitric oxide protects against infection-induced neuroinflammation by preserving the stability of the blood-brain barrier. PLoS Pathogens. 2016;**12**:e1005442

[38] Northington FJ, Tobin JR, Harris AP, Traystman RJ, Koehler RC. Developmental and regional differences in nitric oxide synthase activity and blood flow in the sheep brain. Journal of Cerebral Blood Flow and Metabolism. 1997;**17**:109–115

[39] John CC, Carabin H, Montano SM, Bangirana P, Zunt JR, et al. Global research priorities for infections that affect the nervous system. Nature. 2015;**527**:S178–186

[40] Sankowski R, Mader S, Valdes-Ferrer SI. Systemic inflammation and the brain: Novel roles of genetic, molecular, and environmental cues as drivers of neurodegeneration. Frontiers in Cellular Neuroscience. 2015;**9**:28

[41] Kopitar-Jerala N. Innate immune response in brain, NF-Kappa B signaling and cystatins. Frontiers in Molecular Neuroscience. 2015;**8**:73

[42] Parsons MP, Raymond LA. Extrasynaptic NMDA receptor involvement in central nervous system disorders. Neuron. 2014;**82**:279–293

[43] Fujikawa DG. The role of excitotoxic programmed necrosis in acute brain injury. Computational and Structural Biotechnology Journal. 2015;**13**:212–221

[44] Berg RM, Moller K, Bailey DM. Neuro-oxidative-nitrosative stress in sepsis. Journal of Cerebral Blood Flow and Metabolism. 2011;**31**:1532–1544

[45] Nathan C. Inducible nitric oxide synthase: What difference does it make? The Journal of Clinical Investigation. 1997;**100**:2417–2423

[46] Araujo CV, Estato V, Tibirica E, Bozza PT, Castro-Faria-Neto HC, et al. PPAR gamma activation protects the brain against microvascular dysfunction in sepsis. Microvascular Research. 2012;**84**:218–221

[47] Nisimura LM, Estato V, de Souza EM, Reis PA, Lessa MA, et al. Acute Chagas disease induces cerebral microvasculopathy in mice. PLoS Neglected Tropical Diseases. 2014;**8**:e2998

[48] Reis PA, Estato V, da Silva TI, d'Avila JC, Siqueira LD, et al. Statins decrease neuroinflammation and prevent cognitive impairment after cerebral malaria. PLoS Pathogens. 2012;**8**:e1003099

[49] Ong PK, Melchior B, Martins YC, Hofer A, Orjuela-Sanchez P, et al. Nitric oxide synthase dysfunction contributes to impaired cerebroarteriolar reactivity in experimental cerebral malaria. PLoS Pathogens. 2013;**9**:e1003444

[50] Pesic S, Grbovic L, Jovanovic A. Acetylcholine-induced contractions in the perforating branch of the human internal mammary artery: Protective role of the vascular endothelium. Pharmacology. 2002;**64**:182–188

[51] Bozza FA, D'Avila JC, Ritter C, Sonneville R, Sharshar T, et al. Bioenergetics, mitochondrial dysfunction, and oxidative stress in the pathophysiology of septic encephalopathy. Shock. 2013;**39** (Suppl 1):10–16

[52] Barichello T, Fortunato JJ, Vitali AM, Feier G, Reinke A, et al. Oxidative variables in the rat brain after sepsis induced by cecal ligation and perforation. Critical Care Medicine. 2006;**34**:886–889

[53] Reis PA, Comim CM, Hermani F, Silva B, Barichello T, et al. Cognitive dysfunction is sustained after rescue therapy in experimental cerebral malaria, and is reduced by additive antioxidant therapy. PLoS Pathogens. 2010;**6**:e1000963

[54] Giannopoulos S, Katsanos AH, Tsivgoulis G, Marshall RS. Statins and cerebral hemodynamics. Journal of Cerebral Blood Flow and Metabolism. 2012;**32**:1973–1976

[55] Saha RN, Pahan K. Regulation of inducible nitric oxide synthase gene in glial cells. Antioxidants & Redox Signaling. 2006;**8**:929–947

[56] Radi R. Peroxynitrite, a stealthy biological oxidant. The Journal of Biological Chemistry. 2013;**288**:26464–26472

[57] Szabo C, Ischiropoulos H, Radi R. Peroxynitrite: Biochemistry, pathophysiology and development of therapeutics. Nature Reviews Drug Discovery. 2007;**6**:662–680

[58] Weberpals M, Hermes M, Hermann S, Kummer MP, Terwel D, et al. NOS2 gene deficiency protects from sepsis-induced long-term cognitive deficits. The Journal of Neuroscience. 2009;**29**:14177–14184

[59] Klann E, Dever TE. Biochemical mechanisms for translational regulation in synaptic plasticity. Nature Reviews Neuroscience. 2004;**5**:931–942

[60] Courtney MJ, Li LL, Lai YY. Mechanisms of NOS1AP action on NMDA receptor-nNOS signaling. Frontiers in Cellular Neuroscience. 2014;**8**:252

[61] Lipton SA. NMDA receptor activity regulates transcription of antioxidant pathways. Nature Neuroscience. 2008;**11**:381–382

[62] Lai TW, Zhang S, Wang YT. Excitotoxicity and stroke: Identifying novel targets for neuroprotection. Progress in Neurobiology. 2014;**115**:157–188

[63] Ha HC, Snyder SH. Poly(ADP-ribose) polymerase is a mediator of necrotic cell death by ATP depletion. Proceedings of the National Academy of Sciences of the United States of America. 1999;**96**:13978–13982

[64] Sunico CR, Gonzalez-Forero D, Dominguez G, Garcia-Verdugo JM, Moreno-Lopez B. Nitric oxide induces pathological synapse loss by a protein kinase G-, Rho kinase-dependent mechanism preceded by myosin light chain phosphorylation. The Journal of Neuroscience. 2010;**30**:973–984

[65] Halliwell B. Reactive oxygen species and the central nervous system. Journal of Neurochemistry. 1992;**59**:1609–1623

[66] Finkel T, Holbrook NJ. Oxidants, oxidative stress and the biology of ageing. Nature. 2000;**408**:239–247

[67] Nakamura T, Cho DH, Lipton SA. Redox regulation of protein misfolding, mitochondrial dysfunction, synaptic damage, and cell death in neurodegenerative diseases. Experimental Neurology. 2012;**238**:12–21

[68] Tan XL, Xue YQ, Ma T, Wang X, Li JJ, et al. Partial eNOS deficiency causes spontaneous thrombotic cerebral infarction, amyloid angiopathy and cognitive impairment. Molecular Neurodegeneration. 2015;**10**:24

[69] Levine DA, Galecki AT, Langa KM, Unverzagt FW, Kabeto MU, et al. Trajectory of cognitive decline after incident stroke. The Journal of the American Medical Association. 2015;**314**:41–51

[70] Weiss N, Miller F, Cazaubon S, Couraud PO. The blood-brain barrier in brain homeostasis and neurological diseases. Biochimica et Biophysica Acta. 2009;**1788**:842–857

[71] Eckel B, Ohl F, Bogdanski R, Kochs EF, Blobner M. Cognitive deficits after systemic induction of inducible nitric oxide synthase: A randomised trial in rats. European Journal of Anaesthesiology. 2011;**28**:655–663

[72] Lackner P, Burger C, Pfaller K, Heussler V, Helbok R, et al. Apoptosis in experimental cerebral malaria: Spatial profile of cleaved caspase-3 and ultrastructural alterations in different disease stages. Neuropathology and Applied Neurobiology. 2007;**33**:560–571

[73] Comim CM, Barichello T, Grandgirard D, Dal-Pizzol F, Quevedo J, et al. Caspase-3 mediates in part hippocampal apoptosis in sepsis. Molecular Neurobiology. 2013;**47**:394–398

[74] Anand SS, Babu PP. c-Jun N terminal kinases (JNK) are activated in the brain during the pathology of experimental cerebral malaria. Neuroscience Letters. 2011;**488**:118–122

[75] de Miranda AS, Brant F, Campos AC, Vieira LB, Rocha NP, et al. Evidence for the contribution of adult neurogenesis and hippocampal cell death in experimental cerebral malaria cognitive outcome. Neuroscience. 2015;**284**:920–933

[76] Moraes CA, Santos G, de Sampaio e Spohr TC, D'Avila JC, Lima FR, et al. Activated microglia-induced deficits in excitatory synapses through IL-1beta: Implications for cognitive impairment in sepsis. Molecular Neurobiology. 2015;**52**:653–663

[77] Comim CM, Reis PA, Frutuoso VS, Fries GR, Fraga DB, et al. Effects of experimental cerebral malaria in memory, brain-derived neurotrophic factor and acetylcholinesterase activity [correction for acitivity] in the hippocampus of survivor mice. Neuroscience Letters. 2012;**523**:104–107

[78] Comim CM, Silva NC, Mina F, Dominguini D, Scaini G, et al. Evaluation of NCS-1, DARPP-32, and neurotrophins in hippocampus and prefrontal cortex in rats submitted to sepsis. Synapse. 2014;**68**:474–479

[79] Barichello T, Machado RA, Constantino L, Valvassori SS, Reus GZ, et al. Antioxidant treatment prevented late memory impairment in an animal model of sepsis. Critical Care Medicine. 2007;**35**:2186–2190

Section 2

Nitric Oxide Synthase in Cardiovascular System (CVS)

Chapter 5

Real-Time Monitoring of Nitric Oxide Dynamics in the Myocardium: Biomedical Application of Nitric Oxide Sensor

Gi-Ja Lee, Young Ju Lee and Hun-Kuk Park

Additional information is available at the end of the chapter

http://dx.doi.org/10.5772/67255

Abstract

Nitric oxide (NO) is an important physiological mediator that regulates a wide range of cellular processes in many tissues. Therefore, the accurate and reliable measurement of physiological NO concentration is essential to the understanding of NO signaling and its biological role. Most methods used for NO detection are indirect including spectroscopic approaches such as the Griess assay for nitrite and detection of methemoglobin after NO reaction with oxyhemoglobin. These methods cannot accurately reflect the changes in NO concentration in vivo and in real time. Therefore, direct methods are necessary for investigating biological process and diseases related to NO in biological conditions. There is a growing interest in the development of electrochemically based sensors for direct, in vivo, and real-time monitoring of NO. Electrochemical methods offer simplicity, good sensitivity, high selectivity, fast response times, and long-term calibration stability compared to other techniques including electron paramagnetic resonance, chemiluminescence, and fluorescence. In this article, we present real-time NO dynamics in the myocardium during myocardial ischemia-reperfusion (IR) utilizing electrochemical NO microsensor. And applications of electrochemical NO sensor for the evaluation of cardioprotective effects of therapeutic treatments such as drug administration and ischemic preconditioning are reviewed.

Keywords: nitric oxide, real-time detection, myocardial ischemia-reperfusion, electrochemical sensor, therapeutic treatments

1. Introduction

Nitric oxide (NO) is one of gaseous cellular-signaling molecules, which regulates a wide range of physiological and cellular processes in various tissues. In particular, it plays a vital role in a variety of biological processes including immune defense, neurotransmission, regulation of cell death (apoptosis), and cell motility [1–4]. NO has some key features that make this molecule suited to its cellular-signaling functions. NO is a lipophilic diatomic gas under atmospheric conditions. As it has a relatively small Stokes radius and neutral charge, it can rapidly diffuse the cell membrane. The presence of an unpaired electron in NO supports its high reactivity with oxygen (O_2), superoxide (O_2^-), transition metals, and thiols [5, 6]. The removal of the unpaired electron in NO generates the nitrosonium cation NO^+, while the addition of an electron forms the nitroxyl anion (NO^-). These different forms of NO represent distinct chemical reactivities [6, 7]. And NO reacts with O_2^- to form peroxynitrite ($OONO^-$), a particularly destructive molecule within biological systems [8].

NO is known to play a major role in vascular biology and heart failure. NO is a double-edged sword; NO inhibits ischemia-reperfusion (IR) injury, represses inflammation, and prevents left ventricular remodeling, whereas excess NO and coexistence of reactive oxygen species (ROS) with NO are injurious [6]. In that, low concentration of NO has beneficial effects on heart function, while high concentration of NO has opposite effects. Recently, it was reported that the final action of NO is not only regulated by its concentration and cellular confinement but also strongly depends on the level of oxidative stress in the myocardium [4]. But, the cardioprotective mechanism of NO is not yet clear, and it is not known whether NO effectively acts during ischemia or during reperfusion. Therefore, the accurate and quantitative detection of physiological NO concentration is crucial to the understanding of NO signaling and its biological role. This review focuses on the role of NO in myocardial IR injury. In addition, we will summarize the studies from our laboratory, which evaluates the cardioprotective effects of therapeutic treatments such as drug administration and ischemic preconditioning in the myocardium during myocardial IR utilizing electrochemical NO sensor.

2. Production of nitric oxide

In general, NO is produced from the conversion of L-arginine to L-citrulline, a reaction catalyzed by a family of enzymes called NO synthases (NOSs). Endothelial NOS (eNOS, also known as NOS3) and neuronal NOS (nNOS, also known as NOS1) are constitutive and low-output enzymes, whereas the macrophage-type NOS isoform, known as inducible NOS (iNOS, also known as NOS2), is an inducible and high-output enzyme [9]. NOS is a homodimeric oxidoreductase containing iron protoporphyrin IX (heme), flavin adenine dinucleotide, flavin mononucleotide, and tetrahydrobiopterin (BH_4), which is a cofactor essential for the catalytic activity of all three NOS isoforms [6, 10, 11]. NO biosynthesis by the three NOS isoforms can be suppressed by several small-molecule inhibitors: N^G-methyl-L-arginine (L-NMA) inhibits all NOS isoforms, and L-N^G-nitroarginine methyl ester (L-NAME) has some selectivity for the constitutive NOS isoforms (i.e., nNOS and eNOS), whereas other inhibitors such as aminoguanidine, 1400 W, and many others show selectivity for iNOS [9].

Although NOS had been generally considered to be the primary source of NO in biological systems, nonenzymatic NO synthesis also occurs. NO can be produced in tissues by either direct disproportionation or reduction of nitrite to NO under the acidic and highly reduced conditions which occur in disease states, such as ischemia [12]. Tissue acidosis occurring during ischemia increases NO production independent from eNOS [13], and even at normal pH, xanthine oxidase in the presence of low pO_2 and high nicotinamide adenine dinucleotide (NADH) concentration is capable of producing NO from nitrite [14]. Besides, in the isolated rat heart [15] and in rabbit hindlimb muscle [16], the NO concentration is still increased during ischemia after complete NOS inhibition by Nw-nitro-L-arginine (L-NNA).

3. Nitric oxide measurements

It is difficult to directly measure NO concentration in vivo because NO is present at nanomolar concentrations in the body and highly reactive with numerous endogenous species including free radicals, oxygen, peroxides, transition metals, and metalloproteins. Indeed, the half-life of NO in biological milieu is <10 s [17].

3.1. Indirect method (Griess assay)

Indirect methods measure the stable metabolites of NO such as nitrites (NO_2^-) and nitrates (NO_3^-). The most widely used method for NO detection is based on Griess assay reagents, which can react with nitrite to form a purple azo dye. This method requires that nitrate first be reduced to nitrite and then nitrite determined by the Griess reaction [18]. Briefly, the Griess reaction is a two-step diazotization reaction. First, the NO-derived nitrosating agent, dinitrogen trioxide (N_2O_3), generated from the acid-catalyzed formation of nitrous acid from nitrite (or autoxidation of NO) reacts with sulfanilamide to produce a diazonium ion. And then it is coupled to N-(1-napthyl)ethylenediamine to form a chromophoric azo product that strongly absorbs at 540 nm [19]. This method has some disadvantages including its sensitivity and its ability to detect nitrate. Therefore, nitrate must be converted to nitrite before the total nitrite is detected. The reduction of nitrate to nitrite can be achieved by using bacterial nitrate reductase or reducing metal such as cadmium [20]. However, these methods often fail to accurately reflect the spatial and temporal distributions of NO in biological environments. In addition, the concentration of nitrate and nitrite metabolite may not show the accurate amount of NO in the specific site because other parts of the body can also produce these compounds. Therefore, direct measurement strategies are necessary for investigating the physiological origin and action of endogenously produced NO.

3.2. Direct methods (nitric oxide sensor)

Several methods exist for directly measuring NO including electron paramagnetic resonance spectroscopy [21], chemiluminescence [22], fluorescence [23, 24], and electrochemical sensing [17, 25, 26]. Of these approaches, miniaturized electrochemical (e.g., amperometric and voltammetric) sensors represent the most promising means for determining the spatial and temporal distributions of NO near its physiological source [17]. Electrochemical methods provide simplicity, fast response times, good sensitivity, high selectivity, and long-term calibration stability [27, 28]. The most simple detection scheme to date involves the electrochemical oxidation of NO at a metal (e.g., platinum and gold) or carbon electrode [27, 29]. However, it is necessary to use a relatively high working potential (+0.7 to +0.9 V vs. Ag/AgCl) for the direct electrooxidation of NO. In this condition, several interferences from other readily oxidizable biological species such as nitrite, ascorbic acid, uric acid, and acetaminophen often disturb selective detection of NO [25, 27]. Therefore, further surface modification with permselective membranes is required to achieve the desired selectivity for NO via size exclusion or electrostatic repulsion [4]. Indeed, several polymeric materials have been evaluated as gas-permeable or permselective membranes including nafion, collodion, polycarbazole, o- and m-phenylenediamine, poly(tetrafluoroethylene), cellulose acetate, and multilayer hybrids of these polymers [27, 30–36]. In particular, Shin et al. reported that sol-gel-derived electrochemical sensors showed good sensitivity and selectivity for NO detection [17, 27].

4. Nitric oxide in myocardial ischemia-reperfusion injury

Myocardial infarction (MI) is one of the major causes of morbidity and mortality in industrialized countries, despite the improvement in clinical management of the disease. MI is caused by sudden stoppage of blood supply to the heart that leads to tissue necrosis. The normal myocardium produces more than 90% of its adenosine triphosphate (ATP) by oxidative metabolism and less than 10% by anaerobic glycolysis [4, 37]. After the induction of ischemia, the myocardium can be completely recovered to its normal state if blood supply is adequately restored. But cellular necrosis eventually occurs if ischemia persists [4]. During severe cardiac ischemia, cardiac myocytes must drastically reduce ATP demand or utilization to meet the needs for survival and thus balance the reduced ATP supply with reduced demand during severe ischemia [4, 38].

NO has been extensively studied in the setting of myocardial IR injury. Previous studies demonstrate that the deficiency of eNOS deteriorates myocardial IR injury [39], whereas the overexpression of eNOS [40], the administration of NO donors [41], and inhaled NO gas therapy [42] significantly protect the myocardium. NO possesses several physiological properties that make it a potent cardioprotective-signaling molecule, as follows [43]: First, NO is a potent vasodilator in the ischemic myocardium which enables an essential perfusion of injured tissue. Second, NO reversibly inhibits mitochondrial respiration during early reperfusion. It leads to a decrease in mitochondrial-driven injury by extending the zone of adequate tissue cellular oxygenation away from vessels [43–45]. It is known that restoration of oxygen at reperfusion leads to a lethal burst of reactive oxygen species (ROS) generation. An important source of ROS is the mitochondria. In mitochondria, electrons from intermediary metabolism move down the electron transport chain (ETC) and transferred to oxygen at complex IV [46]. When oxygenation is normal, complex I activity is high because a cysteine residue on its ND3 subunit is protected from modification. During ischemia (without oxygen), electrons accumulate along the ETC [46]. Reperfusion leads to a burst of ROS production from multiple sites which can attack proteins, lipids, and DNA, as well as lethal activation of the mitochondrial permeability transition pore [46]. NO inhibits mitochondrial complex I by S-nitrosation (or S-nitrosylation) of cysteines, which subsequently prevents damage during IR injury [47]. Reversible S-nitrosation of complex I slows the reactivation of mitochondria during the crucial first minutes of the reperfusion, thereby decreasing ROS production, oxidative damage, and tissue necrosis [48]. Third, NO is a potent inhibitor of neutrophil adherence to the vascular endothelium which is a significant event initiating further leukocyte activation and superoxide radical production [43, 49, 50]. Fourth, NO prevents platelet aggregation [51], and this effect attenuates capillary plugging together with the anti-neutrophil actions of NO [52]. Finally, NO inhibits apoptosis either directly or indirectly by inhibiting caspase-3-like activation via a cGMP-dependent mechanism [43, 53] and by direct inhibition of caspase-3-like activity through protein S-nitrosylation [43, 54]. In summary, the release of low concentrations of NO by constitutive NOS played a role in the regulation of coronary blood flow, inhibition of platelet aggregation, adherence to the endothelium, and possibly modulation of myocardial oxygen consumption.

But, excessive generation of NO is detrimental to cardiovascular function as exemplified in septic shock where burst generation of iNOS-derived NO causes hypotension, cardiodepression, and vascular hyporeactivity [55]. The detrimental effect of excess NO is attributed to the action on mitochondria. NO inhibits the mitochondrial respiratory chain, resulting in inhibition of ATP production, increased oxidant production, and increased susceptibility to cell death [56]. Inhibition of mitochondrial respiration by NO and its derivatives stimulates production of reactive oxygen and nitrogen species by mitochondria [56], which contribute to cell death in excess.

In conclusion, NO can preserve blood flow in the ischemic tissues and reduce platelet aggregation and neutrophil-endothelium interaction following IR. Besides, low concentrations of NO improve cardiomyocyte function. On the contrary, higher NO concentrations diminish cardiomyocyte function, mediate inflammatory processes following IR, worsen mitochondrial respiration, and even induce cardiomyocyte death. Therefore, it seems that NO can mediate both protective and detrimental myocardial effects which are crucially dependent upon the experimental conditions. Consensus is being reached in the debate regarding a NO protective effect, with most studies reporting its protective effects. However, the role of their product, NO, in the process of IR is still not well defined mainly because of the difficulty in measuring NO concentration in the body tissue.

In the next section, we summarize real-time NO dynamics in the myocardium during myocardial IR. And applications of electrochemical NO sensor for the evaluation of cardioprotective effects of therapeutic treatments such as hypothermia, drug administration, and ischemic preconditioning are summarized.

5. Real-time monitoring of nitric oxide dynamics in the myocardium during myocardial ischemia-reperfusion

5.1. Hypothermia

In general, hypothermia is thought to reduce the metabolic needs of cells, specifically perhaps by reducing the oxygen demand in the hypothermic tissues [57]. Besides, in isolated heart perfusion system, hearts were placed in ice-cold buffer as quickly as possible to avoid any detrimental effects of hypoxia. Therefore, myocardial hypothermia might be a useful technique to limit ischemic damage during infarction or as adjunctive therapy during minimally invasive cardiac surgery [58]. Lee et al. reported changes in NO dynamics during myocardial IR utilizing a sol-gel-modified electrochemical NO sensor and isolated heart perfusion system [4]. They attempted to clarify the role of endogenous NO release by comparing intact and cardioprotected hearts, in which cardioprotection was conferred by hypothermic treatment of the hearts. For the hypothermic group, hearts were immediately immersed in ice-cold perfusion buffer for 3 min after harvest. In the ischemic myocardium, NO showed a time-dependent change during the 40 min ischemic episode. After myocardial ischemia and early reperfusion, the restoration level of NO was decreased below the pre-ischemic level (**Figure 1**). However, the myocardium with hypothermic treatment (151 ± 37 nM) generated more NO during the ischemic period than that without any treatment (59 ± 15 nM). Besides, the restoration level of NO in the hypothermic group (−57 ± 26 nM) was significantly higher than that of the intact group (−170 ± 50 nM, $p < 0.05$) [4]. As a result, they inferred that hypothermic treatment of the heart would promote endogenous NO production in the ischemic myocardium. It might be a helpful therapeutic strategy for protecting the myocardium from IR injury [4].

5.2. Myocardial oxygen dynamics

Because oxygen plays a critical role in the pathophysiology of myocardial injury during subsequent reperfusion, as well as ischemia, the accurate measurement of myocardial oxygen tension is crucial for the assessment of myocardial viability by IR injury. Lee et al. reported a sol-gel-derived electrochemical oxygen microsensor to monitor changes in oxygen tension (pO_2) during myocardial IR [59]. And they analyzed differences in oxygen tension recovery in the post-ischemic myocardium depending on ischemic time to investigate the correlation between recovery parameters for oxygen tension and the severity of IR injury. **Figure 2** shows the nine parameters for pO_2 dynamics during myocardial IR and the maximum and restoration values of pO_2 at different ischemic times. As a result, they observed that if ischemia was stopped within 20 min, the pO_2 in the myocardium after the onset of reperfusion restored to pre-ischemic levels. However, the pO_2 in the myocardium did not recover to its pre-ischemic state, if the ischemic time was >30 min [59]. These results show that the maximum and restoration values of pO_2 in the post-ischemic myocardium were closely related to the infarct size [59]. In summary, they demonstrated that the degree of reoxygenation in the post-ischemic myocardium was an important index of IR injury and myocardial viability, utilizing a sol-gel-derived electrochemical oxygen microsensor and recovery parameters.

http://dx.doi.org/10.5772/67255

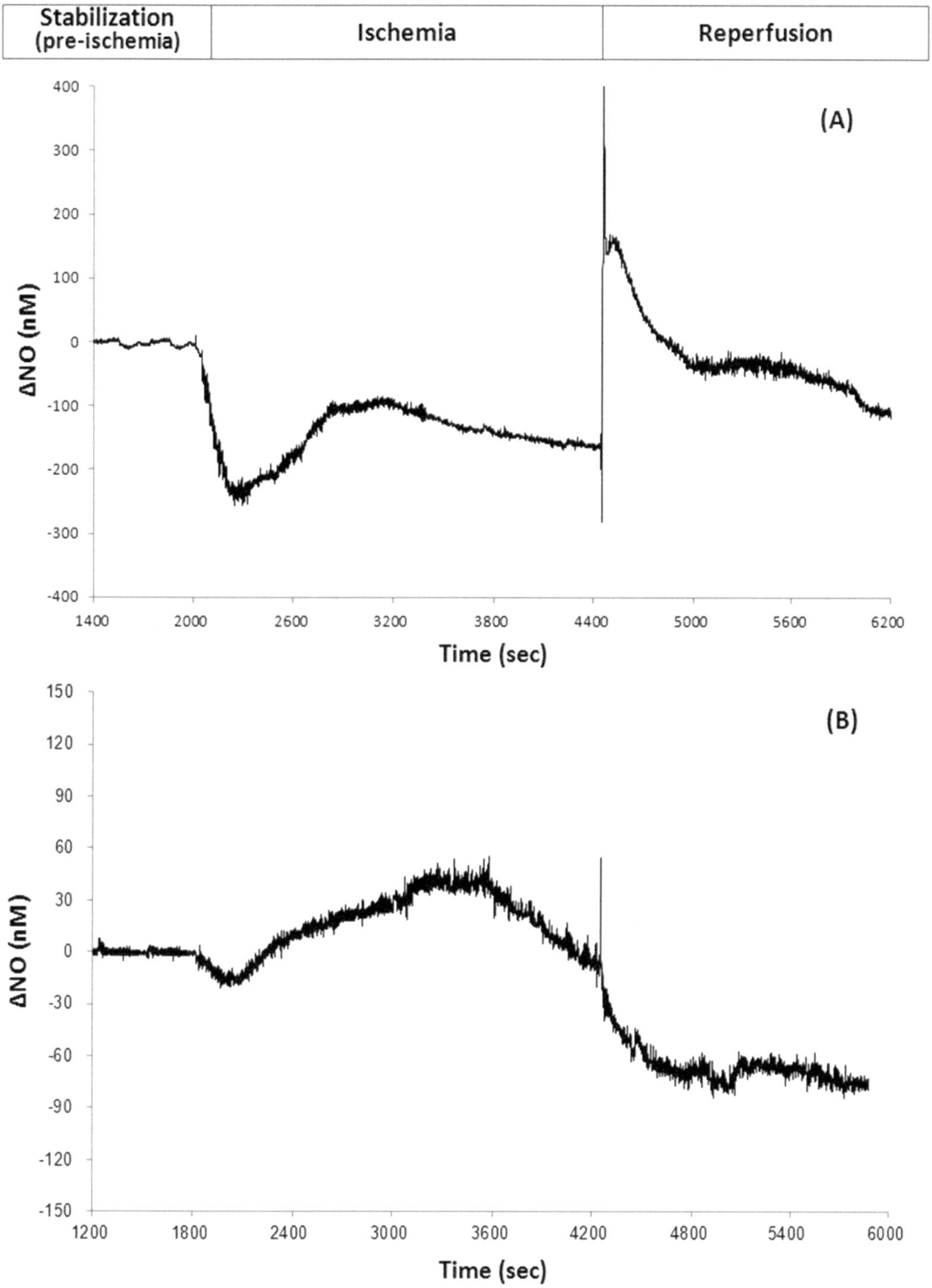

Figure 1. Representative real-time measurement of NO in intact (A) and hypothermic (B) groups during myocardial ischemia-reperfusion of Langendorff-perfused rat hearts. Reproduced with permission from Lee et al. [4]. © 2011 Elsevier B.V.

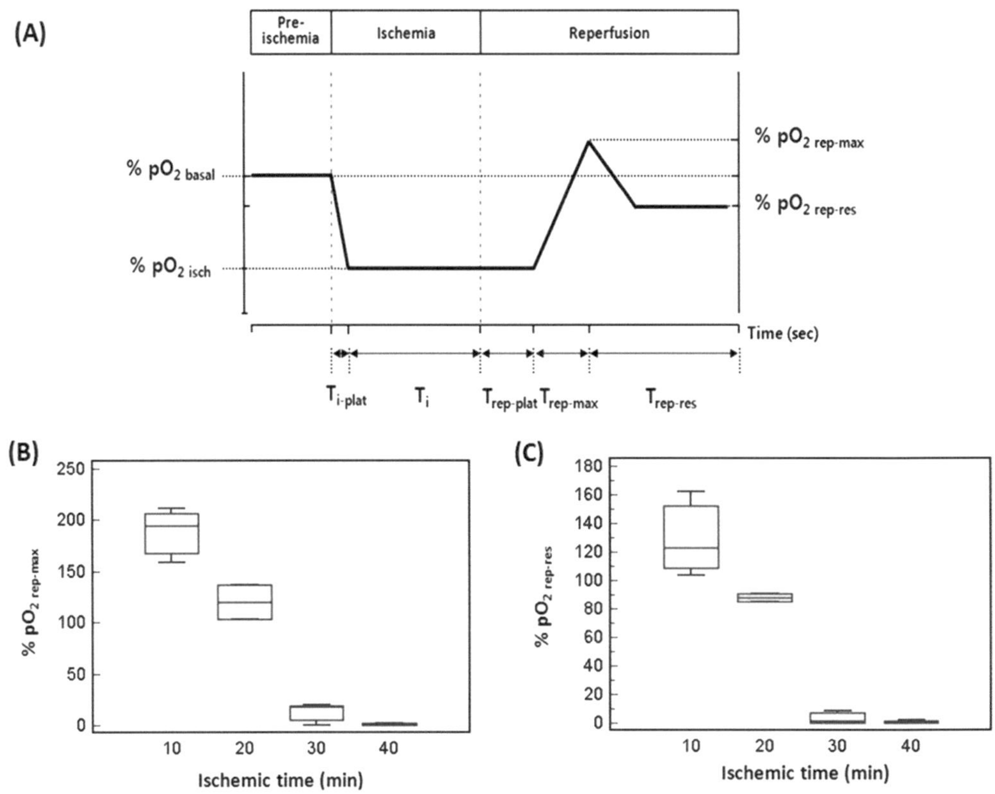

Figure 2. (A) Definition of analysis parameters proposed for changes in oxygen tension throughout the experimental protocol and box and whisker graph depicting maximum (B) and restoration levels (C) of oxygen tension during the reperfusion period as a percentage of pre-ischemic levels at different ischemic times (n = 3 per group). Reproduced from Lee et al. [59]. © The Royal Society of Chemistry 2012.

5.3. Remote ischemic preconditioning

Ischemic preconditioning is an adaptive response of briefly ischemic tissues that serves to protect against subsequent prolonged ischemic insults and reperfusion injury [60]. In particular, remote ischemic preconditioning (RIPC) is a novel method where ischemia followed by reperfusion of one organ is believed to protect remote organs either by the release of biochemical messengers into circulation or by the activation of nerve pathways, resulting in the release of messengers that have a protective effect [60–62]. This preserves the target tissue without trauma to major vessels or direct stress to the target organ [63]. Although some studies have demonstrated that endothelial NO is one of the major contributors to the candidate mechanism of RIPC [60, 64], the mechanism of RIPC-induced cardioprotection has not yet been fully elucidated. Lee et al. simultaneously measured NO and O_2 dynamics in the myocardium during myocardial IR utilizing sol-gel-modified electrochemical NO and O_2 microsensors [65]. By comparing control and RIPC-treated hearts, we attempted to clarify the correlation between NO release in the ischemic period and O_2 restoration in the myocardium after reperfusion. **Figure 3** represents the schematic diagrams of experimental design and experimental setup of an isolated heart perfusion system and a real-time monitoring system for NO and oxygen tension dynamics during myocardial IR of the rat.

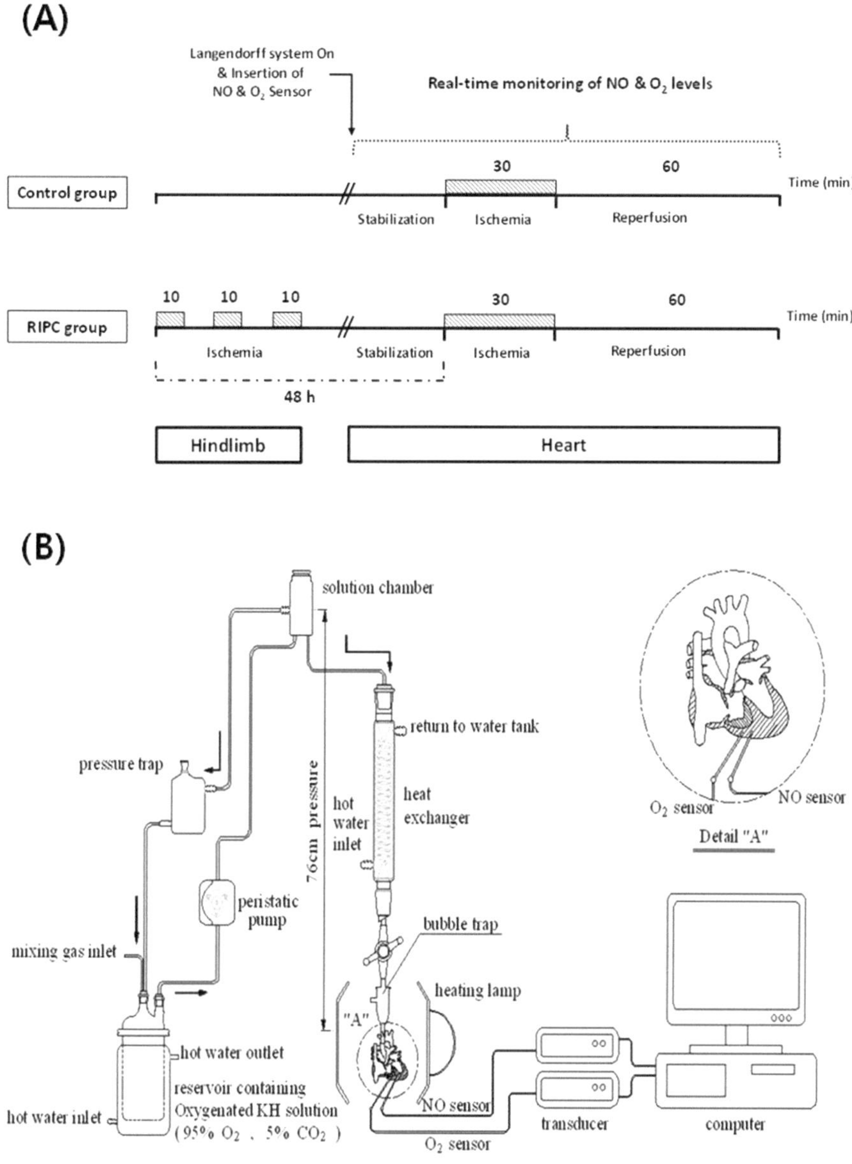

Figure 3. Schematic diagrams of (A) experimental design and (B) experimental setup of an isolated heart perfusion system and a real-time monitoring system for nitric oxide and oxygen tension dynamics during myocardial ischemia-reperfusion of the rat. Reproduced from Kang et al. [65]. © 2013 Elsevier B.V.

As a result, the concentration change of NO in the RIPC group was different from those in the control group during the ischemic period. In the control group, the NO level initially declined but then gradually inclined during the ischemic episode. In contrast, the NO level in the RIPC group rapidly increased after the onset of ischemia and continued to rise throughout the entire ischemic period [65]. When reperfusion was initiated, the pattern of both NO level and pO_2 in the RIPC group was different from that of the control group. As a result, the NO level and the pO_2 of the myocardium in the RIPC group were restored to pre-ischemic levels, unlike those in the control group that did not recover to their pre-ischemic state (**Figures 4** and **5**). In summary, the endogenous production of NO during the ischemic period appears to be correlated with the restoration of NO and pO_2 in the post-ischemic myocardium after early reperfusion. Additionally, RIPC would promote endogenous NO release against ischemic stimuli and subsequently facilitate reoxygenation in post-ischemic myocardia after reperfusion [65].

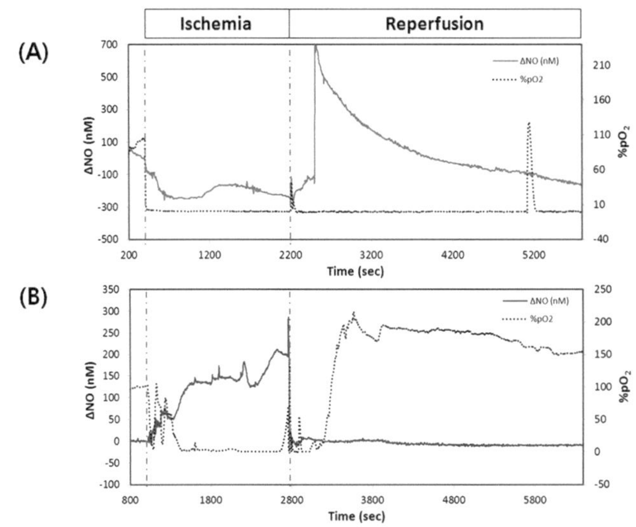

Figure 4. Representative real-time and simultaneous measurement of nitric oxide and oxygen tension in (A) control and (B) remote ischemic preconditioning (RIPC) groups during myocardial ischemia-reperfusion in Langendorff-perfused rat hearts. Reproduced from Kang et al. [65]. © 2013 Elsevier B.V.

5.4. Effect of prostaglandin E1

Prostaglandin E1 (Alprostadil®, PGE1), which is an important member of the prostaglandin (PG) family, is a product of arachidonic acid metabolism by cyclooxygenase [66, 67]. Similar to NO, PGE1 has cardioprotective effects during IR [67, 68], as well as vasodilator effects on the systemic and pulmonary circulation [69]. Fang et al. reported that pretreatment of human umbilical vein endothelial cells with PGE1 significantly protected those cells from H_2O_2-induced cell death [66]. And this effect might depend, at least in part, on the upregulation of NO expression [66]. On the other hand, Huk et al. reported that PGE1 prevents the excessive generation of NO, superoxide, and $ONOO^-$ which trigger a cascade of events leading to IR injury [70]. Though it is known that PGE1 has cardioprotective effects against IR injury, its mechanism and the correlation between NO and PGE1 are not yet clear. Lee et al. monitored the changes in NO and O_2 levels in the myocardium during myocardial IR that were induced by PGE1 pretreatment, utilizing sol-gel-modified amperometric NO and O_2 microsensors [67]. They investigated the correlation between endogenous NO and PGE1 in the ischemic episode, as well as oxygen recovery in the post-ischemic myocardium [67]. For statistical and quantitative analysis, they utilized analytical parameters such as %NO and $\%pO_2$, which are defined as the percentage of normalized NO (Eq. (1)) and pO_2 (Eq. (2)), respectively:

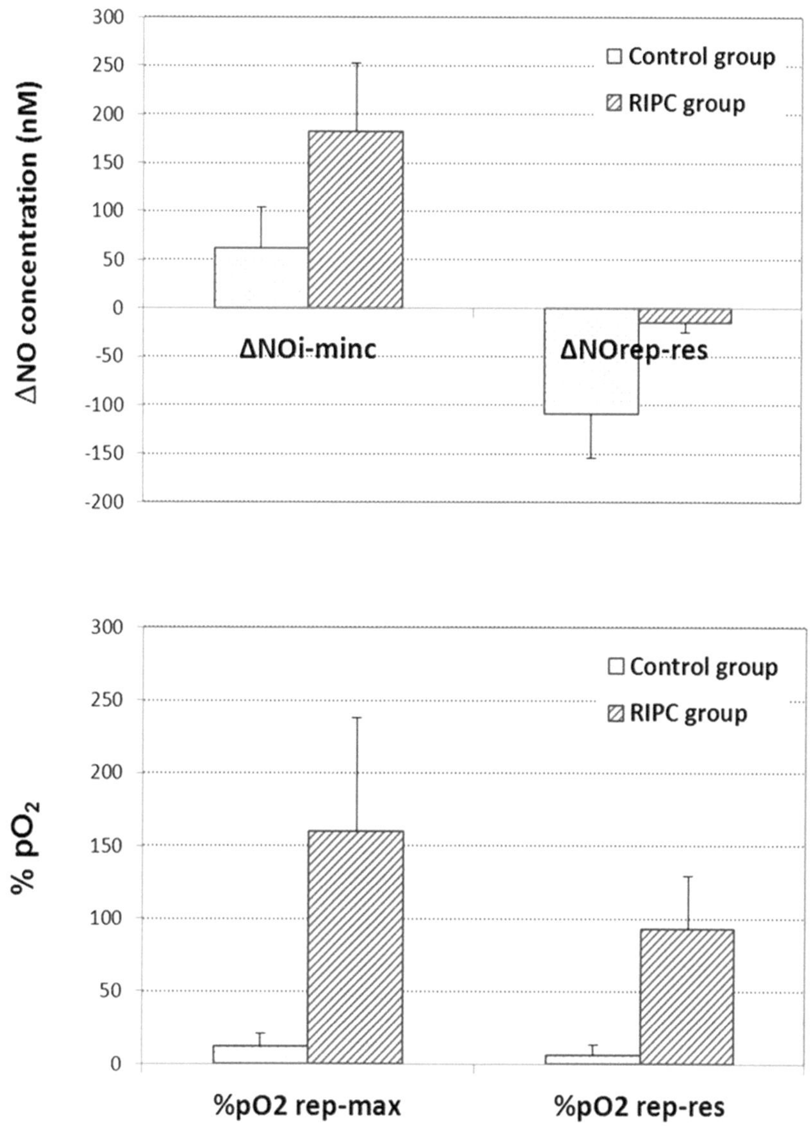

Figure 5. The correlation between ischemia-evoked nitric oxide concentration and reoxygenation parameters of the post-ischemic myocardium in two groups. Error bars represent standard deviation of the mean (n = 5 per group). Reproduced from Ref. [65], Kang SW et al., Anal. Chim. Acta 802, 74 (2013). © 2013 Elsevier B.V.

$$\%NO = \frac{\text{NO level on ischemic or reperfusion period (nM)}}{\text{NO level on pre-ischemic period (mM)}} \times 100 \tag{1}$$

$$\%pO_2 = \frac{pO_2 \text{ level on ischemic or reperfusion period (mmHg)}}{pO_2 \text{ level on pre-ischemic period (mmHg)}} \times 100 \tag{2}$$

As a result, there were significant differences in the NO dynamics during ischemia and reperfusion between the control and PGE1-treated rat hearts (**Table 1** and **Figure 6**) as follows

[67]: In the control group, the initial decrease in %NO was 56.0 ± 12.9, and the maximum NO level was 86.7 ± 18.5 during the ischemic period. However, in the PGE1 group, %NO rapidly declined to 19.9 ± 5.8% of the pre-ischemic levels, and this was maintained throughout the 30 min ischemic episode. In addition, after the onset of reperfusion, NO level inclined to a maximum of 82.0 ± 6.4 but did not exceed the pre-ischemic basal NO level. In the control group, the maximum %NO response (164.9 ± 41.0) to reperfusion was approximately double than that of the PGE1 group ($p < 0.01$, n = 5). They suggest that the cardioprotective effect of PGE1 might be attributed to a reduction in excessive NO production during early reperfusion.

Parameters	Control group (n = 5)	PGE1 group (n = 5)	*p* value
Level of baseline during pre-ischemic period	101.4 ± 0.9	101.0 ± 2.9	0.767
Level of initial decrement after the onset of ischemia	56.0 ± 12.9	19.9 ± 5.8	<0.001
Maximum level during ischemia	86.7 ± 18.5	26.0 ± 13.1	<0.001
Maximum level during the reperfusion period	164.9 ± 41.0	82.0 ± 6.4	0.01
Restoration level after 60 min of reperfusion	98.7 ± 42.1	45.6 ± 10.1	0.046

Reproduced from Ref. [67], Kang et al., Sensor. Actuat. B – Chem. 203, 245 (2014). © 2014 Elsevier B.V.

Table 1. Changes in %NO level during myocardial ischemia-reperfusion of the control and PGE1-treated groups.

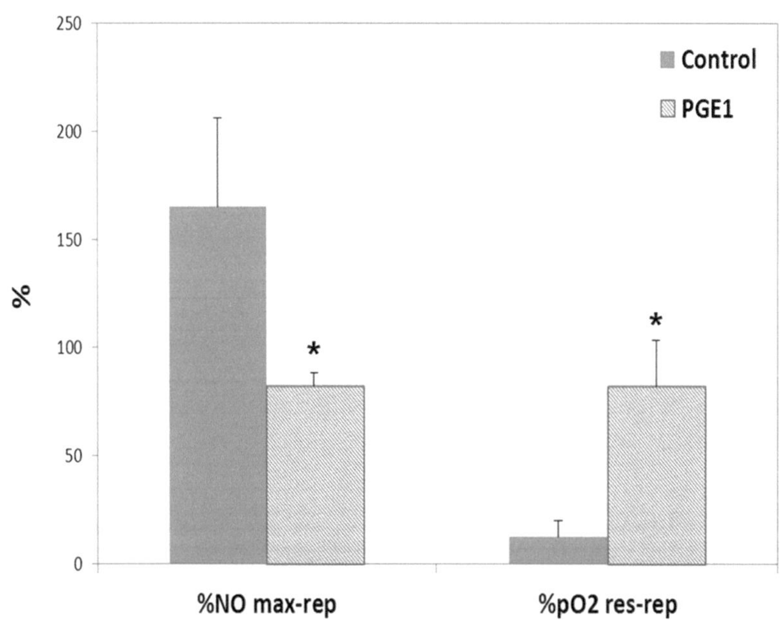

Figure 6. The correlation between the maximum %NO and restoration %pO_2 during 60 min of reperfusion in the two groups. Error bars represent the standard deviation from the mean (n = 5 per group). Reproduced from Kang et al. [67]. © 2014 Elsevier B.V.

http://dx.doi.org/10.5772/67255

6. Conclusions

NO plays important roles in the cardiovascular system by mediating various physiological and pathophysiological processes. From the real-time measurement of endogenous NO dynamics in the myocardium, we summarize as follows: (1) NO concentration was definitely decreased after myocardial ischemia; (2) there was endogenous NO formation as a protective response against ischemia during the ischemic episode, but it was not enough to restore pre-ischemic NO level; (3) the promotion of endogenous formation and inhibition of the time-course alteration of NO during an ischemic episode might be helpful as a therapeutic strategy for protecting the myocardium from ischemic injury; and (4) the reduction of excessive NO production in early reperfusion period might also be helpful as a therapeutic strategy to protect the myocardium from IR injury. And NO permselective microsensors have good sensitivity and specificity for detecting biologically released NO dynamics in vivo and can be applied in real-time monitoring of NO dynamics in various organs.

Acknowledgements

This study was supported by the National Research Foundation of Korea (NRF) funded by the Ministry of Education (NRF-2013R1A1A2065149 and 2016R1D1A1A09919833) and the Ministry of Science, ICT, and Future Planning (NRF-2015M3A9E2029188).

Author details

Gi-Ja Lee[1, 2], Young Ju Lee[1] and Hun-Kuk Park[1, 2]*

*Address all correspondence to: sigmoidus@khu.ac.kr

1 Department of Biomedical Engineering & Healthcare Industry Research Institute, College of Medicine, Kyung Hee University, Seoul, South Korea

2 Department of Medical Engineering, Graduate School, Kyung Hee University, Seoul, South Korea

References

[1] Hibbs JB, Taintor RR, Vavrin Z, Rachelin EM. Nitric oxide: a cytotoxic activated macrophage effector molecule. Biochem. Biophys. Res. Commun. 1988;**157**:87–94. DOI: 10.1016/S0006-291X(88)80015-9

[2] Palmer PMJ, Ferrige AG, Moncada S. Nitric oxide release accounts for the biological activity of endothelium-derived relaxing factor. Nature 1987;**327**:524–526. DOI: 10.1038/327524a0

[3] Szabo C. Physiological and pathophysiological roles of nitric oxide in the central nervous system. Brain Res. Bull. 1996;**44**:131–141. DOI: 10.1016/0361-9230(96)00159-1

[4] Lee GJ, Chae SJ, Kim SK, Lee S, Shin JH, Kim W, Park HK. Real time dynamics of nitric oxide during cardiac ischemia-reperfusion of the rat. Sensor. Actuat. B Chem. 2012;**161**:480–485. DOI: 10.1016/j.snb.2011.10.064

[5] Hong JK, Yun BW, Kang JG, Raja MU, Kwon E, Sorhagen K, Chu C, Wang Y, Loake GJ. Nitric oxide function and signalling in plant disease resistance. J. Exp. Bot. 2008;**59**:147–154. DOI: 10.1093/jxb/erm244

[6] Otani H. The role of nitric oxide in myocardial repair and remodelling. Antioxid. Redox Signal. 2009;**11**:1913–1928. DOI: 10.1089/ARS.2009.2453

[7] Stamler JS, Singel DJ, Loscalzo J. Biochemistry of nitric oxide and its redox-activated forms. Science 1992;**258**:1898–1902. DOI: 10.1126/science.1281928

[8] Bonfoco E, Krainc D, Ankarcrona M, Nicotera P, Lipton SA. Apoptosis and necrosis: two distinct events induced, respectively, by mild and intense insults with N-methyl-D-aspartate or nitric oxide/superoxide in cortical cell cultures. Proc. Natl. Acad. Sci. USA 1995;**92**:7162–7166. DOI: 10.1089/ars.2009.2453

[9] Szabo C. Gasotransmitters in cancer: from pathophysiology to experimental therapy. Nat. Rev. Drug Discov. 2016;**15**:185–203. DOI: 10.1038/nrd.2015.1

[10] Mayer B, Hemmens B. Biosynthesis and action of nitric oxide in mammalian cells. Trends Biochem. Sci. 1997;**22**:477–481. DOI: 10.1016/S0968-0004(97)01147-X

[11] Stuehr D, Pou S, Rosen GM. Oxygen reduction by nitric-oxide synthases. J. Biol. Chem. 2001;**276**:14533–14536. DOI: 10.1074/jbc.R100011200

[12] Zweier JL, Samouilov A, Kuppusamy P. Non-enzymatic nitric oxide synthesis in biological systems. BBA-Bioenergetics 1999;**1411**:250–262. DOI: 10.1016/S0005-2728(99)00018-3

[13] Kitakaze M, Node K, Takashima S, Asanuma H, Asakura M, Sanada S, Shinozaki Y, Mori H, Sato H, Kuzuya T, Hori M. Role of cellular acidosis in production of nitric oxide in canine ischemic myocardium. J. Mol. Cell Cardiol. 2001;**33**:1727–1737. DOI: 10.1006/jmcc.2001.1435

[14] Godber BLJ, Doel JJ, Sapkota GP, Blake DR, Stevens CR, Eisenthal R, Harrison R. Reduction of nitrite to nitric oxide catalyzed by xanthine oxidoreductase. J. Biol. Chem. 2000;**275**:7757–7763. DOI: 10.1074/jbc.275.11.7757

[15] Csonka C, Szilvássy Z, Fülöp F, Páli T, Blasig IE, Tosaki A, Schulz R, Ferdinandy P. Classic preconditioning decreases the harmful accumulation of nitric oxide during ischemia and reperfusion in rat hearts. Circulation 1999;**100**:2260–2266. DOI: 10.1161/01.CIR.100.22.2260

[16] Brovkovych V, Stolarczyk E, Oman J, Tomboulian P, Malinski T. Direct electrochemical measurement of nitric oxide in vascular endothelium. J. Pharm. Biomed. Anal. 1999;**19**:135–143. DOI: 10.1016/S0731-7085(98)00090-9

[17] Shin JH, Privett BJ, Kita JM, Wightman RM, Schoenfisch MH. Fluorinated xerogel-derived microelectrodes for amperometric nitric oxide sensing. Anal. Chem. 2008;**80**:6850–6859. DOI: 10.1021/ac800185x

[18] Bryan NS, Grisham MB. Methods to detect nitric oxide and its metabolites in biological samples. Free Radic. Biol. Med. 2007;**43**:645–657. DOI: 10.1016/j.freeradbiomed.2007.04.026

[19] Grisham MB, Johnson GG, Lancaster JR. Quantitation of nitrate and nitrite in extracellular fluids. Methods Enzymol. 1996;**268**:237–246. DOI: 10.1016/S0076-6879(96)68026-4

[20] Chan KL. Role of nitric oxide in ischemia and reperfusion injury. Curr. Med. Chem. Anti Inflamm. Anti Allergy Agents. 2002;**1**:1–13. DOI: 10.2174/1568014024606511

[21] Takahashi S, Omori Y, Miyazaki H, Yoshino F, Shoji H, Lee MC, Todoki K, Kamibayashi M, Murakami E. Real-time monitoring of nitric oxide in ischemic myocardium using an NO-selective electrode calibrated by electron spin resonance. Life Sci. 2003;**74**:75–85. DOI: 10.1016/j.lfs.2003.06.027

[22] Robinson JK, Bollinger MJ, Birks JW. Luminol/H_2O_2 chemiluminescence detector for the analysis of nitric oxide in exhaled breath. Anal. Chem. 1999;**71**:5131–5136. DOI: 10.1021/ac990646d

[23] Duarte AJ, da Silva JC. Reduced fluoresceinamine as a fluorescent sensor for nitric oxide. Sensors (Basel). 2010;**10**:1661–1669. DOI: 10.3390/s100301661

[24] Gomes A, Fernandes E, Lima JL. Use of fluorescence probes for detection of reactive nitrogen species: a review. J. Fluoresc. 2006;**16**:119–139. DOI: 10.1007/s10895-005-0030-3

[25] Bedioui F, Villeneuve N. Electrochemical nitric oxide sensors for biological samples—principle, selected examples and applications. Electroanalysis 2003;**15**:5–18. DOI: 10.1002/elan.200390006

[26] Ciszewski A, Milczarek G. Electrochemical detection of nitric oxide using polymer modified electrodes. Talanta 2003;**61**:11–26. DOI: 10.1016/S0039-9140(03)00355-2

[27] Shin JH, Weinman SW, Schoenfisch MH. Sol–gel derived amperometric nitric oxide microsensor. Anal. Chem. 2005;**77**:3494–3501. DOI: 10.1021/ac048153i

[28] Shim JH, Lee Y. Glass nanopore-based ion-selective electrodes. Anal. Chem. 2009;**81**:8571–8576. DOI: 10.1021/ac061984z

[29] de Vooys ACA, Beltramo GL, van Riet B, van Veen JAR, Koper MTM. Mechanisms of electrochemical reduction and oxidation of nitric oxide. Electrochim. Acta 2004;**49**:1307–1314. DOI: dx.doi.org/10.1016/j.electacta.2003.07.020

[30] Prakash R, Srivastava RC, Seth PK. Polycarbazole modified electrode; nitric oxide sensor. Polym. Bull. 2001;**46**:487–490. DOI: 10.1007/s002890170036

[31] Friedemann MN, Robinson SW, Gerhardt GA. o-Phenylenediamine-modified carbon fiber electrodes for the detection of nitric oxide. Anal. Chem. 1996;**68**:2621–2628. DOI: 10.1021/ac960093w

[32] Kitamura Y, Uzawa T, Oka K, Komai Y, Takizawa N, Kobayashi H, Tanishita K. Microcoaxial electrode for in vivo nitric oxide measurement. Anal. Chem. 2000;**72**:2957–2962. DOI: 10.1021/ac000165q

[33] Zhang X, Cardosa L, Broderick M, Fein H, Lin J. An integrated nitric oxide sensor based on carbon fiber coated with selective membranes. Electroanalysis 2000;**12**:1113–1117. DOI: 10.1002/1521-4109(200010)12:14<1113::AID-ELAN1113>3.0.CO;2-U

[34] Lee Y, Oh BK, Meyerhoff ME. Improved planar amperometric nitric oxide sensor based on platinized platinum anode. 1. Experimental results and theory when applied for monitoring NO release from diazeniumdiolate-doped polymeric films. Anal. Chem. 2004;**76**:536–544. DOI: 10.1021/ac035064h

[35] Kato D, Sakata M, Hirayama C, Hirata Y, Mizutani F, Kunitake M. Selective permeation of nitric oxide through two dimensional cross-linked polysiloxane LB films. Chem. Lett. 2002:2:1190–1191

[36] Cserey A, Gratzl M. Stationary-state oxidized platinum microsensor for selective and on-line monitoring of nitric oxide in biological preparations. Anal. Chem. 2001;**73**:3965–3974. DOI: 10.1021/ac010123h

[37] Kobayashi K, Neely JR. Control of maximum rates of glycolysis in rat cardiac muscle. Circ. Res. 1979;**44**:166–175. DOI: 10.1161/01.RES.44.2.166

[38] Kawahara K, Hachiro T, Yokokawa T, Nakajima T, Yamauchi Y, Nakayama Y. Ischemia/reperfusion-induced death of cardiac myocytes: possible involvement of nitric oxide in the coordination of ATP supply and demand during ischemia. J. Mol. Cell. Cardiol. 2006;**40**:35–46. DOI: 10.1016/j.yjmcc.2005.06.020

[39] Sharp BR, Jones SP, Rimmer DM, Lefer DJ. Differential response to myocardial reperfusion injury in eNOS-deficient mice. Am. J. Physiol. Heart Circ. Physiol. 2002;**282**:H2422–H2426. DOI: 10.1152/ajpheart.00855.2001

[40] Jones SP, Greer JJ, Kakkar AK, Ware PD, Turnage RH, Hicks M, van Haperen R, de Crom R, Kawashima S, Yokoyama M, Lefer DJ. Endothelial nitric oxide synthase overexpression attenuates myocardial reperfusion injury. Am. J. Physiol. Heart Circ. Physiol. 2004;**286**:H276–H282. DOI: 10.1152/ajpheart.00129.2003

[41] Pabla R, Buda AJ, Flynn DM, Blesse SA, Shin AM, Curtis MJ, Lefer DJ. Nitric oxide attenuates neutrophil-mediated myocardial contractile dysfunction after ischemia and reperfusion. Circ. Res. 1996;**78**:65–72. DOI: 10.1161/01.RES.78.1.65

[42] Hataishi R, Rodrigues AC, Neilan TG, Morgan JG, Buys E, Shiva S, Tambouret R, Jassal DS, Raher MJ, Furutani E, Ichinose F, Gladwin MT, Rosenzweig A, Zapol WM, Picard MH, Bloch KD, Scherrer-Crosbie M. Inhaled nitric oxide decreases infarction size and improves left ventricular function in a murine model of myocardial ischemia-reperfusion injury. Am. J. Physiol. Heart Circ. Physiol. 2006;**291**:H379–H384. DOI: 10.1152/ajpheart.01172.2005

[43] Calvert JW, Lefer DJ. Myocardial protection by nitrite. Cardiovasc. Res. 2009;**83**:195–203. DOI: 10.1093/cvr/cvp079

[44] Loke KE, Laycock SK, Mital S, Wolin MS, Bernstein R, Oz M, Addonizio L, Kaley G, Hintze TH. Nitric oxide modulates mitochondrial respiration in failing human heart. Circulation 1999;**100**:1291–1297. DOI: 10.1161/01.CIR.100.12.1291

[45] Thomas DD, Liu X, Kantrow SP, Lancaster JR Jr. The biological lifetime of nitric oxide: implications for the perivascular dynamics of NO and O_2. Proc. Natl. Acad. Sci. USA 2001;**98**:355–360. DOI: 10.1073/pnas.98.1.355

[46] Schumacker PT. Nitric oxide quenches the fire in heart mitochondria. Nat. Med. 2013;**19**:666–667. DOI: 10.1038/nm.3224

[47] Brown GC, Borutaite V. Inhibition of mitochondrial respiratory complex I by nitric oxide, peroxynitrite and S-nitrosothiols. Biochim. Biophys. Acta. 2004;**1658**:44–49. DOI: 10.1016/j.bbabio.2004.03.016

[48] Chouchani ET, Methner C, Nadtochiy SM, Logan A, Pell VR, Ding S, James AM, Cochemé HM, Reinhold J, Lilley KS, Partridge L, Fearnley IM, Robinson AJ, Hartley RC, Smith RA, Krieg T, Brookes PS, Murphy MP. Cardioprotection by S-nitrosation of a cysteine switch on mitochondrial complex I. Nat. Med. 2013;19:753–759. DOI: 10.1038/nm.3212. Epub 2013 May 26.

[49] Ma XL, Weyrich AS, Lefer DJ, Lefer AM. Diminished basal nitric oxide release after myocardial ischemia and reperfusion promotes neutrophil adherence to coronary endothelium. Circ. Res. 1993;**72**:403–412. DOI: 10.1161/01.RES.72.2.403

[50] Palazzo AJ, Jones SP, Girod WG, Anderson DC, Granger DN, Lefer DJ. Myocardial ischemia-reperfusion injury in CD18- and ICAM-1-deficient mice. Am. J. Physiol. 1998;**275**:H2300–H2307.

[51] Walter U, Gambaryan S. Roles of cGMP/cGMP-dependent protein kinase in platelet activation. Blood 2004;**104**:2609. DOI: 10.1182/blood-2004-06-2389

[52] Calvert JW, Lefer DJ. Statin therapy and myocardial no-reflow. Br. J. Pharmacol. 2006;**149**:229–231. DOI: 10.1038/sj.bjp.0706862

[53] Li J, Bombeck CA, Yang S, Kim YM, Billiar TR. Nitric oxide suppresses apoptosis via interrupting caspase activation and mitochondrial dysfunction in cultured hepatocytes. J. Biol. Chem. 1999;**274**:17325–17333. DOI: 10.1074/jbc.274.24.17325

[54] Kim YM, Talanian RV, Billiar TR. Nitric oxide inhibits apoptosis by preventing increases in caspase-3-like activity via two distinct mechanisms. J. Biol. Chem. 1997;**272**:31138–31148. DOI: 10.1074/jbc.272.49.31138

[55] Titheradge MA. Nitric oxide in septic shock. Biochim. Biophys. Acta 1999;**1411**:437–455. DOI: 10.1016/S0005-2728(99)00031-6

[56] Brown GC, Borutaite V. Nitric oxide and mitochondrial respiration in the heart. Cardiovasc. Res. 2007;**75**:283–290. DOI: 10.1016/j.cardiores.2007.03.022

[57] Olivecrona GK, Götberg M, Harnek J, Van der Pals J, Erlinge D. Mild hypothermia reduces cardiac post-ischemic reactive hyperemia. BMC Cardiovasc. Disord. 2007;**7**:5. DOI: 10.1186/1471-2261-7-5

[58] Hale SL, Kloner RA. Myocardial hypothermia: a potential therapeutic technique for acute regional myocardial ischemia. J. Cardiovasc. Electrophysiol. 1999;**10**:405–413. DOI: 10.1111/j.1540-8167.1999.tb00689.x

[59] Lee GJ, Kim SK, Kang SW, Kim OK, Chae SJ, Choi S, Shi JH, Park HK, Chung JH. Real time measurement of myocardial oxygen dynamics during cardiac ischemia–reperfusion of rats. Analyst 2012;**137**:5312–5319. DOI: 10.1039/c2an35208g

[60] Tapuria N, Kumar Y, Habib MM, Amara MA, Seifalian AM, Davidson BR. Remote ischemic preconditioning: a novel protective method from ischemia reperfusion injury—a review. J. Surg. Res. 2008;**150**:304–330. DOI: 10.1016/j.jss.2007.12.747

[61] Frank A, Bonney M, Bonney S, Weitzel L, Koeppen M, Eckle T. Myocardial ischemia reperfusion injury: from basic science to clinical bedside. Semin. Cardiothorac. Vasc. Anesth. 2012;**16**:123–132. DOI: 10.1177/1089253211436350

[62] Yuan HJ, Zhu XH, Luo Q, Wu YN, Kang Y, Jiao JJ, Gao WZ, Liu YX, Lou JS. Noninvasive delayed limb ischemic preconditioning in rats increases antioxidant activities in cerebral tissue during severe ischemia-reperfusion injury. J. Surg. Res. 2012;**174**:176–183. DOI: 10.1016/j.jss.2010.11.001

[63] Kraemer R, Lorenzen J, Kabbani M, Herold C, Busche M, Vogt PM, Knobloch K. Acute effects of remote ischemic preconditioning on cutaneous microcirculation—a controlled prospective cohort study. BMC Surg. 2011;**11**:32–38. DOI: 10.1186/1471-2482-11-32

[64] Burwell LS, Brookes PS. Mitochondria as a target for the cardioprotective effects of nitric oxide in ischemia–reperfusion injury. Antioxid. Redox Signal. 2008;**10**:579–599. DOI: 10.1089/ars.2007.1845

[65] Kang SW, Kim OK, Seo B, Lee SH, Quan FS, Shin JH, Lee GJ, Park HK. Simultaneous, real-time measurement of nitric oxide and oxygen dynamics during cardiac ischemia–reperfusion of the rat utilizing sol–gel-derived electrochemical microsensors. Anal. Chim. Acta 2013;**802**:74–81. DOI: 10.1016/j.aca.2013.09.029

[66] Fang WT, Li HJ, Zhou LS. Protective effects of prostaglandin E1 on human umbilical vein endothelial cell injury induced by hydrogen peroxide. Acta Pharmacol. Sin. 2010;**31**:485–492. DOI: 10.1038/aps.2010.23

[67] Kang SW, Ahn HJ, Seo B, Kim OK, Lee YJ, Shin JH, Lee GJ, Park HK. Effects of prostaglandin E1 on nitric oxide and oxygen dynamics during rat myocardial ischemia–reperfusion utilizing sol–gel derived microsensors. Sensor. Actuat. B Chem. 2014;**203**:245–251. DOI: 10.1016/j.snb.2014.06.127

[68] Tauber S, Menger MD, Lehr HA. Microvascular in vivo assessment of reperfusion injury: significance of prostaglandin E(1) and I(2) in postischemic "no-reflow" and "reflow-paradox". J. Surg. Res. 2004;**120**:1–11. DOI: 10.1016/S0022-4804(03)00332-9

[69] Locker GJ, Grimm M, Losert H, Stoiser B, Kofler J, Knapp S, Wilfing A, Knoebl P, Kapiotis S, Czerny M, Muhm M, Hiesmayr M, Frass M. Prostaglandin E-1 does not influence plasmatic coagulation, hepatic synthesis, or postoperative blood loss in patients after coronary-artery bypass grafting. J. Clin. Anesth. 2000;**12**:363–370. DOI: 10.1016/S0952-8180(00)00170-7

[70] Huk I, Brovkovych V, Nanobashvili J, Neumayer C, Polterauer P, Prager M, Patton S, Malinski T. Prostaglandin E1 reduces ischemia/reperfusion injury by normalizing nitric oxide and superoxide release. Shock 2000;**14**:234–242.

Chapter 6

Role of Endothelial Nitric Oxide Synthase in Glucocorticoid-Induced Hypertension: An Overview of Experimental Data

Kinga G. Blecharz-Lang and Malgorzata Burek

Additional information is available at the end of the chapter

http://dx.doi.org/10.5772/67025

Abstract

Imbalances in the synthesis or in the bioavailability of nitric oxide (NO), the freely diffusible vasodilator, in myocardial endothelial cells were demonstrated to be crucial in the development of hypertension. Glucocorticoids (GCs) are widely used as immunomodulators. One of the numerous side effects of GC therapy is hypertension arising from reduced release of the endothelium-derived NO. GCs can modulate NO synthesis by targeting the genes involved in it, like nitric oxide synthase (NOS) and guanosine triphosphate (GTP) cyclohydrolase-1 (GTPCH-1). This chapter will give an overview on the impact of GCs on NO synthesis and signalling in animal models as well as in in vitro cell culture models. Moreover, strategies for preventing or neutralizing side effects of long-term GC therapy will be discussed.

Keywords: myocardial endothelial cells, nitric oxide, nitric oxide synthase, glucocorticoids, glucocorticoid receptor

1. Introduction

Hypertension mediated by glucocorticoids (GCs) is a frequent clinical problem. Among others it results from the fact that GCs remain as one of the most commonly prescribed drugs for many conditions, such as inflammatory diseases, asthma, multiple sclerosis, chronic obstructive pulmonary disease (COPD) and many more. Either the pharmacological administration of GCs or intrinsic GC excess in humans (caused that is by Cushing syndrome) may result in hypertension. The problem was originally discussed to originate from sodium excess or from an impairment of water reabsorption by the renal mineralocorticoid receptor [1, 2]. However, extensive experimental and clinical data begun to unravel the complex molecular mechanisms inducing the onset and leading to a persistence of pathologically high blood pressure induced by GCs. In addition, recent research pointed to the role of extra-renal tissues, such as the vascular endothelium, that have not been taken into account so far in the regulation of blood pressure. The importance of glucocorticoid receptor (GR) signalling became more and more obvious in this process [3, 4].

Effects of GCs on vessel vasodilatation and nitric oxide (NO) synthesis as well as imbalances between NO and superoxide were examined extensively in numerous in vitro and in vivo studies. Imbalances in the bioavailability of endothelium-derived NO play a crucial role in diverse cardiovascular diseases including hypertension, arteriosclerosis and hypercholesterolemia [5–11]. NO is also known to influence circulating platelets and white blood cells and modulates various cellular events, such as platelet activation, mitochondrial function, ion transport, inflammation, angiogenesis and cell proliferation, all processes being essential in cardiovascular homeostasis. It also contributes to the control of the heart rate and its contractility. Although its role in these processes is not fully known, the amount of bioactive NO has been suggested to be involved in the control of cardiac contractility. While lower NO levels seem to promote contractility, higher levels have the opposite effects [12]. It limits cardiac remodelling after ischemic injury. NO is known to be constitutively generated in cardiomyocytes by two different NO synthase (NOS) isoforms, the endothelial and neuronal one indicating different roles of these isoforms in cardiac function [13]. In this chapter, we summarized the state of knowledge from available experimental data considering the influence of externally administered GCs on the synthesis of the vasodilatator NO in hypertension. Hypertension arising from intrinsic GC excess will not be discussed here.

2. Glucocorticoids and the glucocorticoid receptor

GCs are steroid hormones synthesized in the adrenal glands. They are essential for life and involved in various cellular processes, such as in the regulation of metabolic and immunological pathways, and the control of cellular homeostasis. Cortisol is the main form of GCs produced in the human body. Multiple naturally occurring as well as synthetic compounds, like dexamethasone, are being used in the treatment of diverse inflammatory and immunological diseases. The cellular effects mediated by GCs, including genomic and non-genomic ones, are exerted by binding to their receptor, the GR. In its inactive state, the GR is kept in the cytosol fraction by chaperon proteins, such as the heat shock protein 90 (HSP90). The lipophilic GCs diffuse through the cell membrane into the cytosol; they bind to the GR which in that moment is released by the HSPs and is then activated (**Figure 1**). After dimerization, the GR-GC complex enters the nucleus where it binds to so-called glucocorticoid response elements (GREs) located in the promoter regions of target genes. By this binding, the GR activates or represses the expression of its specific targets [14]. Moreover, by binding to other transcription factors, the GR can indirectly influence the expression of certain other genes [14].

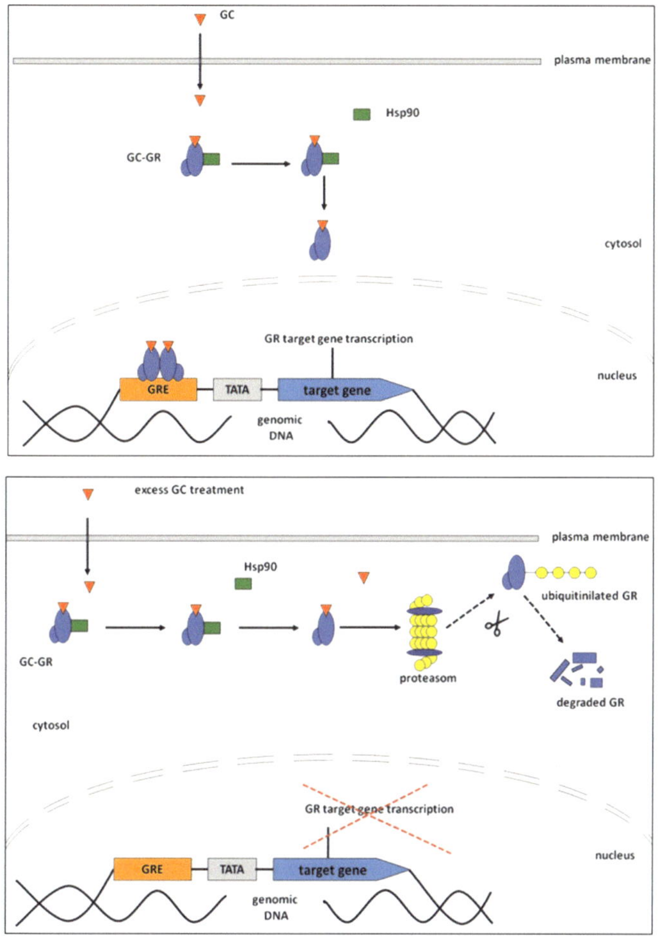

Figure 1. Glucocorticoid receptor signalling. In its inactive state, the glucocorticoid receptor (GR) is kept in the cytosol by chaperon proteins, such as the heat shock protein 90 (HSP90). The lipophilic GCs diffuse through the cell membrane into the cytosol, bind to the GR resulting in release of GR from the HSP complex and its activation. After dimerization, the GR-GC complex enters the nucleus where it binds to so-called glucocorticoid response elements (GREs) located in the promoter regions of target genes. The protein amount of the GR correlates with its transcriptional activity. Under physiological conditions and excess GC treatment, the recycling process of receptor/transcriptional DNA complexes is regulated by the 26S proteasome, GR ubiquitination and degradation.

The GR is conserved among all species and its expression was confirmed in a wide range of tissues including the vascular endothelium. It was also shown to be expressed in microvascular endothelial cells of the myocardium and of the blood-brain barrier [15–19]. The protein amount of the GR correlates with its transcriptional activity. Under physiological conditions, the GC turnover and the recycling process of receptor/transcriptional DNA complexes is regulated by the 26S proteasome as well as by the degradation of the GR (**Figure 1**) [20, 21].

Although GCs are the most widely prescribed medicine, systemic GC administration, especially for a long period, is associated with numerous detrimental side effects [22, 23]. GC-induced hypertension is one of the common undesirable effects. Tissue-specific GR knockout mice shed light on the molecular mechanisms behind GC-induced hypertension. Against expectations, a GR deletion in the distal nephron did not protect mice against hypertension caused by GC administration [24]. Also, the lack of GR in vascular smooth muscle cells has only delayed the onset of hypertension in comparison to wild-type mice [25]. Finally, mice with a tissue-specific GR knockout in the vascular endothelium were resistant to GC-induced hypertension [26]. These facts underlined the pivotal role of the GR expressed in vascular endothelial cells in generation and maintenance of GC-mediated hypertension demonstrating that cell-specific actions of the GR are responsible for the phenotype of a whole organism.

3. Glucocorticoid effects on the NO synthesis and signalling pathway

3.1. Influence on eNOS

At cellular level, GCs negatively influence the synthesis of NO causing endothelial dysfunction. NOS is expressed in three isoforms: neuronal NOS (nNOS, NOS1), inducible NOS (iNOS, NOS2) and endothelial NOS (eNOS, NOS3) [27, 28]. Under physiological circumstances, all NOS isoforms can catalyse the conversion of L-arginine to L-citrulline and NO. Endothelial NOS constitutively produces NO in vascular endothelial cells. Cofactors involved in this process, such as 5,6,7,8,-tetrahydrobiopterin (BH_4), are limiting elements in the synthesis of NO. It has been observed that the absence of BH_4 or L-arginin leads to eNOS uncoupling and to the production of reactive oxygen species (ROS) instead of NO. In consequence, this results in endothelial dysfunction followed by severe vascular disorders, that is, of cardiovascular nature [29]. The pivotal role of eNOS and its product NO in regulating the blood pressure and blood flow was demonstrated in eNOS knockout and eNOS over-expressing mice. While mice lacking eNOS were hypertensive, animals over-expressing eNOS in the vascular tissue became hypotensive and suffered from decreased vasoreactivity [30–32]. These mice models confirmed that blood pressure is regulated by eNOS-derived NO. Therefore, first efforts to explain the molecular mechanism of GC-mediated hypertension focussed on the regulation of eNOS.

GCs lead to a down-regulation of eNOS mRNA and protein in cultured endothelial cells, in the aorta and in organs such as kidney and liver of GC-treated rats [33]. In vivo, disruption of the eNOS gene preserved mice from hypertension [34]. In vitro studies with human umbilical vein endothelial cells(HUVECs) have demonstrated that endogenous GR acts as a negative regulator of both eNOS and iNOS. It was shown that a GR knockdown in HUVECs increased eNOS mRNA and protein levels [35]. An endothelial deletion of the GR in mice resulted in an accelerated expression of eNOS in septic mice [35]. Moreover, other animal studies demonstrated that following GC treatment, the expression of eNOS was significantly down-regulated [33, 36]. The observed eNOS reduction could be further confirmed at promoter level after identification of a potential GRE in the promoter region of this gene [37]. This study which has been performed in HUVECs demonstrated the suppression of the eNOS promoter activity in response to cortisol application. In line with these results, the binding of the GR to the suppressive GRE at −111 to −105 bp located on the NOS promoter was tested

http://dx.doi.org/10.5772/67025

by chromatin immunoprecipitation [37]. Recently, a dexamethasone-mediated regulation of the eNOS promoter in microvascular myocardial endothelial cells derived from mice has been shown. Concordantly with previous results, a GC-mediated suppression of the eNOS promoter could be measured [19]. The GR can inhibit eNOS and other forms of NOS by trans-repressive interactions with other transcription factors, such as NFκB or AP-1 [38, 39]. Moreover, recent reports revealed the meaning of the molar ratio between BH_4 and eNOS rather than the absolute amounts of the cofactor and its enzyme. As we know by now, GCs cause a reduction of both BH_4 and eNOS. They also inhibit NOS phosphorylation without an evident uncoupling of eNOS after GCs treatment [40].

3.2. Effects on L-arginine and BH_4

GCs are known to alter various molecules in the NOS-mediated biosynthesis of NO. L-arginine is an essential substrate for all NOS isoforms. In GC-induced hypertension, systemic levels of L-arginine and L-citrulline have been observed to be diminished reflecting the importance of upstream NOS regulators in this clinical condition [41]. L-arginine supplementation could partially reverse hypertension in GC-treated rats [41, 42]. BH_4 is an essential cofactor of eNOS in the biosynthesis of NO [43]. It can be produced by two different strategies. The first is via the conversion from GTP catalysed by the GTP cyclohydrolase 1 (GTPCH-1), which is the rate-limiting enzyme involved in its synthesis. The second way is the salvage production pathway, where the precursor protein sepiapterin is converted into the intermediate molecule BH_2 and then into BH_4 [44]. The importance of the bioavailability of BH_4 was validated in ex vivo studied vessel pieces. Here, a 30-fold increase of NOS activity in the presence of BH_4 compared with BH_4-free preparations could be determined [45, 46]. Limited amounts of BH_4 can lead to an uncoupling of eNOS and further to endothelial dysfunction [47, 48]. Uncoupled eNOS overproduces ROS which in turn enhance the oxidation of BH_4 to BH_2 resulting in BH_4 deficiency [49–52]. BH_2 can be reconverted to BH_4 by dihydrofolate reductase (DHFR) [53]. A GC supplementation of cultured human endothelial cells was shown to reduce mRNA and protein levels of DHFR [40]. Furthermore, administration of external BH_4 led to an increase of NO levels following GC-treatment, as it was demonstrated in animal studies as well as in coronal endothelial cells [51, 54]. However, BH_4 application or treatment with sepiapterin, a precursor of BH_4, did not alter systolic blood pressure although total concentrations of plasma BH_4 were elevated in GC-induced hypertensive rats [55, 56]. GC effects on BH_4 and NO down-regulation have been proven to be genomic by the usage of specific GR antagonists. One of such antagonistic compounds is RU486, the application of which could reverse the decreasing effects on NO synthesis caused by GC treatment [34, 57].

3.3. Control of GTP cyclohydrolase 1

As the rate-limiting enzyme involved in the de novo biosynthesis of BH_4, GTPCH-1 shows a decreased expression in response to GC treatment. As a result of its decrease, a significantly lowered BH_4 and NO production contributed to impaired endothelium-dependent vaso-relaxation [55, 58, 59]. In vitro studies with human and mouse endothelial cells demonstrated genomic effects of GCs on GTPCH-1 by GC-mediated down-regulation of GTPCH-1 mRNA and promoter activity [19, 40]. It was also shown that dexamethasone mediated a dense down-regulation of both GTPCH-1 and eNOS after a long-term treatment period [19]. Elsewhere it was shown that a gene transfer of human GTPCH-1 restored vascular BH_4 concentrations and positively influenced endothelial function in hypertonic rats [60]. The summary of this data clearly displays that GCs influence the whole multistep cascade of NO production mediated by eNOS.

4. Strategies to reduce glucocorticoid-induced hypertension

Since mechanisms behind GC-induced side reactions, such as hypertension, are complex and in part controversial, a basic knowledge of the pharmacology, clinical guidelines and the adverse effects of these substances is imperative. Once increased blood pressure as an unwanted effect of GC treatment has been detected, the use of appropriate anti-hypertensive compounds, such as NO-generating compounds and direct vasodilators and hybrid drugs possessing a NO-releasing moiety including diuretics, calcium channel blockers, angiotensin-converting enzyme (ACE) inhibitors, angiotensin receptor blockers, β-blockers, α-adrenergic receptor antagonists, centrally acting α-2 receptor agonists, should be considered. There are different strategies and natural as well as synthetic compounds that have been discussed to reduce or prevent GC-induced imbalances occurring in the production of NO. A brief overview on the most important ones that directly influence NO availability or signalling will be given below.

4.1. Reducing glucocorticoid toxicity

First of all, the side effects of a drug administration should be considered in terms of the drug duration/dosage, which in turn should take into account the risk/benefit ratio. GC-induced side effects are usually more severe following systemic than topical application. As an important parameter, a definite indication to start GC treatment should be considered. Moreover, short- and intermediate-acting ones should be preferred over longer-acting GCs and a minimum necessary dose and duration of treatment should be chosen. Furthermore, a once-a-day morning administration should be preferred over a divided dose therapy. Necessarily, the body weight, blood and ocular pressure, the development of cataracts, serum lipids, blood and urine glucose concentration should be monitored during the complete GC therapy.

4.2. NO-generating agents

Several drugs act as NO-releasing agents. Well known are organic nitrates and nitrites, including nitroglycerin acting as anti-anginal drug. However, the use of these agents has been associated with some limitations and some of the compounds are still tested in clinical trials.

Pulmonary rather than cardiac hypertension can be reduced by a direct delivery of NO into the lung though inhalation. Reaching well-ventilated areas of the lungs, inhaled NO increases blood oxygenation and improves ventilation-perfusion. However, its effect on the systemic arterial tone is minor since NO is rapidly scavenged and acutely inactivated after reacting with circulating haemoglobin [61]. Therefore, effects of distal NO rather occur due to the formation of other bioactive nitrogen oxides, including inorganic nitrate and *S*-nitrosothiols acting in an endocrine manner [62].

Nitrite therapy also contributes to the regulation of blood flow and blood pressure through its reduction to NO catalysed by metalloprotein oxidoreductases. The bioconversion of nitrite into NO in the vascular system is regulated by a reductive reaction with deoxyhaemoglobin and couples hypoxic sensing with NO production and therefore augmented under physiological or pathological hypoxia [63–65]. Inhaled nitrite causes pulmonary vasodilatation and induces protective remodelling in animal studies [66]. Moreover, nitrite reduces blood pressure and improves cell and organ viability after ischemia reoxygenation as it was shown in the heart, liver and lungs [67, 68]. Either oral or inhaled delivery of nitrite has been shown to exhibit a rapid and efficient systemic absorption and is still tested for further effects in various clinical studies.

Nitrate can be considered as a pro-drug that is further metabolized to nitrite and other nitrogen oxides after application. Due to its long half-life, the enterosalivary circulation and reformation, nitrite ensures a low-grade NO-like bioactivity over a longer period of time [69]. Nitrate has been shown to possess a robust NO-like activity in the cardiovascular system and can be assimilated from nitrate-rich diet, especially in green leafy vegetables, or as a salt [70]. Animal models revealed compensating effects of dietary nitrite and nitrate on the lack of eNOS expression [71]. Similar to nitrite, nitrate has also cardio- and vascular- and renoprotective properties and both molecules act as a blood pressure-lowering and anti-inflammatory agent in different animal models of cardiovascular diseases [72–74].

Regulatory functions for the cardiovascular system have also been demonstrated for nitrated fatty acids [75]. These electrophilic compounds can be generated in the reaction of unsaturated fatty acids with NO-derived reactive species and act via NO-dependent and NO-independent pathways [76]. Besides positive effects on vascular relaxation, nitrated fatty acids have shown anti-oxidative and anti-inflammatory effects in blood vessels [77, 78]. Despite the cardio-protective action of nitrated fatty acids proven in animal models, the first human clinical trial is still ongoing and safety is being tested.

4.3. L-arginine and L-citrulline administration

A direct and very simple way to enhance NO availability is L-arginine delivery. As a substrate for all NOSs, this amino acid is present in sufficiently high concentrations in most cells. However, L-arginine administration was observed to even increase NO generation [79]. Dietary L-arginine possesses protective effects on the cardiac vascularity, as it was shown in a number of studies and is therefore frequently supplemented in food. However, mixed results of L-arginine supplementation have been shown in multiple clinical trials, that is, no significant effects of L-arginine on NO availability and on various cardiovascular parameters followed by even detrimental effects, such as a higher incidence of myocardial infarction after L-arginine administration, were shown in clinical trials [80]. However, studies focusing on effects on blood pressure have shown blood pressure-lowering effects [81]. The partially converse effects may partly result from the unspecific binding of this amino acid to all three NOS isoforms. Moreover, the molecule mediates many other effects than those associated with

NO metabolism, that is, it is a precursor of other amino acids or is a substrate for arginase, a key enzyme in the urea cycle [82]. In this context, it was identified that L-arginine also interacts with the asymmetric dimethylarginine (ADMA) that acts as an endogenous NOS inhibitor. Since high ADMA levels are generated in cardiovascular diseases, L-arginine supplementation seems to normalize but not improve NO levels and endothelial function in such individuals [83].

A problem arising from L-arginine administration might also be that this amino acid is extensively metabolized by intestinal bacteria and arginase in the liver before reaching the circulation [84]. Compared with L-arginine, L-citrulline has been shown to be effectively transported into endothelial cells where it is enzymatically converted into L-arginine [84]. However, despite potential advantages shown for this molecule, no clinical trial has been initiated to study the effects of L-citrulline on cardiovascular function.

4.4. BH_4 supplementation

BH_4 is an important factor controlling the activity of all NOS isoforms. Concerning the salvage pathway by which this factor can be produced endogenously, the external supplementation of BH_4 has been considered as a strategy improving NO synthesis and reducing eNOS uncoupling. Intracellular levels of BH_4 can be restored by exposing cells to sepiapterin, being converted to BH_4 through sepiapterin reductase. Satisfying results on superoxide formation and vascular function have been achieved by the administration of sepiapterin in various models of hypertension, that is, in spontaneously hypertensive or nephrectomised animals [85, 86]. Endothelial dysfunction has also been restored by sepiapterin in dexamethasone-incubated aorta rings ex vivo [55]. There are also in vitro studies applying aortic endothelial cells showing an improvement of NO release and decrease of ROS production [58]. There is one main difference between a direct BH_4 and a sepiapterin administration. While sepiapterin is absorbed by cells and converted to intracellular BH_4, a direct BH_4 administration acts extracellularly [43]. Despite this seemingly important difference, there is only one report describing sepiapterin administration, whereas a direct BH_4 administration has not been described in the context of GC-induced hypertension [56]. The low number of experimental data might be due to the fact that sepiapterin did not affect dexamethasone-mediated hypertension in rats although plasma levels of BH_4 were significantly increased. As the authors explained, the uncoupling of eNOS seems not to play a major role in the pathogenesis of GC-induced hypertension [56].

4.5. Response to antioxidants and anti-inflammatory agents

Antioxidants and anti-inflammatory agents such as tempol, apocynin, N-acetylcysteine, folic acid and atorvastatin can prevent and partially reverse GCs-induced hypertension in vivo [87–92]. Therefore, these substances represent promising candidates for prevention and treatment of increased systolic blood pressure induced by long-term and high-dosage GC treatment.

Tempol, 4-hydroxy-2,2,6,6-tetramethyl piperidinoxyl, is a superoxide scavenger with either water-soluble or membrane-permeable characteristics. It is known to normalize blood pressure, as it was shown in numerous animal models of differently induced hypertension [90, 93–95].

Interesting findings have been reported by Zhang et al. They showed that tempol did not have an influence on oxidative stress measured in rat plasma. However, they reported tempol to have prevented and partially reversed dexamethasone-induced hypertension independently of systemic ROS.

The hydrophilic antioxidant N-acetylcysteine is already widely used in the clinic, that is, as a mucolytic agent or in ischemia-reperfusion injury [96, 97]. In dexamethasone-induced hypertension among rats, this agent was shown to lower systolic blood pressure as well as restore plasma NO levels, without any effect on oxidative markers in the plasma. Similar experimental settings revealed N-acetylcysteine treatment having partially preventing but not reversing properties on GC-induced hypertension in comparison to what has been shown for tempol [91].

The same laboratory tested folic acid in hypertension induced by synthetic GCs [87]. Folic acid is a B group vitamin that is known from its ROS decreasing properties [98]. It also lowers the concentration of homocysteine that reduces intracellular BH_4 levels, thereby increasing BH_4 bioavailability [99]. Patients suffering from Cushing syndrome tend to have higher systemic homocysteine and reduced folate levels than healthy controls [100]. It is therefore not surprising that the supplementation of folic acid might prevent and partially revers GC-induced hypertension by increasing BH_4 amounts, although the detailed mechanism remains elusive [87].

The same group has demonstrated that antioxidant apocynin, which is a specific NAD(P)H oxidase inhibitor, is able either to reverse or to prevent dexamethasone-caused hypertension [101]. Interestingly, the same study has shown that the NO precursor L-arginine did not have the same effects as those being observed for apocynin in dexamethasone-induced hypertensive rats.

Among its pleiotropic functions, atorvastatin has been reported to improve endothelial function through increased bioavailability of NO and ameliorated ROS and pro-inflammatory cytokine production in different models of hypertension [92, 102, 103]. Atorvastatin could lower the blood pressure of dexamethasone-treated hypertensive rats due to its positive influence on the expression of eNOS measured in aortic samples and due to the lowered levels of superoxide in plasma. However, there was no significant effect of this compound on vasorelaxation although the endothelial function was markedly improved [92].

4.6. Influencing the GC signalling pathway

As it was shown in a transgenic animal model, an endothelial GR knockout was protected against GC-induced hypertension [26]. Targeting GR in endothelial cells could therefore constitute a preventive treatment strategy directed against GC-induced hypertension. A recent study by Blecharz et al. provided new insights into the GC-mediated disturbances of NO production in the mouse microvascular endothelial cell line MyEND [19]. They observed negative effects on NO-synthesis either after a short treatment period of 24 and 48 h or after a long-term GC administration persisting 30 days. These effects were accompanied by lowered NO concentrations and eNOS activity in response to dexamethasone application. However, instead of reduced eNOS protein expression, a decrease of GTPCH-1 protein they documented a decrease of GTPCH-1 protein. Interestingly, the observed changes in enzymes and cofactors involved in NO synthesis were accompanied by a significantly lowered immune reactivity of the GR in MyEND cells. Ligand-dependent down-regulation of the GR, associated with the loss of functional GC responsiveness, was referred in numerous publications [20, 21, 104, 105]. It was also shown that the expression of the GR is lower in endothelial cells derived from the mouse myocardium than in those isolated from the blood-brain barrier [106]. Although the degradation of the receptor by the proteasome can be observed in capillaries from both organs, brain endothelial cells remain responsive to GCs due to a sufficient expression of the GR [17]. The complete loss of GC responsiveness verified in MyEND cells in contrast to endothelium of the central nervous system hinders the translocation of the GC-GR complex into the nucleus, thereby hindering the transactivation of the GC target gene GTPCH-1 (as depicted in **Figure 1**) [19]. As it was shown, blocking the proteasomal GR degradation by calpain inhibitor I or by over-expressing a nondegradable GR isoform rescued either NO or BH_4 levels, as well as eNOS activity in MyEND cells after GC treatment. Hence, these data provide the biochemical evidence that GC-induced disturbed vasodilatation may result from the involvement of the proteasome in the failure of GC responsiveness rather than from GC-mediated trans-repression of NO-synthesizing molecules, as it was thought before [19, 57]. On the other hand, together with data gained in brain endothelial cells, the results support the endocrine heterogeneity of different vascular beds and a blockage of the proteasomal signalling pathway may be a potential future option for neutralizing or reducing unfavourable effects of long-term GC therapy [17, 107].

5. Conclusions

GC-induced hypertension remains a severe clinical problem, since GCs are the widest prescribed drugs by clinicians. Despite numerous basic and clinical studies, the complexity of the induction and persistence of hypertension remains not entirely understood. A robust explanation for GC hypertension is that GCs reduce the activity and expression of eNOS. Moreover, enzymes and cofactors upstream of eNOS are negatively influenced. The effects mediated by GCs on these molecules can be genomic or non-genomic. Therefore, further examinations are necessary to learn about this very frequent and serious clinical issue. On one hand, a detailed comprehension of the multistep GC/GR signalling cascade in the vasculature as well as in other non-renal tissues is required. Moreover, the regulation of molecules upstream and downstream of eNOS by GCs in a genomic and non-genomic manner should be clarified, given the fact that these effects might differ depending on the tissue. This might be explained by the evidence that the expression of the GR differs in a tissue-specific manner. In addition, it will be necessary to develop new in vivo models mimicking this complex disease and helping to understand the interplay between organs such as the kidney, the liver, the cardiovascular and the central nervous system. These animal models should be supported by single- or multi-tissue culture models in vitro. They would accelerate and promote the development of new compounds, such as synthetic and natural antioxidants lowering GC-induced pathologically increased blood pressure, and enable the identification of potential therapeutic targets.

Author details

Kinga G. Blecharz-Lang[1] and Malgorzata Burek[2]*

*Address all correspondence to: Burek_M@ukw.de

1 Department of Neurosurgery, Charité University Hospital Berlin, Berlin, Germany

2 Department of Anaesthesia and Critical Care, University of Würzburg, Würzburg, Germany

References

[1] Whitworth, J.A., G.J. Mangos, and J.J. Kelly, *Cushing, cortisol, and cardiovascular disease.* Hypertension, 2000. **36**(5): pp. 912–6.

[2] Baid, S. and L.K. Nieman, *Glucocorticoid excess and hypertension.* Curr Hypertens Rep, 2004. **6**(6): pp. 493–9.

[3] Ullian, M.E., *The role of corticosteriods in the regulation of vascular tone.* Cardiovasc Res, 1999. **41**(1): pp. 55–64.

[4] Pimenta, E., M. Wolley, and M. Stowasser, *Adverse cardiovascular outcomes of corticosteroid excess.* Endocrinology, 2012. **153**(11): pp. 5137–42.

[5] Gounarides, J.S., et al., *Effect of dexamethasone on glucose tolerance and fat metabolism in a diet-induced obesity mouse model.* Endocrinology, 2008. **149**(2): pp. 758–66.

[6] Hadoke, P.W., J. Iqbal, and B.R. Walker, *Therapeutic manipulation of glucocorticoid metabolism in cardiovascular disease.* Br J Pharmacol, 2009. **156**(5): pp. 689–712.

[7] Panoulas, V.F., et al., *Long-term exposure to medium-dose glucocorticoid therapy associates with hypertension in patients with rheumatoid arthritis.* Rheumatology (Oxford), 2008. **47**(1): pp. 72–5.

[8] Shinozaki, K., et al., *Oral administration of tetrahydrobiopterin prevents endothelial dysfunction and vascular oxidative stress in the aortas of insulin-resistant rats.* Circ Res, 2000. **87**(7): pp. 566–73.

[9] Laursen, J.B., et al., *Endothelial regulation of vasomotion in apoE-deficient mice: implications for interactions between peroxynitrite and tetrahydrobiopterin.* Circulation, 2001. **103**(9): pp. 1282–8.

[10] Zerr-Fouineau, M., et al., *Progestins overcome inhibition of platelet aggregation by endothelial cells by down-regulating endothelial NO synthase via glucocorticoid receptors.* FASEB J, 2007. **21**(1): pp. 265–73.

[11] Widder, J.D., et al., *Regulation of tetrahydrobiopterin biosynthesis by shear stress.* Circ Res, 2007. **101**(8): pp. 830–8.

[12] Tang, L., H. Wang, and M.T. Ziolo, *Targeting NOS as a therapeutic approach for heart failure.* Pharmacol Ther, 2014. **142**(3): pp. 306–15.

[13] Seddon, M., A.M. Shah, and B. Casadei, *Cardiomyocytes as effectors of nitric oxide signalling.* Cardiovasc Res, 2007. **75**(2): pp. 315–26.

[14] Oakley, R.H. and J.A. Cidlowski, *The biology of the glucocorticoid receptor: new signaling mechanisms in health and disease.* J Allergy Clin Immunol, 2013. **132**(5): pp. 1033–44.

[15] Tischner, D. and H.M. Reichardt, *Glucocorticoids in the control of neuroinflammation.* Mol Cell Endocrinol, 2007. **275**(1–2): pp. 62–70.

[16] Schweingruber, N., et al., *Mechanisms of glucocorticoids in the control of neuroinflammation.* J Neuroendocrinol, 2012. **24**(1): pp. 174–82.

[17] Forster, C., et al., *Occludin as direct target for glucocorticoid-induced improvement of blood-brain barrier properties in a murine in vitro system.* J Physiol, 2005. **565**(Pt 2): pp. 475–86.

[18] Forster, C., et al., *Differential effects of hydrocortisone and TNFalpha on tight junction proteins in an in vitro model of the human blood-brain barrier.* J Physiol, 2008. **586**(7): pp. 1937–49.

[19] Blecharz, K.G., et al., *Inhibition of proteasome-mediated glucocorticoid receptor degradation restores nitric oxide bioavailability in myocardial endothelial cells in vitro.* Biol Cell, 2014. **106**(7): pp. 219–35.

[20] Wallace, A.D. and J.A. Cidlowski, *Proteasome-mediated glucocorticoid receptor degradation restricts transcriptional signaling by glucocorticoids.* J Biol Chem, 2001. **276**(46): pp. 42714–21.

[21] Bellingham, D.L., M. Sar, and J.A. Cidlowski, *Ligand-dependent down-regulation of stably transfected human glucocorticoid receptors is associated with the loss of functional glucocorticoid responsiveness.* Mol Endocrinol, 1992. **6**(12): pp. 2090–102.

[22] Ciriaco, M., et al., *Corticosteroid-related central nervous system side effects.* J Pharmacol Pharmacother, 2013. **4**(Suppl 1): p. S94–8.

[23] Poetker, D.M. and D.D. Reh, *A comprehensive review of the adverse effects of systemic corticosteroids.* Otolaryngol Clin North Am, 2010. **43**(4): pp. 753–68.

[24] Goodwin, J.E., et al., *The glucocorticoid receptor in the distal nephron is not necessary for the development or maintenance of dexamethasone-induced hypertension.* Biochem Biophys Res Commun, 2010. **394**(2): pp. 266–71.

[25] Goodwin, J.E., J. Zhang, and D.S. Geller, *A critical role for vascular smooth muscle in acute glucocorticoid-induced hypertension.* J Am Soc Nephrol, 2008. **19**(7): pp. 1291–9.

[26] Goodwin, J.E., et al., *Knockout of the vascular endothelial glucocorticoid receptor abrogates dexamethasone-induced hypertension.* J Hypertens, 2011. **29**(7): pp. 1347–56.

[27] Tsutsui, M., et al., *Significance of nitric oxide synthases: Lessons from triple nitric oxide synthases null mice.* J Pharmacol Sci, 2015. **127**(1): pp. 42–52.

[28] Ignarro, L.J., C. Napoli, and J. Loscalzo, *Nitric oxide donors and cardiovascular agents modulating the bioactivity of nitric oxide: an overview.* Circ Res, 2002. **90**(1): pp. 21–8.

[29] Feletou, M. and P.M. Vanhoutte, *Endothelial dysfunction: a multifaceted disorder (The Wiggers Award Lecture).* Am J Physiol Heart Circ Physiol, 2006. **291**(3): p. H985–1002.

[30] Huang, P.L., et al., *Hypertension in mice lacking the gene for endothelial nitric oxide synthase.* Nature, 1995. **377**(6546): pp. 239–42.

[31] Shesely, E.G., et al., *Elevated blood pressures in mice lacking endothelial nitric oxide synthase.* Proc Natl Acad Sci U S A, 1996. **93**(23): pp. 13176–81.

[32] Ohashi, Y., et al., *Hypotension and reduced nitric oxide-elicited vasorelaxation in transgenic mice overexpressing endothelial nitric oxide synthase.* J Clin Invest, 1998. **102**(12): pp. 2061–71.

[33] Wallerath, T., et al., *Down-regulation of the expression of endothelial NO synthase is likely to contribute to glucocorticoid-mediated hypertension.* Proc Natl Acad Sci U S A, 1999. **96**(23): pp. 13357–62.

[34] Wallerath, T., et al., *Dexamethasone lacks effect on blood pressure in mice with a disrupted endothelial NO synthase gene.* Nitric Oxide, 2004. **10**(1): pp. 36–41.

[35] Goodwin, J.E., et al., *Endothelial glucocorticoid receptor is required for protection against sepsis.* Proc Natl Acad Sci U S A, 2013. **110**(1): pp. 306–11.

[36] Schafer, S.C., et al., *Dexamethasone suppresses eNOS and CAT-1 and induces oxidative stress in mouse resistance arterioles.* Am J Physiol Heart Circ Physiol, 2005. **288**(1): pp. H436–44.

[37] Liu, Y., et al., *Glucocorticoid response elements and 11 beta-hydroxysteroid dehydrogenases in the regulation of endothelial nitric oxide synthase expression.* Cardiovasc Res, 2009. **81**(1): pp. 140–7.

[38] Biddie, S.C. and G.L. Hager, *Glucocorticoid receptor dynamics and gene regulation.* Stress, 2009. **12**(3): pp. 193–205.

[39] Radomski, M.W., R.M. Palmer, and S. Moncada, *Glucocorticoids inhibit the expression of an inducible, but not the constitutive, nitric oxide synthase in vascular endothelial cells.* Proc Natl Acad Sci U S A, 1990. **87**(24): pp. 10043–7.

[40] Tobias, S., et al., *Dexamethasone, tetrahydrobiopterin and uncoupling of endothelial nitric oxide synthase.* J Geriatr Cardiol, 2015. **12**(5): pp. 528–39.

[41] Wen, C., et al., *L-arginine partially reverses established adrenocorticotrophin-induced hypertension and nitric oxide deficiency in the rat.* Blood Press, 2000. **9**(5): pp. 298–304.

[42] Turner, S.W., et al., *L-arginine prevents corticotropin-induced increases in blood pressure in the rat.* Hypertension, 1996. **27**(2): pp. 184–9.

[43] Thony, B., G. Auerbach, and N. Blau, *Tetrahydrobiopterin biosynthesis, regeneration and functions.* Biochem J, 2000. **347**(Pt 1): pp. 1–16.

[44] Nichol, C.A., G.K. Smith, and D.S. Duch, *Biosynthesis and metabolism of tetrahydrobiopterin and molybdopterin.* Annu Rev Biochem, 1985. **54**: pp. 729–64.

[45] Schmidt, K., et al., *Tetrahydrobiopterin-dependent formation of endothelium-derived relaxing factor (nitric oxide) in aortic endothelial cells.* Biochem J, 1992. **281** (Pt 2): pp. 297–300.

[46] Werner-Felmayer, G., G. Golderer, and E.R. Werner, *Tetrahydrobiopterin biosynthesis, utilization and pharmacological effects.* Curr Drug Metab, 2002. **3**(2): pp. 159–73.

[47] Kuzkaya, N., et al., *Interactions of peroxynitrite, tetrahydrobiopterin, ascorbic acid, and thiols: implications for uncoupling endothelial nitric-oxide synthase.* J Biol Chem, 2003. **278**(25): pp. 22546–54.

[48] Thum, T., et al., *Endothelial nitric oxide synthase uncoupling impairs endothelial progenitor cell mobilization and function in diabetes.* Diabetes, 2007. **56**(3): pp. 666–74.

[49] Li, H. and U. Forstermann, *Uncoupling of endothelial NO synthase in atherosclerosis and vascular disease.* Curr Opin Pharmacol, 2013. **13**(2): pp. 161–7.

[50] Xia, Y., et al., *Superoxide generation from endothelial nitric-oxide synthase. A Ca2+/calmodulin-dependent and tetrahydrobiopterin regulatory process.* J Biol Chem, 1998. **273**(40): pp. 25804–8.

[51] Cosentino, F. and T.F. Luscher, *Tetrahydrobiopterin and endothelial nitric oxide synthase activity.* Cardiovasc Res, 1999. **43**(2): pp. 274–8.

[52] Cosentino, F. and Z.S. Katusic, *Tetrahydrobiopterin and dysfunction of endothelial nitric oxide synthase in coronary arteries.* Circulation, 1995. **91**(1): pp. 139–44.

[53] Crabtree, M.J., et al., *Critical role for tetrahydrobiopterin recycling by dihydrofolate reductase in regulation of endothelial nitric-oxide synthase coupling: relative importance of the de novo biopterin synthesis versus salvage pathways.* J Biol Chem, 2009. **284**(41): pp. 28128–36.

[54] Meininger, C.J., et al., *Impaired nitric oxide production in coronary endothelial cells of the spontaneously diabetic BB rat is due to tetrahydrobiopterin deficiency.* Biochem J, 2000. **349**(Pt 1): pp. 353–6.

[55] Johns, D.G., et al., *Glucocorticoids inhibit tetrahydrobiopterin-dependent endothelial function.* Exp Biol Med (Maywood), 2001. **226**(1): pp. 27–31.

[56] Thida, M., et al., *Effects of sepiapterin supplementation and NOS inhibition on glucocorticoid-induced hypertension.* Am J Hypertens, 2010. **23**(5): pp. 569–74.

[57] Mitchell, B.M., et al., *Glucocorticoids decrease GTP cyclohydrolase and tetrahydrobiopterin-dependent vasorelaxation through glucocorticoid receptors.* J Cardiovasc Pharmacol, 2004. **43**(1): pp. 8–13.

[58] Mitchell, B.M., A.M. Dorrance, and R.C. Webb, *GTP cyclohydrolase 1 downregulation contributes to glucocorticoid hypertension in rats.* Hypertension, 2003. **41(3 Pt 2)**: pp. 669–74.

[59] Simmons, W.W., et al., *Glucocorticoids regulate inducible nitric oxide synthase by inhibiting tetrahydrobiopterin synthesis and L-arginine transport.* J Biol Chem, 1996. **271**(39): pp. 23928–37.

[60] Zheng, J.S., et al., *Gene transfer of human guanosine 5′-triphosphate cyclohydrolase I restores vascular tetrahydrobiopterin level and endothelial function in low renin hypertension.* Circulation, 2003. **108**(10): pp. 1238–45.

[61] Frostell, C., et al., *Inhaled nitric oxide. A selective pulmonary vasodilator reversing hypoxic pulmonary vasoconstriction.* Circulation, 1991. **83**(6): pp. 2038–47.

[62] Cannon, R.O., 3rd, et al., *Effects of inhaled nitric oxide on regional blood flow are consistent with intravascular nitric oxide delivery.* J Clin Invest, 2001. **108**(2): pp. 279–87.

[63] van Faassen, E.E., et al., *Nitrite as regulator of hypoxic signaling in mammalian physiology.* Med Res Rev, 2009. **29**(5): pp. 683–741.

[64] Cosby, K., et al., *Nitrite reduction to nitric oxide by deoxyhemoglobin vasodilates the human circulation.* Nat Med, 2003. **9**(12): pp. 1498–505.

[65] Liu, C., et al., *Mechanisms of human erythrocytic bioactivation of nitrite.* J Biol Chem, 2015. **290**(2): pp. 1281–94.

[66] Hunter, C.J., et al., *Inhaled nebulized nitrite is a hypoxia-sensitive NO-dependent selective pulmonary vasodilator.* Nat Med, 2004. **10**(10): pp. 1122–7.

[67] Shiva, S., et al., *Nitrite augments tolerance to ischemia/reperfusion injury via the modulation of mitochondrial electron transfer.* J Exp Med, 2007. **204**(9): pp. 2089–102.

[68] Webb, A., et al., *Reduction of nitrite to nitric oxide during ischemia protects against myocardial ischemia-reperfusion damage.* Proc Natl Acad Sci U S A, 2004. **101**(37): pp. 13683–8.

[69] Omar, S.A., E. Artime, and A.J. Webb, *A comparison of organic and inorganic nitrates/nitrites.* Nitric Oxide-Biol Chem, 2012. **26**(4): pp. 229–240.

[70] Lundberg, J.O., E. Weitzberg, and M.T. Gladwin, *The nitrate-nitrite-nitric oxide pathway in physiology and therapeutics.* Nat Rev Drug Discov, 2008. **7**(2): pp. 156–67.

[71] Carlstrom, M., et al., *Dietary inorganic nitrate reverses features of metabolic syndrome in endothelial nitric oxide synthase-deficient mice.* Proc Natl Acad Sci U S A, 2010. **107**(41): pp. 17716–20.

[72] Carlstrom, M., et al., *Dietary nitrate attenuates oxidative stress, prevents cardiac and renal injuries, and reduces blood pressure in salt-induced hypertension.* Cardiovasc Res, 2011. **89**(3): pp. 574–85.

[73] Bryan, N.S., et al., *Dietary nitrite supplementation protects against myocardial ischemia-reperfusion injury.* Proc Natl Acad Sci U S A, 2007. **104**(48): pp. 19144–9.

[74] Hendgen-Cotta, U.B., et al., *Dietary nitrate supplementation improves revascularization in chronic ischemia.* Circulation, 2012. **126**(16): pp. 1983–92.

[75] Trostchansky, A., et al., *Nitro-fatty acids: formation, redox signaling, and therapeutic potential.* Antioxid Redox Signal, 2013. **19**(11): pp. 1257–65.

[76] O'Donnell, V.B., et al., *Nitration of unsaturated fatty acids by nitric oxide-derived reactive species.* Methods Enzymol, 1999. **301**: pp. 454–70.

[77] Delmastro-Greenwood, M., B.A. Freeman, and S.G. Wendell, *Redox-dependent anti-inflammatory signaling actions of unsaturated fatty acids.* Annu Rev Physiol, 2014. **76**: pp. 79–105.

[78] Rudolph, V., et al., *Endogenous generation and protective effects of nitro-fatty acids in a murine model of focal cardiac ischaemia and reperfusion.* Cardiovasc Res, 2010. **85**(1): pp. 155–66.

[79] Wu, G. and S.M. Morris, Jr., *Arginine metabolism: nitric oxide and beyond.* Biochem J, 1998. **336**(Pt 1): pp. 1–17.

[80] Schulman, S.P., et al., *L-arginine therapy in acute myocardial infarction: the vascular interaction with age in myocardial infarction (VINTAGE MI) randomized clinical trial.* JAMA, 2006. **295**(1): pp. 58–64.

[81] Wilson, A.M., et al., *L-arginine supplementation in peripheral arterial disease: no benefit and possible harm.* Circulation, 2007. **116**(2): pp. 188–95.

[82] Pernow, J. and C. Jung, *Arginase as a potential target in the treatment of cardiovascular disease: reversal of arginine steal?* Cardiovasc Res, 2013. **98**(3): pp. 334–43.

[83] Boger, R.H., *The pharmacodynamics of L-arginine.* J Nutr, 2007. **137**(6 Suppl 2): pp. 1650S–1655S.

[84] Solomonson, L.P., et al., *The caveolar nitric oxide synthase/arginine regeneration system for NO production in endothelial cells.* J Exp Biol, 2003. **206**(Pt 12): pp. 2083–7.

[85] Podjarny, E., et al., *Effect of chronic tetrahydrobiopterin supplementation on blood pressure and proteinuria in 5/6 nephrectomized rats.* Nephrol Dial Transplant, 2004. **19**(9): pp. 2223–7.

[86] Hong, H.J., et al., *Supplemention with tetrahydrobiopterin suppresses the development of hypertension in spontaneously hypertensive rats.* Hypertension, 2001. **38**(5): pp. 1044–8.

[87] Miao, Y., et al., *Folic acid prevents and partially reverses glucocorticoid-induced hypertension in the rat.* Am J Hypertens, 2007. **20**(3): pp. 304–10.

[88] Krug, S., et al., *N-Acetylcysteine prevents but does not reverse dexamethasone-induced hypertension.* Clin Exp Pharmacol Physiol, 2008. **35**(8): pp. 979–81.

[89] Zhang, Y., et al., *Apocynin but not allopurinol prevents and reverses adrenocorticotropic hormone-induced hypertension in the rat.* Am J Hypertens, 2005. **18**(7): pp. 910–6.

[90] Zhang, Y., Y. Miao, and J.A. Whitworth, *Aspirin prevents and partially reverses adrenocorticotropic hormone-induced hypertension in the rat.* Am J Hypertens, 2007. **20**(11): pp. 1222–8.

[91] Zhang, Y., et al., *The anti-oxidant Tempol reverses and partially prevents adrenocorticotrophic hormone-induced hypertension in the rat.* J Hypertens, 2003. **21**(8): pp. 1513–8.

[92] Mondo, C.K., et al., *Anti-oxidant effects of atorvastatin in dexamethasone-induced hypertension in the rat.* Clin Exp Pharmacol Physiol, 2006. **33**(11): pp. 1029–34.

[93] Beswick, R.A., et al., *Long-term antioxidant administration attenuates mineralocorticoid hypertension and renal inflammatory response.* Hypertension, 2001. **37**(2 Pt 2): pp. 781–6.

[94] Ortiz, M.C., et al., *Antioxidants block angiotensin II-induced increases in blood pressure and endothelin.* Hypertension, 2001. **38**(3 Pt 2): pp. 655–9.

[95] Schnackenberg, C.G. and C.S. Wilcox, *Two-week administration of tempol attenuates both hypertension and renal excretion of 8-Iso prostaglandin f2alpha.* Hypertension, 1999. **33**(1 Pt 2): pp. 424–8.

[96] Arstall, M.A., et al., *N-acetylcysteine in combination with nitroglycerin and streptokinase for the treatment of evolving acute myocardial infarction. Safety and biochemical effects.* Circulation, 1995. **92**(10): pp. 2855–62.

[97] Smilkstein, M.J., et al., *Efficacy of oral N-acetylcysteine in the treatment of acetaminophen overdose. Analysis of the national multicenter study (1976 to 1985).* N Engl J Med, 1988. **319**(24): pp. 1557–62.

[98] Das, U.N., *Folic acid says NO to vascular diseases.* Nutrition, 2003. **19**(7–8): pp. 686–92.

[99] Topal, G., et al., *Homocysteine induces oxidative stress by uncoupling of NO synthase activity through reduction of tetrahydrobiopterin.* Free Radic Biol Med, 2004. **36**(12): pp. 1532–41.

[100] Terzolo, M., et al., *Hyperhomocysteinemia in patients with Cushing's syndrome.* J Clin Endocrinol Metab, 2004. **89**(8): pp. 3745–51.

[101] Hu, L., et al., *Apocynin but not L-arginine prevents and reverses dexamethasone-induced hypertension in the rat.* Am J Hypertens, 2006. **19**(4): pp. 413–8.

[102] Wassmann, S., et al., *Cellular antioxidant effects of atorvastatin in vitro and in vivo.* Arterioscler Thromb Vasc Biol, 2002. **22**(2): pp. 300–5.

[103] Sarath, T.S., et al., *Atorvastatin ameliorates arsenic-induced hypertension and enhancement of vascular redox signaling in rats.* Toxicol Appl Pharmacol, 2014. **280**(3): pp. 443–54.

[104] Silva, C.M., et al., *Regulation of the human glucocorticoid receptor by long-term and chronic treatment with glucocorticoid.* Steroids, 1994. **59**(7): pp. 436–42.

[105] Wallace, A.D., et al., *Lysine 419 targets human glucocorticoid receptor for proteasomal degradation.* Steroids, 2011. **75**(12): pp. 1016–23.

[106] Forster, C., et al., *Glucocorticoid effects on mouse microvascular endothelial barrier permeability are brain specific.* J Physiol Lond, 2006. **573**(2): pp. 413–425.

[107] Aird, W.C., *Endothelial cell heterogeneity.* Crit Care Med, 2003. **31**(4 Suppl): pp. S221–30.

Section 3

Nitric Oxide Synthase in Reproductive System

Chapter 7

Nitric Oxide Synthase in Male Urological and Andrologic Functions

Qingfeng Yu, Tieqiu Li, Jingping Li, Liren Zhong and Xiangming Mao

Additional information is available at the end of the chapter

http://dx.doi.org/10.5772/67562

Abstract

Nitric oxide (NO), a crucial signaling molecule, is synthesized by the nitric oxide synthase (NOS) enzyme. The significant effects of NOS are under exploration, and the roles of potential therapy targets for diseases of NOS are widely accepted. In this chapter, we summarized the important roles of NOS mainly on pathogenesis of prostate diseases, male infertility, erectile dysfunction and, addition, the potential therapeutic efficacies of NOS for those diseases.

Keywords: nitric oxide synthase, nitric oxide, prostate cancer, male infertility, erectile dysfunction, male reproduction

1. Introduction

Urology and andrology are the branches of medicine that focus on urinary tract system and male reproductive organs. In recent years, incidences of diseases in urology and andrology system such as prostate cancer and male infertility are increasing and causing heavy burden to our society. Growing studies have been demonstrating that nitric oxide synthase (NOS), which synthesized nitric oxide (NO) by converting L-arginine to L-citrulline, locates in tissues of urinary and male reproductive system and acts as key regulators for sexual function, male reproduction, cancer progression and so on [1–3]. The aims of this chapter are to present the roles of NOS and the recent advances of regulation and therapy function with regard to sexual function, male infertility, prostate carcinoma, Peyronie disease, priapism and cryptorchidism.

2. NOS and male sexual function

2.1. NOS and erectile dysfunction

Erectile dysfunction (ED) refers to the symptom that the penis cannot reach and (or) maintain the adequate erection to complete the satisfaction of sexual intercourse, and the course of disease will last at least 6 months or more. Penile erection is an integrated process of artery blood supply and cavernous blood storage launched by nerve, and during this process, neurotransmitter plays an important role [4]. NO is a main messenger, which involves in the induction and maintenance of erection through hemangiectasis and corpus cavernosum relaxation [5]. It has been clear that NO penetrates the smooth muscle cell membrane and catalyzes the formation of cGMP after combining with the ornithine enzyme on the iron ring and then changing the intracellular calcium concentration of smooth muscle to cause relaxation.

2.1.1. NOS in penile tissue

The nNOS and iNOS were found in the central nervous system, especially the hypothalamic area, such as paraventricular nucleus and the medial optic zone, that control the erectile and sexual behavior and also regulate penile erection through spinal nerve centers [6, 7] (**Figure 1**). Specially, nNOS mainly distributes in the penile and pelvic nerve plexus in adult rats, whereas eNOS is in the penis and pelvic area of the urethra but less in the body part of the penis [8, 9].

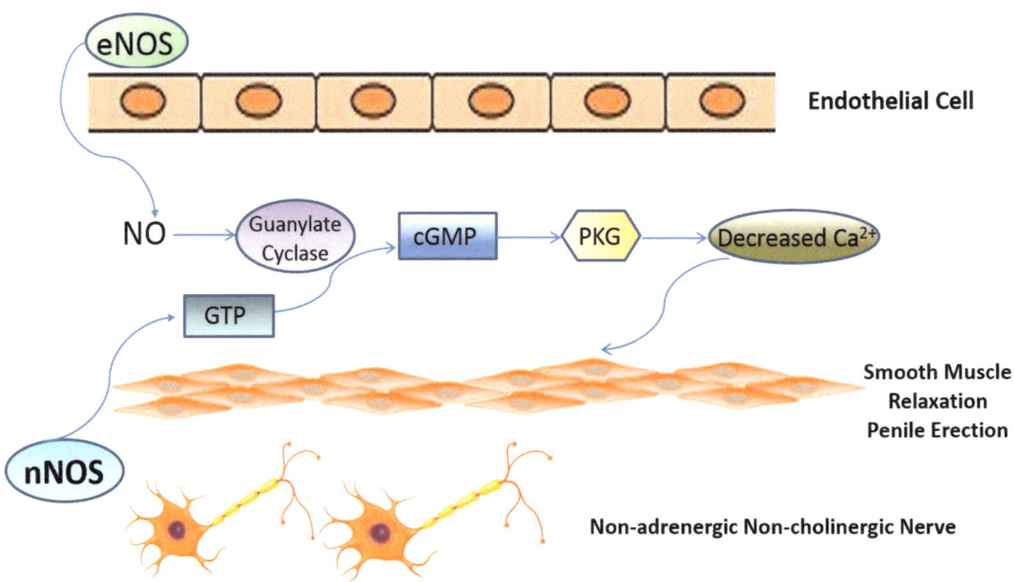

Figure 1. Role of NOS and NO in penile erection. The nNOS and iNOS regulate penile erection through NO/cGMP/PKG pathway. Abbreviations: eNOS, endothelial nitric oxide synthase; nNOS, neuronal nitric oxide synthase; NO, nitric oxide; GTP, guanosine triphosphate; cGMP, cyclic guanosine monophosphate; PKG, protein kinase G.

http://dx.doi.org/10.5772/67562

2.1.2. Current reviews on the effects of NOS on disease-related ED

2.1.2.1. Diabetes-related ED

A possible percentage ranges from 50% to 90% of diabetes patients suffer from ED [10], which counts around three times more than that in healthy cohort. Also, previous report showed that diabetes patients who firstly suffered from ED had a younger age in average compared with people without diabetes and more severe appeared of their symptoms [11]. It has now reached a consensus on the relationship between diabetes and ED, the damage of endothelial cell, ultrastructure changes in cavernous smooth muscle and matrix fibrosis are the common factors affecting the cavernous diastolic function and resulting in ED [12], and damage of eNOS-NO-cGMP pathway was considered to be the main molecular mechanism [13]. The expression of Recombinant Nitric Oxide Synthase Trafficker (NOSTRIN) in Dulbecco's modified eagle medium (DMED) corpus cavernosum increased while the expression of eNOS decreased. It is theorized that increased NOSTRIN may be an important mechanism for the reduction of eNOS expression, while further study is still needed [14].

2.1.2.2. Benign prostatic hyperplasia-related ED

Approximately 49% of BPH patients suffer from ED. A significant correlation was reported between BPH/lower urinary tract obstruction (LUTS) and ED after excluding the effects of age and other etiologies on ED [15, 16]. BPH/LUTS seems to be one of the most harmful factors contribute to ED compared with diabetes, hypertension and (or) heart disease [17]. In fact, both parasympathetic innervation of prostate and cavernous nerve of penis are from the pelvic plexus [18]. Pathophysiology studies also showed that the mechanisms of BPH are similar to those of sexual dysfunction which include decrease of the ratio of endothelial NOS/NO, enhancement of endothelial presenilin-1 contraction effect, overreaction of autonomic nervous of bladder, prostate and penis, enhancement of signaling pathway Rho kinase expression/activity and (or) pelvic vascular sclerosis [19]. Since NOS has been found to play significant effects on BPH patients with ED, new treatment by exogenous NO donor and NOS activating enzyme would be promising [20]. However, only few animal experiments have been under exploration up to date, and the mechanisms for the occurrence of ED in BPH patients are still needed to be identified.

2.1.2.3. Hypertension-related ED

Hypertension is another important risk factor for ED. Jensen et al. reported that nearly 27% of hypertensive patients suffered from ED [21]. The increase of plasma asymmetrical dimethylarginine (ADMA) concentration caused reduction of the NO expression in penile tissue by inhibiting the NOS activity, which might be a possible mechanism for hypertension-related ED [22].

2.1.3. Possible therapy strategies of NOS on ED

Increasing evidence has been indicating that L-Arg-NO-cGMP pathway might be a crucial mechanism of penile erection [23]. As a key enzyme in the synthesis of NO, NOS has always been one of the research focuses. Specially, nNOS acts as a key role in erection launch, whereas eNOS enables cavernous body dilate and maintains the status of erection [24]. Although the effect of iNOS was absent in the direct regulation of penile erection, a special "double effect" in the elderly and the pathological state was reported [25]. Since the reduction of NOSs or the decrease of its activity might contribute to ED, the treatment on L-Arg-NO-cGMP for ED might be revolutionary breakthrough, as phosphodiesterase type 5 inhibitors (PDE5Is) was found to improve the erectile function by increasing the NO concentration but reducing the eGMP degradation [26]. However, nearly 20% of patients with ED still showed little benefit after receiving PDE5Iss, especially in patients with diabetes or prostate cancer (Prostate carcinoma) after radical mastectomy [27]. Future NOSs gene transfer therapy from the molecular level would be another choice [28], which might have long-term curative effect, little side effect to the body and, even, completely cure ED. Therapies including increasing expression of NOSs (nNOS, iNOS, eNOS) or inhibiting the expression of protein inhibitor of NOS (PIN) might be promising and worthwhile exploration [29]. However, shortcomings such as short effect duration, possibility of inducing abnormal erection and other potential unknown side effects from the long-term excessive expression of NO are addressed.

2.2. NOS and libido

ED may cause low sexual desire or loss of libido in men [30]. Previous study reported that treatment for ED could somehow retrieve sexual desire [31]. It is believed that NO and NOS are beneficial for penile erection, and consequently, NO and NOS may enhance sexual motivation in indirect ways. NO could also affect libido in the direct ways.

Areas for male sexual behavior in brain distribute NO responsive guanylyl cyclase, which involves cellular events of NO [32], and previous studies showed that the NOS inhibitor N^G-nitro-L-arginine methyl ester (NAME) administered to medial preoptic area by reverse dialysis caused reduced mounting of male rat [32, 33]. Chu et al. further reported that Impaza, a stimulator of eNOS, could raise the sexual incentive motivation rates of male rats through the NO-guanylyl cyclase pathway [34], and nNOS was also considered to affect the male sexual behavior by activating the cyclic guanosine monophosphate (cGMP) [35]. However, adverse result was reported by other researchers, and the conclusion was still controversial [36].

3. NOS and male infertility

Approximately 15% of couples suffer from infertility while male cause contributed to nearly 50% in these infertile couples [37]. Male reproduction is known to involve complicated aspects such as spermatogenesis, sperm dynamics, sperm morphology and acrosome reaction. Increasing evidences have been indicating that NOS and NO are associated with male infertility [38].

3.1. NOS and male reproductive system

The hypothalamic-pituitary axis plays core roles in reproduction and steroid hormone production in man. Gonadotropin-releasing hormone (GnRH), also known as luteinizing hormone-releasing hormone (LHRH), which is produced and secreted by the arcuate nucleus

http://dx.doi.org/10.5772/67562

of the hypothalamus, could stimulate the anterior pituitary to episodically release follicle-stimulating hormone (FSH) and luteinizing hormone (LH). LH stimulates the Leydig cells to produce testosterone, and FSH exerts its effect directly on the Sertoli cells to promote spermatogenesis (**Figure 2**).

In vitro studies have shown that NO stimulates LHRH secretion from the hypothalamus and modulates LH release from the pituitary [39, 40]. Ceccatelli [41] reported that sodium nitroprusside, a NO donor, suppressed GnRH-stimulated LH release from pituitaries in male rats. Chatterjee et al. [42] showed that NOS inhibitor p-nitro-L-arginine methyl ester (L-NAME) enhanced GnRH-induced LH release from pituitaries in rats. Decreased level of GnRH and gonadotropin in chronic NO deficiency rats were also observed [43].

3.1.1. Testis

3.1.1.1. Testicular microcirculation

The testis has a rich vascular system that plays a very important role in maintaining the normal functions and stable inner environment of the testis [44]. The regulation of testicular blood microcirculation is very complex, including self-regulation, neural regulation and humoral regulation [45]. NO is the major physiological regulator of basal blood vessel tone and is continually released from endothelium of testicular arteries [46]. A study showed that the regulation effect of NO on testicular blood flow was limited under basal conditions, but this limitation could be significant reversed after HCG treatment; in this case, NO showed

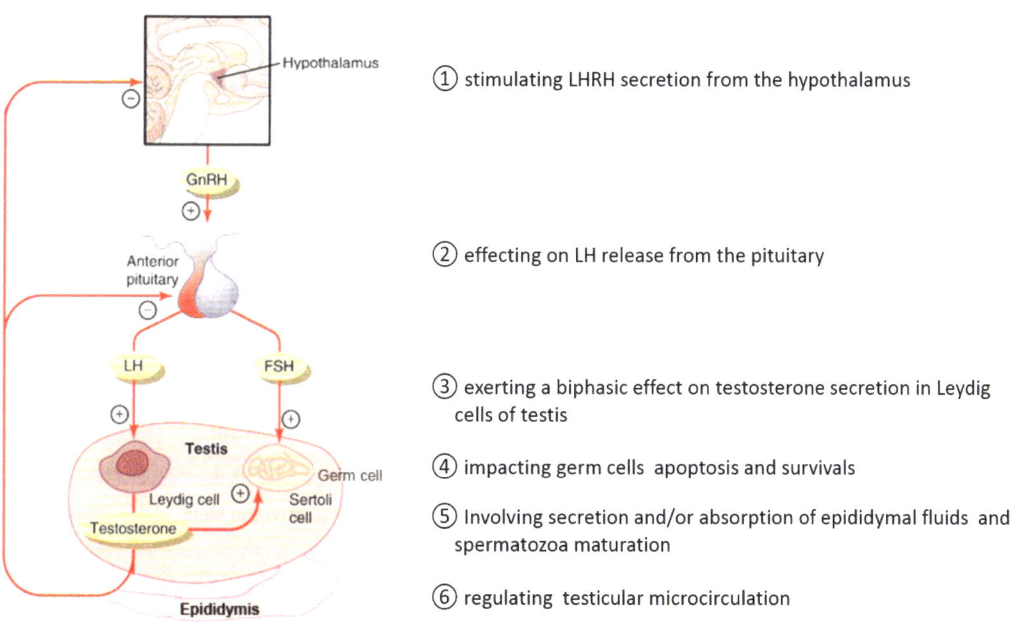

Figure 2. Regulation of hypothalamic-pituitary axis. Abbreviations: LHRH, luteinizing hormone-releasing hormone; GnRH, gonadotropin-releasing hormone; LH, luteinizing hormone; FSH, follicle-stimulating hormone.

the effects of increasing blood flow and inhibiting leukocyte accumulation on rat testicular arteries [46]. NO is also an important factor in regulating testicular vessel tension at different temperatures, at 34–37°C, disturbance of testicular arteries reaction appeared after L-NAME treatment [47]. Interestingly, NO content and NOS activity could be significantly increased at abnormal high temperatures caused by varicose spermatic veins in varicocele patients [48].

3.1.1.2. Leydig cells

Leydig cells, also known as interstitial cells, are adjacent to the seminiferous tubules in the testicle. Leydig cells produce and release testosterone under the control of LH and act as autocrine and/or paracrine hormones in gonad under the modulation of NO [49]. An immunohistochemistry study demonstrated that eNOS, nNOS and iNOS all expressed in cytoplasm of Leydig cells in rat testis [50]. Interestingly, a testis-specific subclass of nNOS, known as the truncated form of nNOS (TnNOS), has been recently identified as a major contributor to the formation of NO [51]. TnNOS has been found to be localized solely in the Leydig cells of the testes but neither in the Sertoli nor germ cells [51], which enable us to predict that NO may associate with functions of Leydig cells. Kozieł et al. found that NOS was able to act directly within the male gonad by means of regulating androgen secretion though Leydig cells [52]. Another study showed that stress-induced stimulation of the testicular NO signaling pathway leaded to the inhibition of steroidogenic enzymes [53]. But NOS seemingly exerted a biphasic effect on testosterone secretion [54]. At low concentrations, NO exerted a transient stimulatory effect on testosterone secretions mediated by cyclic GMP, whereas at high concentrations, it inhibited steroidogenesis by Leydig cells.

3.1.1.3. Sertoli cells

Spermatogenesis is a complex process in which Sertoli cells closely involve, and NOS also plays a crucial role in this process through Sertoli cells. Zini et al. [55] showed that eNOS protein located in Sertoli cells and some parts of germ cells in seminiferous tubules, especially in degenerating germ cells and spermatids in histologically normal testes. The iNOS was also found in Sertoli cells, as well as a small subset of pachytene spermatocytes and elongated spermatids in the normal testis [56]. However, iNOS expression could be very intense in Sertoli cells in pathological conditions, for example, the absence of seminiferous tubules [57]. iNOS involves in germ cell death in testicular ischemia-reperfusion injury model, and inhibition of iNOS could improve impaired spermatogenesis [58]. In cryptorchidism model, the transgene expression of eNOS increased testicular germ cell apoptosis. In iNOS⁄mice, the numbers of spermatocytes, spermatids and Sertoli cells per tubule were significantly more than those with wild-type testes [59]. A possible conclusion could be drawn that NO plays an important role in both numerical and functional regulation of key somatic cells in the testis, which in turn impacts on germ cells and their survivals during the process of daily sperm production.

3.1.1.4. Epididymis

The main functions of the epididymis are promoting spermatozoa mature and storing spermatozoa [60]. An immunostaining study in human epididymis showed that NOS almost exclusively located in the epithelium [55], and the greatest concentration was in the adluminal region [55]. It suggests that NOS may involve secretion and/or absorption of epididymal fluids, or in another way diffuse into the tubule lumen to affect nearby spermatozoa. Another study showed a similar distribution of NOS protein in rat epididymis, speculating that epididymal NOS protein might contribute to spermatozoa maturation [61].

3.2. NOS and sperm function

Approximately 15% of couples suffer from infertility while male cause contributed to nearly 50% in these infertile couples [37]. Male reproduction is known to involve complicated aspects, such as spermatogenesis, sperm dynamics, sperm morphology and acrosome reaction. Increasing evidences have been indicating that NOS and NO are associated with male infertility [38].

3.2.1. Sperm motility, morphology and viability

Sperm motility is an essential factor for male fertility. Low sperm motility, also referred as asthenozoospermia, is one of the major causes to male infertility [62]. Previous study indicated that nearly 80% of semen samples from infertile males were defective in sperm motility [63]. Hellstrom et al. reported for the first time that sodium nitroprusside, a NO releaser, was beneficial for maintenance of thaw-sperm motility by reducing lipid peroxidative damage to sperm membranes. Significantly improved motion parameters of sperm were observed in semen samples treated with sodium nitroprusside in concentrations of 50 and 100 nM compared to control samples, and this beneficial effect maintained for 5–6 hours after thaw [64]. However, NO concentration in normozoospermic fertile men was observed to be significantly lower than those of asthenospermia infertile men [65]. In fact, the effect of NO seems to be double-sided, low concentration of NO improves sperm motility, while high concentration contributes to adverse effect [66]. Herrero et al. reported that a significant decrease on sperm motility was observed in semen samples treated with sodium nitroprusside in a higher concentration of 300 mM, and this effect could be blocked by hemoglobin, a scavenger of NO, as sperm motility in samples furtherly treated with hemoglobin was significantly higher than those without. While when the incubating concentration of sodium nitroprusside reduced to 150 mM, no modifications of sperm motility were found [67]. Besides, the other NO releaser, S-nitroso-N-acetylpenicillamine (0.012–0.6 mM), along with sodium nitroprusside (0.25–2.5 mM), was found to decrease percentage of forward progressive sperm motility and straight line velocity in a concentration-dependent manner [68].

As to sperm morphology and viability, the effects of NO reveal controversial contributions. a positive correlation between NO with defects in sperm morphology has been found in male with normal sperm rate ≥14% but a negative correlation with defects in sperm morphology in male with normal sperm rate <14% [69]. However, later study failed to find any significant association between NO production and sperm morphology [70]. Researchers reported that semen treated with 0.25–2.5 mM sodium nitroprusside revealed significantly less sperm bound to the zona pellucida compared with the control group which treated without NO [71], whereas some other researchers reported no any significant effect of NO on sperm viability [66, 69, 72]. And meanwhile, low concentration of NO also plays a role in the maintenance of sperm viability after cryopreservation and post-thaw sperm [65, 66, 68].

3.2.2. *Capacitation, hyperactivation and acrosome reaction*

Capacitation is a process in which spermatozoa acquire the ability to bind to the egg's zone pellucida and fertilize an oocyte during their transit in the female genital tract [73, 74]. Capacitation involves in some molecular events, and it was clear that low level of NO from NO-releasing agents induces human sperm capacitation [75]. Indeed, it has been reported that NO-releasing compounds significantly benefit the capacitation, whereas NO inhibitors decrease this process [76]. In fact, NO produced by spermatozoa involves in a cascade of molecular events of capacitation, which is needed over the course of this process [77–79].

Hyperactivation can be treated as a subcategory of capacitation. Hyperactivation of spermatozoa exhibits high amplitude and asymmetric flagellar movement, non-linear motility and penetrate the oocyte with strong propulsive force [70, 80]. The effect of NO on hyperactivation was found to be similar to which on sperm motility, little concentrations of NO increased spermatozoa hyperactivation, whereas excessive concentrations decreased the hyperactivated spermatozoa motility [81, 82].

Acrosome reaction denotes the process that capacitated and hyperactivated motility sperm binds to the zona pellucida and continues to pass through the exocytotic release of proteolytic enzymes from the acrosome so that to bind to the mature ovum. The amount of NO influenced the acrosome reaction, and increased amount of sperm was observed to undergo the acrosome reaction with the presence of NO donor compound [83]. Meanwhile, significantly increased amount of sperm was found to bind to membrane of the ovum with their plasma membrane [84].

3.2.3. *Sperm mitochondria*

Mitochondria in sperm activate as a generator which supply sperm with energy for the process of motility, acrosome reaction, oocyte fusion, fertilization and so on [85]. NO has been reported to involve in functions of mitochondrial that include biogenesis, remodeling and mitochondrial respiration [86–88]. Specially, different levels of NO could cause different sperm mitochondrial functions, low concentrations of NO enhanced the sperm motility, while NO with higher levels cause mitochondrial hyperpolarization and sperm apoptosis [64, 89]. This might explain the adverse effects of various concentrations NO on sperm motility.

3.3. Single-nucleotide polymorphisms of NOS and male infertility

Genetic variations are crucial etiological factors contribute to male infertility. Up to date, some single-nucleotide polymorphisms (SNPs) have been identified to involve in sperm defects and male infertility in ethnic populations. Polymorphisms T786C and G894T of eNOS were reported to decrease sperm motility and quantity by increasing the seminal oxidative stress in Egyptian infertile male population [90]. Similar results of G894T were reported in Italian and Iranian infertile male populations [91, 92], and this SNP also found to be associated with higher level of sperm DNA fragmentation in Chinese infertile males [93]. The polymorphism 4a4b, which refers to a sequence variant with variable sequence of tandem 4a4b repeats in intron 4, was found to be associated with poor sperm morphology and male infertility in a

Korean and Chinese population [94, 95]. Associations between SNPs of NOS and male infertility are under exploration, which would be promising tools for diagnosis or further curing male infertility.

3.4. Possible therapy strategies of NOS on male infertility

Increasing evidences has been showing that inappropriate concentration level of NO may contribute to male infertility in some extend by means of decreasing sperm motility and normal sperm morphology, reducing efficiency of capacitation and acrosome action. It is reasonable to consider possible therapy strategies to the utilization of NOS donors or inhibitors so that to adjust the concentration of NO to the "right" level. In fact, significantly higher fertile rate was observed in animal experiment in vitro, and further researches would be needed to warrant the potential benefits for human beings.

4. NOS and prostate carcinoma

Prostate carcinoma is one of the most common cancers among men and second in cancer-related deaths in the United States. An estimated study predicted that there will be 180, 890 new prostate carcinoma cases and 26, 120 deaths due to the disease in the country in 2016 [96]. Etiological studies implicated that multiple reasons involved in prostate carcinoma susceptibility, such as dietary, environment, hormone status and genetic factors [97]. Growing studies indicated that NOS and NO system play crucial roles in progression of human prostate carcinoma [98–100].

The physiological functions of NO are dependent primarily on concentrations. Low concentration of NO acted as a signal transducer and affects many physiological processes including blood flow regulation, platelet activity, iron homeostasis, cell proliferation and neurotransmission, whereas, in high concentrations, it exerted a cytotoxic protective effect, for example, to against pathogens and perhaps tumors [101, 102].

4.1. Role of nitric oxide synthase in cancer biology

The roles of NOS and NO on DNA damage, apoptosis, cell cycle, enhancement of cell proliferation, angiogenesis and metastasis are currently viewed, and NO was found to be associated with tumor environment, for example, the vasculature cells and other stromal cells [103–105]. Research also indicated that NOS2 expression was correlated with tumor vascularization, accumulations of p53 mutations and activation of epidermal growth factor receptor, even could be treated as an independent predictor of poor survival in women with estrogen receptor (ER)-negative breast tumors [106]. Low concentrations of NO acted as a promotional role in angiogenesis which stimulates tumor progression by providing blood flow access to the tumor and subsequently resulting in cell proliferation. On the contrary, high levels of NO tend to be cytotoxic to cancer cells [107]. While in animal models, iNOS overexpression produced either pro-tumor or anti-tumor effect on tumor growth, these alterable effects seem to be dependent on the tumor microenvironment and the tumor type itself [104, 108]. The effects of NO possibly differ in expression level of iNOS, duration and timing of NO delivery, the microenvironment, the genetic background and the cell type (**Figure 3**) [109].

4.2. NOS and proliferation of prostate carcinoma

NO generated by eNOS or iNOS might be involved in prostate proliferation. At low concentrations, NO acted as a signaling molecule regulating smooth muscle relaxation and blood flow, neurotransmission, platelet activity, iron homeostasis, cell survival and proliferation, while at high concentrations acted as modulating immune-mediated anti-tumor activities [110, 111]. Concentration of NO less than 100nM had an effect of preventing certain cell types from apoptosis and thereby favors tumorigenesis and progression [112]. Higher expression of iNOS was detected in cancer specimens than that in normal tissues of prostate carcinoma patient. Aaltoma et al. also demonstrated a positive association between expression level of iNOS and rapid cancer cell proliferation rate, dedifferentiation and advanced stage cancer [113]. A recent study has shown that NO also regulated cell proliferation in a pathway of CPD-Arg-NO [114].

4.3. Nitric oxide synthase and angiogenesis of prostate carcinoma

Angiogenesis is a critical molecular event in tumor progression [115, 116]. Epidermal growth factor receptor (EGFR) signaling pathway, tumor suppressor p53 and VEGF, which are collective mediators that exacerbate angiogenesis can be stimulated by NO [115, 117]. The involvement of eNOS in the NO-induced human endothelial and prostate carcinoma cell migration was further warranted [116]. Recent research also reported that NO played vital roles in maintaining blood supply for prostate carcinoma, and an anti-tumor vascular activity effect revealed with presence in inhibition of NOS [115].

4.4. Single-nucleotide polymorphisms of NOS and susceptibility of prostate carcinoma

Several studies suggested that polymorphisms of some NOS genes were genetic susceptibility factors for prostate carcinoma, especially for aggressive diseases [118, 119]. A plethora of meta-analyses has identified eNOS gene polymorphisms as strong susceptibility factors for the progression toward prostate carcinoma [120]. Another study also reported that NOS3 gene

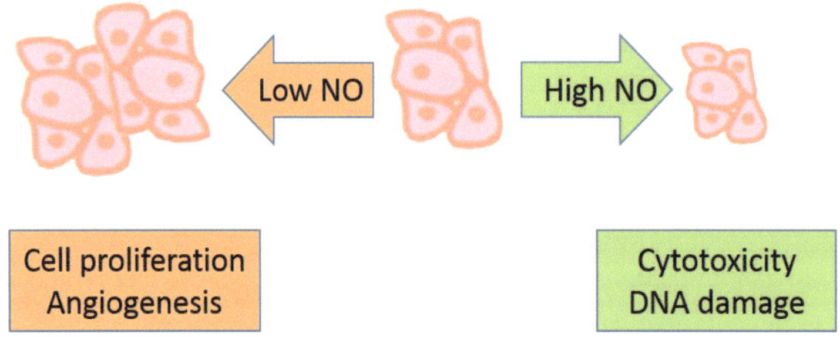

Figure 3. Roles of NO in prostate cancer. Abbreviation: NO, nitric oxide.

polymorphisms were genetic susceptibility factors for the progression of prostate carcinoma and poor patient outcomes [121]. A meta-analysis conducted by Zhao et al. suggested that eNOS gene 894G > T polymorphism contributed to aggravate the onset of prostate carcinoma in males [122]. Nikolic et al [123] also corroborated the involvement of eNOS or NOS3 gene in the pathogenesis of prostate carcinoma. NOS3 rs1799983 polymorphism augmented the risk of prostate carcinoma in various populations. As one of the possible mechanisms, the involvement of NO receptor component, sGC-1, in mediating the proliferation of prostate carcinogenesis, has been surmised [124].

4.5. Possible therapy strategies of NOS on prostate carcinoma

Anti-cancer agents such as gold lotion have successfully demonstrated their anti-carcinogenic potential through the regulation of both iNOS and eNOS [125–127]. Yu et al. [128] also elucidated the significance of eNOS as a seemingly promising strategy for targeting anti-androgen resistant prostate carcinoma. Arginine-releasing compounds such as carboxypeptidase-D increased NO production, which slackened progression of prostate carcinoma so that prolonged survival time [129]. NO-donor drugs also have been under increasing explorations. A few NO-donor drugs have been confirmed to have favorable anticancer activity and could be potential anticancer therapies [3, 130]. GIT-27NO, a novel NO donor, inhibited the growth of PC3 and LnCap prostate carcinoma cells xenografted into nude mice in a concentration-dependent manner [131]. And DETA-NONOate was revealed to inhibit epithelial-mesenchymal transition (EMT) and invasion of human prostate metastatic cells by producing large amount of NO [132]. It is sensible that novel NOS-based therapeutics may prove valuable in the future treatment of prostate carcinoma.

5. NOS and other urinary and male reproductive diseases

5.1. Peyronie disease

Peyronie disease (PD) is an intractable, sexually dysfunctional disease resulting in penile curvature, penile pain, penile deformity, difficulty with coitus, shortening, hinging, narrowing and ED. Mechanisms of PD have not been fully elucidated. A recent hypothesis was that the recurrent microtrauma of the tunica albuginea caused small damages that activated processes of wound healing and fibrotic plaque development during sexual intercourse [133]. Inflammatory cells and iNOS accumulated in the process of wound healing, the increased NO then led to the myofibroblasts and proliferation of fibroblasts and redundant collagen between the layers of the tunica albuginea (penile plaque) [134]. Although surgical therapy is now the first-option for PD patients, researchers are focusing on the nonsurgical treatments of PD, and NOs inhibitors might be a promising choice [135].

5.2. Priapism

Priapism is defined as a persistent and painful erection that lasts longer than 4 hours without sexual stimulation and can lead to ED [136]. The relation between penile erection and production of NOS has been well investigated: nNOS and eNOS were the causes of both the initiation and maintenance phases of penile erection [137]. However, decreased function in NO generated by decreased activation of eNOS resulted in PDE5 downregulation that was thought to be a derivate of NO and, therefore, reduced basal levels of PDE5 and caused priapism [138].

5.3. Cryptorchidism

Cryptorchidism denotes failure of the movement of the testis to the scrotum, and in most cases, it raises risk of testicular germ cell cancer and subfertility later in patients' life course. Testicular germ cell apoptosis which causes by exposure of testicular in elevated temperature and oxidative stress is the primary etiology of infertility. Animal models with cryptorchidism induced by surgery revealed that eNOS played a significant role in mouse spermatogenesis in cryptorchidism-induced apoptosis [139]. Contemporaneously, reduced rate of testicular atrophy was observed in heterozygous Hoxa 11 knockout mice which had congenital bilateral cryptorchidism when early treated with Nomega-nitro-L-arginine methyl ester (L-NAME), a NOS inhibitor [140].

6. Conclusion

It has been becoming evident that redox regulation driven by NOS and NO represents a promising tool for exploring fundamental diseases process and new development of strategies to treat urinary, male reproductive and sexual diseases.

Acknowledgements

This work was supported by Foundation of Science and Technology Projects of Shenzhen (Grant Number: JCYJ20140415162542992).

Author details

Qingfeng Yu[1,2], Tieqiu Li[3], Jingping Li[4], Liren Zhong[5] and Xiangming Mao[1]*

*Address all correspondence to: dr.xiangmingmao@gmail.com

1 Department of Urology, Zhujiang Hospital, Southern Medical University, Guangzhou, PR China

2 Department of Urology, Ludwig-Maximilians University, Munich, Germany

3 Department of Urology, Hunan Provincial People's Hospital, The First Affiliated Hospital of Hunan Normal University, Changsha, PR China

4 Department of Reproductive Endocrinology, Women's Hospital, School of Medicine, Reproductive Medicine Center, Zhejiang University, Hangzhou, Zhejiang, PR China

5 Department of Urology, Southern Medical University affiliated Nanfang Hospital, Guangzhou, PR China

References

[1] Buzadzic B, Vucetic M, Jankovic A, Stancic A, Korac A, Korac B, Otasevic V. New insights into male (in)fertility: The importance of NO. British Journal of Pharmacology. 2015;**172**:1455–1467.

[2] Toque HA, Caldwell RW. New approaches to the design and discovery of therapies to prevent erectile dysfunction. Expert Opinion on Drug Discovery. 2014;**9**:1447–1469.

[3] Hirst D, Robson T. Nitric oxide in cancer therapeutics: Interaction with cytotoxic chemotherapy. Current Pharmaceutical Design. 2010;**16**:411–420.

[4] Patel CK, Bennett N. Advances in the treatment of erectile dysfunction: What's new and upcoming? F1000Research. 2016;**5**.

[5] Gur S, Kadowitz PJ, Sikka SC, Peak TC, Hellstrom WJ. Overview of potential molecular targets for hydrogen sulfide: A new strategy for treating erectile dysfunction. Nitric oxide: Biology and Chemistry/official Journal of the Nitric Oxide Society. 2015;**50**:65–78.

[6] Bivalacqua TJ, Champion HC, Usta MF, Cellek S, Chitaley K, Webb RC, Lewis RL, Mills TM, Hellstrom WJ, Kadowitz PJ. Rhoa/rho-kinase suppresses endothelial nitric oxide synthase in the penis: A mechanism for diabetes-associated erectile dysfunction. Proceedings of the National Academy of Sciences of the United States of America. 2004;**101**:9121–9126.

[7] Azadzoi KM, Master TA, Siroky MB. Effect of chronic ischemia on constitutive and inducible nitric oxide synthase expression in erectile tissue. Journal of Andrology. 2004;**25**:382–388.

[8] Seyam RM, Huynh HT, Brock GB. Neuronal and endothelial nitric oxide synthase isoforms: Quantification of protein and mRNA in the normal rat penis. International Journal of Impotence Research. 1999;**11**:301–308.

[9] Mizusawa H, Hedlund P, Brioni JD, Sullivan JP, Andersson KE. Nitric oxide independent activation of guanylate cyclase by yc-1 causes erectile responses in the rat. The Journal of Urology. 2002;**167**:2276–2281.

[10] Schramek P, Holzmann RM, Floth A. Disorders of erectile potency in diabetes mellitus. Wiener klinische Wochenschrift. 1988;**100**:460–463.

[11] Chu NV, Edelman SV. Erectile dysfunction and diabetes. Current Diabetes Reports. 2002;**2**:60–66.

[12] Pelliccione F, D'Angeli A, D'Andrea S, Barbonetti A, Pezzella A, Necozione S, Falone S, Amicarelli F, Francavilla F, Francavilla S. Tadalafil treatment had a modest effect on endothelial cell damage and repair ability markers in men with erectile dysfunction and vascular risk. Asian Journal of Andrology. 2014;**16**:290–294.

[13] Nunes KP, Wynne BM, Cordeiro MN, Borges MH, Richardson M, Leite R, DeLima ME, Webb RC. Increased cavernosal relaxation by *Phoneutria nigriventer* toxin, pntx2-6, via activation at no/cgmp signaling. International Journal of Impotence Research. 2012;**24**:69–76.

[14] Mookerjee RP, Wiesenthal A, Icking A, Hodges SJ, Davies NA, Schilling K, Sen S, Williams R, Novelli M, Muller-Esterl W, Jalan R. Increased gene and protein expression of the novel eNOS regulatory protein nostrin and a variant in alcoholic hepatitis. Gastroenterology. 2007;**132**:2533–2541.

[15] Dogan Y, Uruc F, Aras B, Sahin A, Kivrak M, Urkmez A, Guner ND, Aydin S. The relationships between metabolic syndrome, erectile dysfunction and lower urinary tract symptoms associated with benign prostatic hyperplasia. Turkish Journal of Urology. 2015;**41**:7–12.

[16] Lee LK, Goren A, Boytsov NN, Donatucci CF, McVary KT. Treatment satisfaction among men with concurrent benign prostatic hyperplasia and erectile dysfunction treated with tadalafil or other phosphodiesterase type-5 inhibitor combinations. Patient Preference and Adherence. 2016;**10**:1205–1215.

[17] Mevcha A, Gulur DM, Gillatt D. Diagnosing urological disorders in ageing men. The Practitioner. 2010;**254**:25–26, 28–29, 22–23.

[18] Ganzer R, Stolzenburg JU, Wieland WF, Brundl J. Anatomic study of periprostatic nerve distribution: Immunohistochemical differentiation of parasympathetic and sympathetic nerve fibres. European Urology. 2012;**62**:1150–1156.

[19] Fusco F, D'Anzeo G, Sessa A, Pace G, Rossi A, Capece M, d'Emmanuele di Villa Bianca R. BPH/LUTS and ED: Common pharmacological pathways for a common treatment. The Journal of Sexual Medicine. 2013;**10**:2382–2393.

[20] McVary K. Lower urinary tract symptoms and sexual dysfunction: Epidemiology and pathophysiology. BJU International. 2006;**97** Suppl 2:23–28; discussion 44–25.

[21] Jensen J, Lendorf A, Stimpel H, Frost J, Ibsen H, Rosenkilde P. The prevalence and etiology of impotence in 101 male hypertensive outpatients. American Journal of Hypertension. 1999;**12**:271–275.

[22] Maas R, Wenske S, Zabel M, Ventura R, Schwedhelm E, Steenpass A, Klemm H, Noldus J, Boger RH. Elevation of asymmetrical dimethylarginine (adma) and coronary artery disease in men with erectile dysfunction. European Urology. 2005;**48**:1004–1011; discussion 1011–1002.

[23] Wessells H, Hruby VJ, Hackett J, Han G, Balse-Srinivasan P, Vanderah TW. Ac-nle-c[asp-his-dphe-arg-trp-lys]-nh2 induces penile erection via brain and spinal melanocortin receptors. Neuroscience. 2003;**118**:755–762.

[24] Burnett AL, Nelson RJ, Calvin DC, Liu JX, Demas GE, Klein SL, Kriegsfeld LJ, Dawson VL, Dawson TM, Snyder SH. Nitric oxide-dependent penile erection in mice lacking neuronal nitric oxide synthase. Molecular Medicine (Cambridge, Mass). 1996;**2**:288–296.

[25] Gonzalez-Cadavid NF, Rajfer J. The pleiotropic effects of inducible nitric oxide synthase (iNOS) on the physiology and pathology of penile erection. Current Pharmaceutical Design. 2005;**11**:4041–4046.

[26] Chen L, Staubli SE, Schneider MP, Kessels AG, Ivic S, Bachmann LM, Kessler TM. Phosphodiesterase 5 inhibitors for the treatment of erectile dysfunction: A trade-off network meta-analysis. European Urology. 2015;**68**:674–680.

[27] Wang X, Wang X, Liu T, He Q, Wang Y, Zhang X. Systematic review and meta-analysis of the use of phosphodiesterase type 5 inhibitors for treatment of erectile dysfunction following bilateral nerve-sparing radical prostatectomy. PLoS One. 2014;**9**:e91327.

[28] Gonzalez-Cadavid NF, Rajfer J. Molecular pathophysiology and gene therapy of aging-related erectile dysfunction. Experimental Gerontology. 2004;**39**:1705–1712.

[29] Magee TR, Kovanecz I, Davila HH, Ferrini MG, Cantini L, Vernet D, Zuniga FI, Rajfer J, Gonzalez-Cadavid NF. Antisense and short hairpin RNA (shRNA) constructs targeting pin (protein inhibitor of NOS) ameliorate aging-related erectile dysfunction in the rat. The Journal of Sexual Medicine. 2007;**4**:633–643.

[30] Lewis RW, Fugl-Meyer KS, Corona G, Hayes RD, Laumann EO, Moreira ED, Jr., Rellini AH, Segraves T. Definitions/epidemiology/risk factors for sexual dysfunction. The Journal of Sexual Medicine. 2010;**7**:1598–1607.

[31] Jannini EA, Screponi E, Carosa E, Pepe M, Lo Giudice F, Trimarchi F, Benvenga S. Lack of sexual activity from erectile dysfunction is associated with a reversible reduction in serum testosterone. International Journal of Andrology. 1999;**22**:385–392.

[32] De Vente J, Hopkins DA, Markerink-Van Ittersum M, Emson PC, Schmidt HH, Steinbusch HW. Distribution of nitric oxide synthase and nitric oxide-receptive, cyclic GMP-producing structures in the rat brain. Neuroscience. 1998;**87**:207–241.

[33] Ratnasooriya WD, Dharmasiri MG, Wadsworth RM. Reduction in libido and fertility of male rats by administration of the nitric oxide (no) synthase inhibitor n-nitro-l-arginine methyl ester. International Journal of Andrology. 2000;**23**:187–191.

[34] Chu X, Zhavbert ES, Dugina JL, Kheyfets IA, Sergeeva SA, Epstein OI, Agmo A. Sildenafil and a compound stimulating endothelial no synthase modify sexual incentive motivation and copulatory behavior in male Wistar and Fisher 344 rats. The Journal of Sexual Medicine. 2008;**5**:2085–2099.

[35] Chu X, Agmo A. Sexual incentive motivation in old male rats: The effects of sildenafil and a compound (impaza) stimulating endothelial no synthase. Pharmacology, Biochemistry, and Behavior. 2008;**89**:209–217.

[36] Ratnasooriya WD, Jayakody JR, Dharmasiri MG. Sodium nitroprusside impairs sexual competence of male rats. Human & Experimental Toxicology. 2004;**23**:187–192.

[37] Schlegel PN. Evaluation of male infertility. Minerva Ginecologica. 2009;**61**:261–283.

[38] Lee NP, Cheng CY. Nitric oxide and cyclic nucleotides: Their roles in junction dynamics and spermatogenesis. Advances in Experimental Medicine and Biology. 2008;**636**:172–185.

[39] McCann SM, Mastronardi C, Walczewska A, Karanth S, Rettori V, Yu WH. The role of nitric oxide in reproduction. Brazilian Journal of Medical and Biological Research. 1999;**32**:1367–1379.

[40] Rettori V, Belova N, Dees WL, Nyberg CL, Gimeno M, McCann SM. Role of nitric oxide in the control of luteinizing hormone-releasing hormone release in vivo and in vitro. Proceedings of the National Academy of Sciences of the United States of America. 1993;**90**:10130–10134.

[41] Ceccatelli S. Expression and plasticity of no synthase in the neuroendocrine system. Brain Research Bulletin. 1997;**44**:533–538.

[42] Chatterjee S, Collins TJ, Yallampalli C. Inhibition of nitric oxide facilitates lh release from rat pituitaries. Life Sciences. 1997;**61**:45–50.

[43] Barnes MJ, Lapanowski K, Rafols JA, Lawson DM, Dunbar JC. Gnrh and gonadotropin release is decreased in chronic nitric oxide deficiency. Experimental Biology and Medicine (Maywood, NJ). 2001;**226**:701–706.

[44] Damber JE, Bergh A. Testicular microcirculation—A forgotten essential in andrology? International Journal of Andrology. 1992;**15**:285–292.

[45] Gu L, Jin B, Zhang X. Progress in testicular microcirculation structure and regulation. Chinese Journal of Microcirculation. 2015;**24**:65–67.

[46] Lissbrant E, Lofmark U, Collin O, Bergh A. Is nitric oxide involved in the regulation of the rat testicular vasculature. Biology of Reproduction. 1997;**56**:1221–1227.

[47] Sabanegh E, Dewire D, Inman S. The effect of nitric oxide blocked on rat testicular microcirculation. Fertility and Sterility. 1994;**62**:1–17.

[48] Xu Y, Xu QY, Yang BH, Zhu XM, Peng YF. Relationship of nitric oxide and nitric oxide synthase with varicocele infertility. Zhonghua nan ke xue [National Journal of Andrology]. 2008;**14**:414–417.

[49] Lamanna C, Assisi L, Vittoria A, Botte V, Di Fiore MM. D-aspartic acid and nitric oxide as regulators of androgen production in boar testis. Theriogenology. 2007;**67**:249–254.

[50] Li MX, He LP, Guo ZQ, Liu YS, Long ZF. The expression of nitric oxide synthase in testes of male rat. Zhonghua nan ke xue [National Journal of Andrology]. 2002;**8**:250–252.

[51] Wang Y, Newton DC, Miller TL, Teichert AM, Phillips MJ, Davidoff MS, Marsden PA. An alternative promoter of the human neuronal nitric oxide synthase gene is expressed specifically in Leydig cells. American Journal of Pathology. 2002;**160**:369–380.

[52] Koziel E, Kotula M, Andronowska A, Pierscinski A, Bilinska B. Correlation between nadph-diaphorase and iNOS in bank vole Leydig cells in vitro and in testicular sections with the use of histochemistry and immunocytochemistry. Folia Histochemica et Cytobiologica. 2000;**38**:71–78.

[53] Kostic TS, Andric SA, Maric D, Kovacevic RZ. Inhibitory effects of stress-activated nitric oxide on antioxidant enzymes and testicular steroidogenesis. The Journal of Steroid Biochemistry and Molecular Biology. 2000;**75**:299–306.

[54] Valenti S, Cuttica CM, Fazzuoli L, Giordano G, Giusti M. Biphasic effect of nitric oxide on testosterone and cyclic GMP production by purified rat Leydig cells cultured in vitro. International Journal of Andrology. 1999;**22**:336–341.

[55] Zini A, O'Bryan MK, Magid MS, Schlegel PN. Immunohistochemical localization of endothelial nitric oxide synthase in human testis, epididymis, and vas deferens suggests a possible role for nitric oxide in spermatogenesis, sperm maturation, and programmed cell death. Biology of Reproduction. 1996;**55**:935–941.

[56] O'Bryan MK, Schlatt S, Gerdprasert O, Phillips DJ, de Kretser DM, Hedger MP. Inducible nitric oxide synthase in the rat testis: Evidence for potential roles in both normal function and inflammation-mediated infertility. Biology of Reproduction. 2000;**63**: 1285–1293.

[57] Costur P, Filiz S, Gonca S, Culha M, Gulecen T, Solakoglu S, Canberk Y, Caliskan E. Expression of inducible nitric oxide synthase (iNOS) in the azoospermic human testis. Andrologia. 2012;**44** Suppl 1:654–660.

[58] Shiraishi K, Naito K, Yoshida K. Nitric oxide promotes germ cell necrosis in the delayed phase after experimental testicular torsion of rat. Biology of Reproduction. 2001;**65**:514–521.

[59] Auharek SA, Avelar GF, Lara NL, Sharpe RM, Franca LR. Sertoli cell numbers and spermatogenic efficiency are increased in inducible nitric oxide synthase mutant mice. International Journal of Andrology. 2011;**34**:e621–629.

[60] Jones RC. To store or mature spermatozoa? The primary role of the epididymis. International Journal of Andrology. 1999;**22**:57–67.

[61] Burnett AL, Ricker DD, Chamness SL, Maguire MP, Crone JK, Bredt DS, Snyder SH, Chang TS. Localization of nitric oxide synthase in the reproductive organs of the male rat. Biology of Reproduction. 1995;**52**:1–7.

[62] Luconi M, Forti G, Baldi E. Pathophysiology of sperm motility. Frontiers in Bioscience: A Journal and Virtual Library. 2006;**11**:1433–1447.

[63] Curi SM, Ariagno JI, Chenlo PH, Mendeluk GR, Pugliese MN, Sardi Segovia LM, Repetto HE, Blanco AM. Asthenozoospermia: Analysis of a large population. Archives of Andrology. 2003;**49**:343–349.

[64] Hellstrom WJ, Bell M, Wang R, Sikka SC. Effect of sodium nitroprusside on sperm motility, viability, and lipid peroxidation. Fertility and Sterility. 1994;**61**:1117–1122.

[65] Balercia G, Moretti S, Vignini A, Magagnini M, Mantero F, Boscaro M, Ricciardo-Lamonica G, Mazzanti L. Role of nitric oxide concentrations on human sperm motility. Journal of Andrology. 2004;**25**:245–249.

[66] Doshi SB, Khullar K, Sharma RK, Agarwal A. Role of reactive nitrogen species in male infertility. Reproductive Biology and Endocrinology: RB&E. 2012;**10**:109.

[67] Herrero MB, Cebral E, Boquet M, Viggiano JM, Vitullo A, Gimeno MA. Effect of nitric oxide on mouse sperm hyperactivation. Acta Physiologica, Pharmacologica et Therapeutica Latinoamericana. organo de la Asociacion Latinoamericana de Ciencias Fisiologicas y [de] la Asociacion Latinoamericana de Farmacologia. 1994;**44**:65–69.

[68] Rosselli M, Dubey RK, Imthurn B, Macas E, Keller PJ. Effects of nitric oxide on human spermatozoa: Evidence that nitric oxide decreases sperm motility and induces sperm toxicity. Human Reproduction (Oxford, England). 1995;**10**:1786–1790.

[69] Wu TP, Huang BM, Tsai HC, Lui MC, Liu MY. Effects of nitric oxide on human spermatozoa activity, fertilization and mouse embryonic development. Archives of Andrology. 2004;**50**:173–179.

[70] Miraglia E, Rullo ML, Bosia A, Massobrio M, Revelli A, Ghigo D. Stimulation of the nitric oxide/cyclic guanosine monophosphate signaling pathway elicits human sperm chemotaxis in vitro. Fertility and Sterility. 2007;**87**:1059–1063.

[71] Bolanos JP, Delgado-Esteban M, Herrero-Mendez A, Fernandez-Fernandez S, Almeida A. Regulation of glycolysis and pentose-phosphate pathway by nitric oxide: Impact on neuronal survival. Biochimica et Biophysica Acta. 2008;**1777**:789–793.

[72] Kruger TF, Menkveld R, Stander FS, Lombard CJ, Van der Merwe JP, van Zyl JA, Smith K. Sperm morphologic features as a prognostic factor in in vitro fertilization. Fertility and Sterility. 1986;**46**:1118–1123.

[73] Chang MC. Fertilizing capacity of spermatozoa deposited into the fallopian tubes. Nature. 1951;**168**:697–698.

[74] Austin CR. The capacitation of the mammalian sperm. Nature. 1952;**170**:326.

[75] Zini A, De Lamirande E, Gagnon C. Low levels of nitric oxide promote human sperm capacitation in vitro. Journal of Andrology. 1995;**16**:424–431.

[76] Herrero MB, de Lamirande E, Gagnon C. Nitric oxide regulates human sperm capacitation and protein-tyrosine phosphorylation in vitro. Biology of Reproduction. 1999;**61**:575–581.

[77] O'Flaherty C, de Lamirande E, Gagnon C. Positive role of reactive oxygen species in mammalian sperm capacitation: Triggering and modulation of phosphorylation events. Free Radical Biology & Medicine. 2006;**41**:528–540.

[78] de Lamirande E, O'Flaherty C. Sperm activation: Role of reactive oxygen species and kinases. Biochimica et Biophysica Acta. 2008;**1784**:106–115.

[79] de Lamirande E, Lamothe G. Reactive oxygen-induced reactive oxygen formation during human sperm capacitation. Free Radical Biology and Medicine. 2009;**46**:502–510.

[80] Kothari S, Thompson A, Agarwal A, du Plessis SS. Free radicals: Their beneficial and detrimental effects on sperm function. Indian Journal of Experimental Biology. 2010;**48**:425–435.

[81] Otasevic V, Korac A, Vucetic M, Macanovic B, Garalejic E, Ivanovic-Burmazovic I, Filipovic MR, Buzadzic B, Stancic A, Jankovic A, Velickovic K, Golic I, Markelic M, Korac B. Is manganese (ii) pentaazamacrocyclic superoxide dismutase mimic beneficial for human sperm mitochondria function and motility. Antioxidants and Redox Signaling. 2013;**18**:170–178.

[82] Miraglia E, De Angelis F, Gazzano E, Hassanpour H, Bertagna A, Aldieri E, Revelli A, Ghigo D. Nitric oxide stimulates human sperm motility via activation of the cyclic GMP/protein kinase g signaling pathway. Reproduction (Cambridge, England). 2011; **141**:47–54.

[83] Donnelly ET, Lewis SE, Thompson W, Chakravarthy U. Sperm nitric oxide and motility: The effects of nitric oxide synthase stimulation and inhibition. Molecular Human Reproduction. 1997;**3**:755–762.

[84] Sikka SC. Relative impact of oxidative stress on male reproductive function. Current Medicinal Chemistry. 2001;**8**:851–862.

[85] Agarwal A, Makker K, Sharma R. Clinical relevance of oxidative stress in male factor infertility: An update. American Journal of Reproductive Immunology (New York, NY: 1989). 2008;**59**:2–11.

[86] Petrovic V, Buzadzic B, Korac A, Vasilijevic A, Jankovic A, Korac B. No modulates the molecular basis of rat interscapular brown adipose tissue thermogenesis. Comparative Biochemistry and Physiology. Toxicology and Pharmacology: CBP. 2010;**152**:147–159.

[87] Vasilijevic A, Vojcic L, Dinulovic I, Buzadzic B, Korac A, Petrovic V, Jankovic A, Korac B. Expression pattern of thermogenesis-related factors in interscapular brown adipose tissue of alloxan-treated rats: Beneficial effect of l-arginine. Nitric Oxide: Biology and Chemistry/Official Journal of the Nitric Oxide Society. 2010;**23**:42–50.

[88] Vucetic M, Otasevic V, Korac A, Stancic A, Jankovic A, Markelic M, Golic I, Velickovic K, Buzadzic B, Korac B. Interscapular brown adipose tissue metabolic reprogramming during cold acclimation: Interplay of hif-1alpha and ampkalpha. Biochimica et Biophysica Acta. 2011;**1810**:1252–1261.

[89] Mishra DP, Shaha C. Estrogen-induced spermatogenic cell apoptosis occurs via the mitochondrial pathway: Role of superoxide and nitric oxide. The Journal of Biological Chemistry. 2005;280:6181–6196.

[90] Mostafa T, Rashed LA, Nabil N, Fouad H, Sabry D, El-Saied DM. Endothelial nitric oxide synthase gene polymorphism relationship with semen parameters and oxidative stress in infertile oligoasthenoteratozoospermic men. Urology. 2015;**85**:1058–1061.

[91] Buldreghini E, Mahfouz RZ, Vignini A, Mazzanti L, Ricciardo-Lamonica G, Lenzi A, Agarwal A, Balercia G. Single nucleotide polymorphism (SNP) of the endothelial nitric oxide synthase (eNOS) gene (glu298asp variant) in infertile men with asthenozoospermia. Journal of Andrology. 2010;**31**:482–488.

[92] Safarinejad MR, Shafiei N, Safarinejad S. The role of endothelial nitric oxide synthase (eNOS) t-786c, g894t, and 4a/b gene polymorphisms in the risk of idiopathic male infertility. Molecular Reproduction and Development. 2010;**77**:720–727.

[93] Yan L, Guo W, Wu S, Liu J, Zhang S, Shi L, Ji G, Gu A. Genetic variants in nitric oxide synthase genes and the risk of male infertility in a Chinese population: A case-control study. PLoS One. 2014;**9**:e115190.

[94] Yun YJ, Park JH, Song SH, Lee S. The association of 4a4b polymorphism of endothelial nitric oxide synthase (eNOS) gene with the sperm morphology in Korean infertile men. Fertility and Sterility. 2008;**90**:1126–1131.

[95] Ying HQ, Pu XY, Liu SR, A ZC. Genetic variants of eNOS gene may modify the susceptibility to idiopathic male infertility. Biomarkers: Biochemical Indicators of Exposure, Response, and Susceptibility to Chemicals. 2013;**18**:412–417.

[96] Siegel RL, Miller KD, Jemal A. Cancer statistics, 2016. CA: A Cancer Journal for Clinicians. 2016;**66**:7–30.

[97] Gronberg H. Prostate cancer epidemiology. Lancet. 2003;**361**:859–864.

[98] Tong X, Li H. Enos protects prostate cancer cells from trail-induced apoptosis. Cancer Letters. 2004;**210**:63–71.

[99] Nanni S, Grasselli A, Benvenuti V, Aiello A, Pantisano V, Re A, Gaetano C, Capogrossi MC, Bacchetti S, Pontecorvi A, Farsetti A. The role of nuclear endothelial nitric oxide synthase in the endothelial and prostate microenvironments. Hormone Molecular Biology and Clinical Investigation. 2011;**5**:91–96.

[100] Nanni S, Aiello A, Re A, Guffanti A, Benvenuti V, Colussi C, Castro-Vega LJ, Felsani A, Londono-Vallejo A, Capogrossi MC, Bacchetti S, Gaetano C, Pontecorvi A, Farsetti A. Estrogen-dependent dynamic profile of eNOS-DNA associations in prostate cancer. PLoS One. 2013;**8**:e62522

[101] Estrada C, Murillo-Carretero M. Nitric oxide and adult neurogenesis in health and disease. The Neuroscientist: A Review Journal Bringing Neurobiology, Neurology and Psychiatry. 2005;**11**:294–307.

[102] Burke AJ, Sullivan FJ, Giles FJ, Glynn SA. The yin and yang of nitric oxide in cancer progression. Carcinogenesis. 2013;**34**:503–512.

[103] Hiraku Y, Kawanishi S, Ichinose T, Murata M. The role of iNOS-mediated DNA damage in infection- and asbestos-induced carcinogenesis. Annals of the New York Academy of Sciences. 2010;**1203**:15–22.

[104] Korde Choudhari S, Sridharan G, Gadbail A, Poornima V. Nitric oxide and oral cancer: A review. Oral Oncology. 2012;**48**:475–483.

[105] Vannini F, Kashfi K, Nath N. The dual role of iNOS in cancer. Redox Biology. 2015;**6**: 334–343.

[106] Glynn SA, Boersma BJ, Dorsey TH, Yi M, Yfantis HG, Ridnour LA, Martin DN, Switzer CH, Hudson RS, Wink DA, Lee DH, Stephens RM, Ambs S. Increased nos2 predicts poor survival in estrogen receptor-negative breast cancer patients. The Journal of Clinical Investigation. 2010;**120**:3843–3854.

[107] Lee SY, Rim Y, McPherson DD, Huang SL, Kim H. A novel liposomal nanomedicine for nitric oxide delivery and breast cancer treatment. Bio-Medical Materials and Engineering. 2014;**24**:61–67.

[108] Bogdan C. Nitric oxide synthase in innate and adaptive immunity: An update. Trends in Immunology. 2015;**36**:161–178.

[109] Vahora H, Khan MA, Alalami U, Hussain A. The potential role of nitric oxide in halting cancer progression through chemoprevention. Journal of Cancer Prevention. 2016;**21**:1–12.

[110] Yu H, Payne TJ, Mohanty DK. Effects of slow, sustained, and rate-tunable nitric oxide donors on human aortic smooth muscle cells proliferation. Chemical Biology & Drug Design. 2011;**78**:527–534.

[111] Ambs S, Hussain SP, Harris CC. Interactive effects of nitric oxide and the p53 tumor suppressor gene in carcinogenesis and tumor progression. FASEB Journal: Official Publication of the Federation of American Societies for Experimental Biology. 1997;**11**:443–448.

[112] Lechner M, Lirk P, Rieder J. Inducible nitric oxide synthase (iNOS) in tumor biology: The two sides of the same coin. Seminars in Cancer Biology. 2005;**15**:277–289.

[113] Aaltoma SH, Lipponen PK, Kosma VM. Inducible nitric oxide synthase (iNOS) expression and its prognostic value in prostate cancer. Anticancer Research. 2001;**21**:3101–3106.

[114] Thomas LN, Merrimen J, Bell DG, Rendon R, Too CK. Prolactin- and testosterone-induced carboxypeptidase-d correlates with increased nitrotyrosines and ki67 in prostate cancer. The Prostate. 2015;**75**:1726–1736.

[115] Ng QS, Goh V, Milner J, Stratford MR, Folkes LK, Tozer GM, Saunders MI, Hoskin PJ. Effect of nitric-oxide synthesis on tumour blood volume and vascular activity: A phase i study. The Lancet Oncology. 2007;**8**:111–118.

[116] Polytarchou C, Hatziapostolou M, Poimenidi E, Mikelis C, Papadopoulou A, Parthymou A, Papadimitriou E. Nitric oxide stimulates migration of human endothelial and prostate cancer cells through up-regulation of pleiotrophin expression and its receptor protein tyrosine phosphatase beta/zeta. International Journal of Cancer. 2009;**124**: 1785–1793.

[117] Cianchi F, Cortesini C, Fantappie O, Messerini L, Sardi I, Lasagna N, Perna F, Fabbroni V, Di Felice A, Perigli G, Mazzanti R, Masini E. Cyclooxygenase-2 activation mediates the proangiogenic effect of nitric oxide in colorectal cancer. Clinical Cancer Research: An Official Journal of the American Association for Cancer Research. 2004;**10**:2694–2704.

[118] Ozturk E, Dikensoy E, Balat O, Ugur MG, Balci SO, Aydin A, Kazanci U, Pehlivan S. Association of endothelial nitric oxide synthase gene polymorphisms with endometrial carcinoma: A preliminary study. Journal of the Turkish German Gynecological Association. 2011;**12**:229–233.

[119] Lee KM, Kang D, Park SK, Berndt SI, Reding D, Chatterjee N, Chanock S, Huang WY, Hayes RB. Nitric oxide synthase gene polymorphisms and prostate cancer risk. Carcinogenesis. 2009;**30**:621–625.

[120] Safarinejad MR, Safarinejad S, Shafiei N, Safarinejad S. Effects of the t-786c, g894t, and intron 4 vntr (4a/b) polymorphisms of the endothelial nitric oxide synthase gene on the risk of prostate cancer. Urologic Oncology. 2013;**31**:1132–1140.

[121] Brankovic A, Brajuskovic G, Nikolic Z, Vukotic V, Cerovic S, Savic-Pavicevic D, Romac S. Endothelial nitric oxide synthase gene polymorphisms and prostate cancer risk in Serbian population. International Journal of Experimental Pathology. 2013;**94**:355–361.

[122] Zhao C, Yan W, Zu X, Chen M, Liu L, Zhao S, Liu H, Hu X, Luo R, Xia Y, Qi L. Association between endothelial nitric oxide synthase 894g>t polymorphism and prostate cancer risk: A meta-analysis of literature studies. Tumour Biology: The Journal of the International Society for Oncodevelopmental Biology and Medicine. 2014;**35**:11727–11733.

[123] Nikolic ZZ, Pavicevic D, Romac SP, Brajuskovic GN. Genetic variants within endothelial nitric oxide synthase gene and prostate cancer: A meta-analysis. Clinical and Translational Science. 2015;**8**:23–31.

[124] Wu JH, Yang K, Ma HS, Xu Y. Association of endothelia nitric oxide synthase gene rs1799983 polymorphism with susceptibility to prostate cancer: A meta-analysis. Tumour Biology: The Journal of the International Society for Oncodevelopmental Biology and Medicine. 2014;**35**:7057–7062.

[125] Cai C, Hsieh CL, Gao S, Kannan A, Bhansali M, Govardhan K, Dutta R, Shemshedini L. Soluble guanylyl cyclase alpha1 and p53 cytoplasmic sequestration and down-regulation in prostate cancer. Molecular Endocrinology. 2012;**26**:292–307.

[126] Lai CS, Li S, Miyauchi Y, Suzawa M, Ho CT, Pan MH. Potent anti-cancer effects of citrus peel flavonoids in human prostate xenograft tumors. Food and Function. 2013;**4**:944–949.

[127] Oktem G, Bilir A, Selvi N, Yurtseven ME, Vatansever S, Ates U, Uysal A, Omay SB. Chemotherapy influences inducible nitric oxide synthase (iNOS) and endothelial nitric oxide synthase (eNOS) activity on 3d breast cancer cell line. Oncology Research. 2006; **16**:195–203.

[128] Yu S, Jia L, Zhang Y, Wu D, Xu Z, Ng CF, To KK, Huang Y, Chan FL. Increased expression of activated endothelial nitric oxide synthase contributes to antiandrogen resistance in prostate cancer cells by suppressing androgen receptor transactivation. Cancer Letters. 2013;**328**:83–94.

[129] Thomas LN, Morehouse TJ, Too CK. Testosterone and prolactin increase carboxypeptidase-d and nitric oxide levels to promote survival of prostate cancer cells. The Prostate. 2012;**72**:450–460.

[130] Gao Y, Li L, Zhang J, Su F, Gong Z, Lai Y, Zhang Y. Physicochemical characterization and an injection formulation study of water insoluble zcvi(4)-2, a novel no-donor anticancer compound. Archives of Pharmacal Research. 2012;**35**:1177–1186.

[131] Donia M, Mijatovic S, Maksimovic-Ivanic D, Miljkovic D, Mangano K, Tumino S, Biondi A, Basile F, Al-Abed Y, Stosic-Grujicic S, Nicoletti F. The novel no-donating compound git-27no inhibits in vivo growth of human prostate cancer cells and prevents murine immunoinflammatory hepatitis. European Journal of Pharmacology. 2009; **615**: 228–233.

[132] Baritaki S, Huerta-Yepez S, Sahakyan A, Karagiannides I, Bakirtzi K, Jazirehi A, Bonavida B. Mechanisms of nitric oxide-mediated inhibition of emt in cancer: Inhibition of the metastasis-inducer snail and induction of the metastasis-suppressor RKIP. Cell Cycle. 2010;**9**:4931–4940.

[133] Bivalacqua TJ, Musicki B, Kutlu O, Burnett AL. New insights into the pathophysiology of sickle cell disease-associated priapism. The Journal of Sexual Medicine. 2012;**9**:79–87.

[134] B. GPT. Inflammatory mechanisms and oxidative stress in Peyronie's disease therapeutic rationale and related emerging treatment strategies. Inflammation and Allergy—Drug Targets. 2012;**11**:48–57.

[135] Shaw EJ, Mitchell GC, Tan RB, Sangkum P, Hellstrom WJ. The non-surgical treatment of Peyronie disease: 2013 update. The World Journal of Men's Health. 2013;**31**:183–192.

[136] Shigehara K, Namiki M. Clinical management of priapism: A review. The World Journal of Men's Health. 2016;**34**:1–8.

[137] Hurt KJ, Musicki B, Palese MA, Crone JK, Becker RE, Moriarity JL, Snyder SH, Burnett AL. Akt-dependent phosphorylation of endothelial nitric-oxide synthase mediates penile erection. Proceedings of the National Academy of Sciences of the United States of America. 2002;**99**:4061–4066.

[138] Anele UA, Morrison BF, Burnett AL. Molecular pathophysiology of priapism: Emerging targets. Current Drug Targets. 2015;**16**:474–483.

[139] Ishikawa T, Kondo Y, Goda K, Fujisawa M. Overexpression of endothelial nitric oxide synthase in transgenic mice accelerates testicular germ cell apoptosis induced by experimental cryptorchidism. Journal of Andrology. 2005;**26**:281–288.

[140] DeFoor WR, Kuan CY, Pinkerton M, Sheldon CA, Lewis AG. Modulation of germ cell apoptosis with a nitric oxide synthase inhibitor in a murine model of congenital cryptorchidism. The Journal of Urology. 2004;**172**:1731–1735; Discussion 1735.

Chapter 8

Nitric Oxide: Key Features in Spermatozoa

Florentin-Daniel Staicu and Carmen Matas Parra

Additional information is available at the end of the chapter

http://dx.doi.org/10.5772/67551

Abstract

Several *in vitro* studies have pointed to the importance of nitric oxide (NO) in the female and male reproductive system in mammals. Its functions vary from preventing oocyte aging, improving the integrity of the microtubular spindle apparatus in aged oocytes, modulating the contraction of the oviduct, to regulating sperm physiology by affecting the motility, inducing chemotaxis in spermatozoa, regulating tyrosine phosphorylation, enhancing the sperm-zona pellucida binding ability, and modulating the acrosomal reaction. In spermatozoa, NO exerts its functions in different ways, which involve key elements such as the soluble isoform of guanylate cyclase, cyclic guanosine monophosphate (cGMP), cyclic adenosine monophosphate (cAMP), protein kinase A (PKA), adenylate cyclase, and the extracellular signal-regulated kinase (ERK) pathway. Furthermore, NO leads to the S-nitrosylation of several sperm proteins, among them a substantial group associated with energy generation and cell movement, but also with signal transduction, suggesting a role for S-nitrosylation in sperm motility and in modulating the sperm function, respectively. In this chapter, an overview of how NO modulates the sperm physiology is presented, based on the knowledge acquired to this day.

Keywords: nitric oxide, nitric oxide synthase, S-nitrosylation, spermatozoa, fertilization

1. Introduction

NO is a small hydrophobic molecule which can easily diffuse through biological membranes [1]. *In vivo*, it is synthesized during the conversion of L-arginine to L-citrulline by nitric oxide synthase, with the help of co-factors such as the reduced form of nicotinamide adenine dinucleotide phosphate, flavin mononucleotide, flavin adenine dinucleotide, and tetrahydrobiopterin [2].

The nitric oxide synthase (NOS) may be found in three different isoforms. Two of them, the endothelial and the neuronal NOS (eNOS, nNOS), require calcium/calmodulin to be activated and are responsible for the continuous basal release of NO. The third isoform, known as inducible NOS (iNOS), is calcium independent [3, 4]. Since NOS activity depends on the availability of its substrate and its co-factors, all these elements jointly determine the cellular rates of NO synthesis [5].

Substantial evidence indicates that NO is a crucial biological messenger involved in a wide variety of physiological and pathological processes in different systems in mammals, including the vascular, nervous, and reproductive system [1, 6].

2. NOS/NO duo in the reproductive system

The NOS/NO duo regulates key functions in both the female and male reproductive systems [6].

All three isoforms of NOS have been identified in the oviduct [7, 8], oocytes, and cumulus and corona cells [9, 10] of several species [7, 11, 12]. The expression of NOS isoforms differs during the estrous cycle in the follicles as well as in the oviduct [13]. Tao *et al.* [10] showed that the immunoreactivity of eNOS in early antral follicles was restricted to the oocyte and increased from small and medium to large follicle-enclosed oocytes. In contrast, no immunoreactivity for iNOS was found in primordial, early antral follicle, or the cumulus-oocyte complexes aspirated from small and medium follicles.

In the oviduct, the endogenous basal release of NO regulates its contraction and the ciliary beating of the ciliated epithelial cells and induces chemotaxis in human spermatozoa via activation of the nitric oxide/soluble isoform of guanylate cyclase/cyclic guanosine monophosphate pathway (NO/sGC/cGMP) [14–16]. NO forms a vital component of the oocyte microenvironment and has been positively implicated in meiotic resumption [17], in preventing oocyte aging and improving the integrity of the microtubular spindle apparatus in aged oocytes [18]. It may also contribute as an anti-platelet agent during implantation [19].

As far as the male gamete is concerned, research was first concentrated on determining the effects of NO-releasing compounds on sperm motility and viability. Low concentrations of sodium nitroprusside (SNP), an NO-releasing compound, stimulated sperm hyperactivation in mouse, fish, and hamster [20–22] and were beneficial to the maintenance of post-thaw human sperm motility [23]. On the other hand, high concentrations of NO-releasing compounds decreased sperm motility [20, 24–26].

Numerous studies have also been conducted to determine the presence and localization of NOS in sperm from several species (**Table 1**). For example, Herrero *et al.* [27] located nNOS in the head of freshly ejaculated human spermatozoa, with a more concentrated fluorescent staining toward the equatorial region. O'Bryan *et al.* [28] described the pattern of eNOS expression in human spermatozoa, finding that morphologically normal spermatozoa exhibited post-acrosomal and equatorial eNOS immunostaining. Interestingly, though, abnormally shaped sperm cells exhibited aberrant staining, especially in the midpiece and/or head region, which correlated negatively with the percentage of motile sperm.

Species	Authors	Techniques	Identified isoforms
Human	Herrero *et al.* [27]	Immunofluorescence	nNOS
	O'Bryan *et al.* [28]	Immunocytochemistry	eNOS
Mouse	Herrero *et al.* [29]	Kinetic assays measuring the conversion of L-arginine to L-citrulline	nNOS, iNOS, eNOS
		Western blot	
Bull	Meiser and Schulz [30]	Modified griess reaction [92]	nNOS, eNOS
		Western blot	
		Immunofluorescence	
Boar	Hou [31]	NO assay kit with the Griess reagent	nNOS, iNOS, eNOS
	Aquila *et al.* [32]	Western blot	
Stallion	Ortega Ferrusola *et al.* [33]	Flow cytometry	nNOS, eNOS
		Western blot	
Cat	Liman and Alan [34]	Histochemistry	nNOS, iNOS, eNOS
		Immunohistochemistry	
		Western blot	

Table 1. Summary of *in vitro* studies and the techniques used to identify NOS isoforms in different species.

NOSs were revealed in mature mouse spermatozoa by means of biochemical techniques and Western blot. Herrero *et al.* [29] showed that mouse spermatozoa can synthesize L-citrulline, depending on the concentration of L-arginine present in the incubation medium while different concentrations of N(G)-nitro-L-arginine methyl ester (L-NAME) inhibit the formation of the amino acid. Furthermore, when sperm protein extracts were incubated under denaturing and nonreducing conditions and then subjected to immunoblotting assay, a protein fraction of 140 kDa was recognized by the three anti-NOS antibodies.

Bull spermatozoa were examined for the presence of constitutive NOS [30]. NO generation seemed to be enhanced by L-arginine and abolished by the NOS-inhibitor, L-NAME. In addition, Meiser and Schulz [30] verified the presence of NOS in bull sperm cells by immunohistochemistry, which was confirmed by Western blot. Confocal laser microscopy localized nNOS-related immunofluorescence at the acrosome cap and the main part of the flagellum. The same technique also identified eNOS staining spread over the spermatozoan head. Moreover, when these findings were confirmed by Western blot, immunoreactive bands at 161 kDa (nNOS) and 133 kDa (eNOS) were identified.

Hou *et al.* [31] investigated whether boar sperm can generate NO, finding that porcine spermatozoa synthesized low levels of NO under noncapacitating conditions, but that the NO concentration almost doubled when sperms were capacitated. Furthermore, NO production was significantly inhibited when capacitated sperms were treated with L-NAME. In another study [32], Western blot analysis was performed to identify NOS enzymes in boar sperm samples. The immunoblots showed three distinct bands: ~160, ~130, and ~135 kDa, corresponding to nNOS, iNOS, and eNOS, respectively.

NO production was evaluated in stallion spermatozoa before and after freezing/thawing [33] by means of flow cytometry, after loading the sperm suspension with an NO detection probe. NO synthesis was positively correlated with sperm motility after thawing and, interestingly, the presence of egg yolk in the semen extender radically reduced the amount of NO produced. The authors further investigated in fresh and frozen/thawed stallion sperm the presence of NOS enzymes by Western blot, using anti-nNOS, anti-eNOS, and anti-universal NOS antibodies. Two bands of approximately 83 kDa and 96 kDa were labeled by the antibodies anti-nNOS and anti-eNOS, respectively. Moreover, the other antibody, which recognized an epitope present in all the NOS isoforms described so far, showed two similar bands of 84 and 92 kDa.

Recently, Liman and Alan [34] investigated the localization of NOS isoforms in spermatozoa within the intratesticular and excurrent duct systems of adult domestic cats. Overall, the spermatozoa head did not exhibit immunoreactivity. On the other hand, immunoreactivity for all three isoforms was observed in the flagellum, in the proximal cytoplasmic droplets of spermatozoa (located in the neck region) within the lumen of the intratesticular and efferent ducts, in the epididymal duct of the caput epididymis, and in the distal cytoplasmic droplets of spermatozoa (located at the mid-principal piece junction of the tail) within the lumen of corpus and cauda epididymis and the vas deferens.

3. Role of NO on sperm functionality

Several *in vitro* studies were conducted in order to determine the effects that NO has on sperm physiology (**Figure 1**). It has been shown that NO affects sperm motility [28, 35, 36], acts as chemoattractant [16, 37], regulates the tyrosine phosphorylation of different sperm proteins [38, 39], enhances the sperm-zona pellucida binding ability [40], and modulates the acrosomal reaction [41, 42].

In detail, NO seems to play an important role in the maintenance of sperm motility at physiological levels. A study [36] showed that the basal release of NO by spermatozoa from normozoospermic samples tended to be greater than that from asthenozoospermic samples, suggesting a physiological and beneficial role for endogenous NO in the preservation of sperm motility. These observations agree with a previous report that normozoospermic spermatozoa express more NOS and generate more nitrite than asthenozoospermic spermatozoa [35]. On the other hand, as previously mentioned, it has been shown that spermatozoa with an abnormal morphology show aberrant staining for eNOS, which was negatively correlated with the motility [28]. A detrimental effect on motility has also been reported by Rosselli *et al.* [24] and Weinberg *et al.* [25] when millimolar concentrations of exogenous NO donors were added to sperm samples.

http://dx.doi.org/10.5772/67551

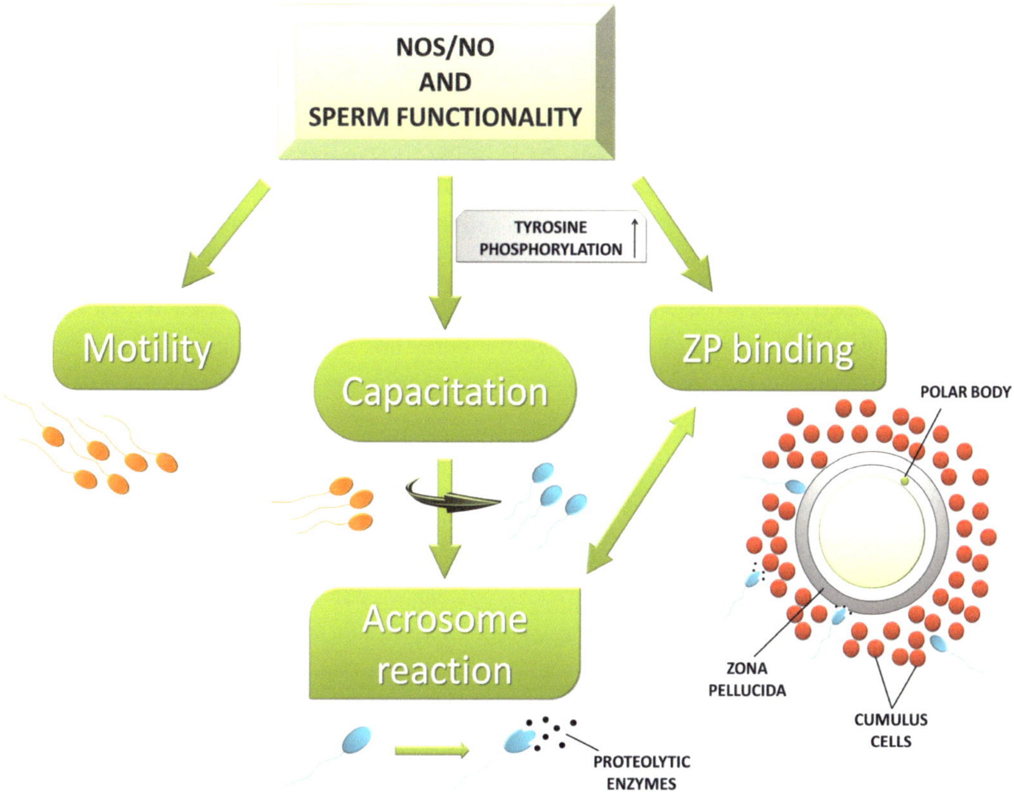

Figure 1. Some aspects of the sperm physiology modulated by the NOS/NO system. At physiological levels, endogenous NO has a beneficial role in maintaining sperm motility, enhances tyrosine phosphorylation, which, in turn, promotes the capacitation process. NO also increases the sperm-zona pellucida binding ability and leads to a rise in the percentage of reacted spermatozoa, especially in the presence of follicular fluid or protein-enriched extracts of follicular fluid.

It has been suggested that, upon approaching and entering the cumulus oophorus, both NO and progesterone, which are synthesized by the cumulus cells [9, 10, 43–45], provide a synergistic stimulus to mobilize stored calcium in the sperm neck/midpiece [46]. As a consequence, they can modulate flagellar activity and contribute to the hyperactivation that is vital for penetration of the oocyte vestments [47].

Interestingly, it has also been suggested that NO may exert a chemoattractant effect on spermatozoa. In fact, the percentage of mouse sperm migrating toward the medium containing an NO donor increased significantly [37]. Similar results were obtained when human spermatozoa were exposed to an NO donor [16]. In the latter case, the signal transduction pathway was also studied. It was proposed that NO exerts its chemoattractant effect through the activation of the NO/sGC/cGMP pathway, since the use of an NO scavenger and/or an sGC and cGMP-dependent protein kinase inhibitor reverted the NO donor-induced migration of sperm.

Since tyrosine phosphorylation in different sperm proteins is associated with the capacitation process [48], this aspect was investigated in order to further define the involvement of

NO in capacitation. Herrero *et al.* [38] observed an increase in tyrosine phosphorylation when human sperm capacitation was accelerated by an NO-releasing compound. On the other hand, when sperm capacitation was inhibited by L-NAME, there was an attenuation in the tyrosine phosphorylation of sperm proteins. In addition, Thundathil *et al.* [39] reported that L-NAME prevented, and a NO donor promoted, the increase in threonine, glutamine, and tyrosine phosphorylation in human spermatozoa. Furthermore, the addition of L-arginine reversed the inhibitory effect of L-NAME on the capacitation and the associated increase in phosphorylation.

The correlation between NO and sperm-zona pellucida binding ability was investigated by Sengoku *et al.* [40], who reported that when treated with low concentrations of a NO donor, the number of spermatozoa which binds to the hemizona is higher than in sperm treated with a higher concentration. Additionally, a NO quencher lowered the enhancement of sperm binding by the NO donor.

NO also seems to modulate the acrosome reaction. The percentage of acrosome loss induced by human follicular fluid or by calcium ionophore was studied when human spermatozoa were capacitated in the presence/absence of NO-releasing compounds or NOS inhibitors [38]. NO donors induced sperm cells to respond faster to human follicular fluid, whereas NOS inhibitors decreased the percentage of acrosome reaction. Similar results were obtained by Revelli *et al.* [41], who showed that different NO-releasing compounds were able to increase the percentage of reacted spermatozoa in the presence of protein-enriched extracts of human follicular fluid. Also, hemoglobin, a NO scavenger, inhibited the follicular fluid–induced acrosomal reaction. In an in-depth analysis of the signaling pathway of the nitric oxide–induced acrosome reaction in human spermatozoa [42], the authors suggested that the acrosome reaction-inducing effect of exogenous NO on capacitated human spermatozoa is accomplished via the NO/sGC/cGMP pathway, which leads to the activation of cGMP-dependent protein kinase (PKG). In fact, both the intracellular cGMP levels and the percentage of reacted spermatozoa were significantly increased after incubation with SNP. Furthermore, the SNP-induced acrosome reaction was significantly reduced in the presence of sGC inhibitors, a reduction that was reversed by the addition of a cell-permeating cGMP analogue to the incubation medium. Finally, PKG inhibition reduced the SNP-induced acrosome reaction.

4. NOS-activating molecules

As previously stated, NOS activity depends on the availability of its substrate and co-factors [5]. However, the scientific literature does not include many studies on the molecules present in the female reproductive tract which may activate, in one way or another, NOS enzymes in spermatozoa.

Starting from follicular fluid samples, Revelli *et al.* [41] obtained a protein-enriched follicular fluid solution (PFF), which was then used to study its effects on NOS activity, citrulline synthesis, and acrosome reaction in human sperm. Interestingly, this study showed for the first time that the endogenous NOS activity of human sperm may be increased by PFF. Moreover, the authors demonstrated that PFF-mediated induction of sperm NOS activity leads to acrosome reaction in the same cells, thereby, establishing a link between follicle-derived substances, the activation of NO synthesis in sperm and biological responses.

Furthermore, the increase in NO synthesis mediated by PFF was not associated with a rise in the expression of NOS catalytic units, which is not surprising since specialized cells possess very poor, if any, transcriptional activity [41]. The authors hypothesized that PFF first determines the transient enzyme activation of sperm NOS, which is subsequently strengthened by a more stable modification of the enzyme.

However, more studies should be performed in order to identify the NOS-activating molecule(s) in the follicular fluid.

5. NO pathway in spermatozoa

In spermatozoa, NO acts via three main pathways (**Figure 2**) [13]. First, NO is able to activate sGC, leading to a rise in the intracellular levels of cGMP [49]. The latter activates the cyclic nucleotide-gated channels (CNG) localized in the flagellum of mammalian spermatozoa [50, 51]. These channels seem to play an important role in the sperm motility control, by allowing the entry of Ca^{2+} ions to the cytoplasm during the capacitation process of mammal sperm [50]. Their activation is one of the first events that occur during capacitation in the mouse spermatozoa [52]. cGMP also activates PKG [53, 54], which is involved in the serine/threonine phosphorylation of proteins that promote sperm capacitation and the acrosome reaction [55, 56]. Furthermore, since cGMP and cAMP compete for the catalytic sites of phosphodiesterases [57, 58], an increase in the intracytoplasmic cGMP concentration may inhibit cAMP degradation via cyclic nucleotide phosphodiesterase type 3 [59], thus increasing cAMP intracellular levels and activating protein kinase A (PKA). The latter leads to an increase in protein tyrosine phosphorylation [60].

Second, NO is directly involved in tyrosine phosphorylation by modulating the cAMP/PKA and the extracellular signal-regulated kinase (ERK) pathways. The cAMP/PKA pathway can be influenced by NO via activation of sGC (as described above), but it can also be regulated directly. In fact, S-nitrosylation of adenylate cyclase (AC) has been suggested as a possible mechanism of action of NO [61]. Low levels of NO may activate AC, consequently increasing the cAMP concentration and activating PKA [62]. However, high levels of NO can inhibit AC [61]. As far as the ERK pathway is concerned, NO reacts with the cysteine residues of the RAS protein, inducing its activation [35]. In turn, RAS triggers the RAF, MEK, and ERK1/2 complex, necessary for tyrosine phosphorylation [63].

Third, NO regulates the post-translational protein modification in spermatozoa via S-nitrosylation [64], a process similar to phosphorylation and acetylation [65, 66]. S-nitrosylation consists of the covalent incorporation of NO into thiol groups (-SH) to form S-nitrosothiols (S-NO), a modification that is selective and reversible [13].

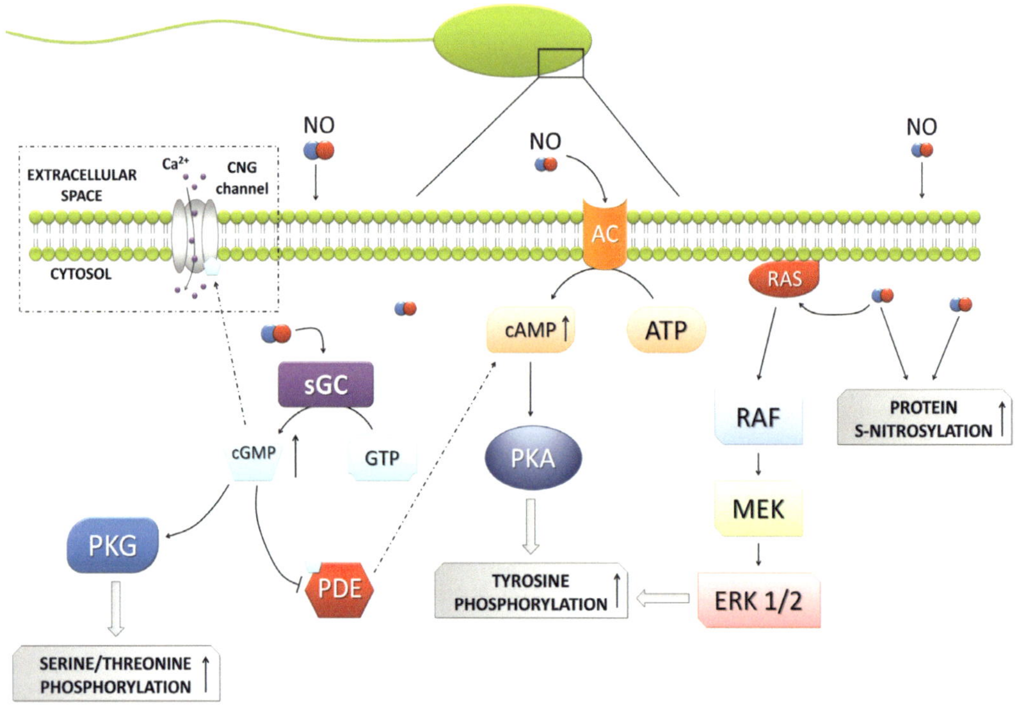

Figure 2. Representation of the main pathways through which NO acts in spermatozoa. NO leads to an increase in the intracellular levels of cyclic guanosine monophosphate (cGMP) by activating the soluble isoform of guanylate cyclase (sGC). The cGMP can activate the cyclic nucleotide-gated channels (CNG) localized in the flagellum of mammalian spermatozoa, which regulate the influx of Ca^{2+} ions to the cytoplasm during the capacitation process and also activates the cGMP-dependent protein kinase (PKG), leading to the serine/threonine phosphorylation of different proteins. It can also inhibit cyclic adenosine monophosphate (cAMP) degradation via cyclic nucleotide phosphodiesterase (PDE), which leads to the activation of cAMP-dependent protein kinase A (PKA) and tyrosine phosphorylation. Furthermore, NO is involved in the tyrosine phosphorylation process in a direct manner, by activating adenylate cyclase (AC) and the extracellular signal-regulated kinase (ERK) pathway. Finally, NO determines post-translational protein modification in spermatozoa via S-nitrosylation.

6. Function of S-nitrosoproteins in spermatozoa

An extensive study by Lefièvre *et al.* [64] described a large number of proteins present in the sperm of normozoospermic men, which can be subjected to S-nitrosylation in the presence of NO donors. Although the function of some nitrosylated proteins remains to be discovered, a considerable group of them are known to be metabolic proteins and proteins associated with energy generation and cell movement, suggesting a role for S-nitrosylation in sperm motility. This agrees with a previous proteomic analysis [67], in which the most abundant group was also involved in energy production.

Other considerable groups of proteins were those involved in signal transduction, which agrees with a role for S-nitrosylation in modulating the sperm function [64]. Interestingly, since sperms are generally assumed to be transcriptionally inactive, a small percentage of the

S-nitrosylated proteins identified by Lefièvre *et al.* [64] was related to transcription. Previous proteomic studies in sperm also observed the presence of proteins involved in transcription [67, 68]. However, when comparing the human sperm S-nitrosoproteome with proteins identified during a proteomic study of sperm-oocyte interaction, only three proteins were found in common, suggesting that S-nitrosylation is not a regulatory mechanism employed during fertilization [64, 69].

It is known that the mobilization of Ca^{2+} stored in the sperm neck/midpiece is necessary for the hyperactivation process [46]. The Ca^{2+} store in the neck of the sperm coincides with the region occupied by the redundant nuclear envelope (RNE) [70] and in order to mobilize Ca^{2+} from this site, ryanodine receptors (RyRs), which are intracellular Ca^{2+}-release channels involved in regulation of cytosolic calcium levels [71], need to be activated. These proteins contain a large number of thiol groups and are thus prone to S-nitrosylation by NO [64, 72, 73]. S-nitrosylation can potentiate the opening of RyRs [74–79], probably through the generation of the membrane permanent product S-nitrosocysteine [80]. It has been shown that an increase in Ca^{2+} induced by NO is accompanied by an increase in S-nitrosylation levels of endogenous RyRs [81, 82] while these Ca^{2+} channels may be inhibited under strongly nitrosylating conditions or at high doses of NO (**Figure 3**) [76, 79, 82]. Furthermore, progesterone acts synergistically with NO to mobilize Ca^{2+} in the sperm neck/midpiece by activation of RyRs [47], contributing to the hyperactivation process.

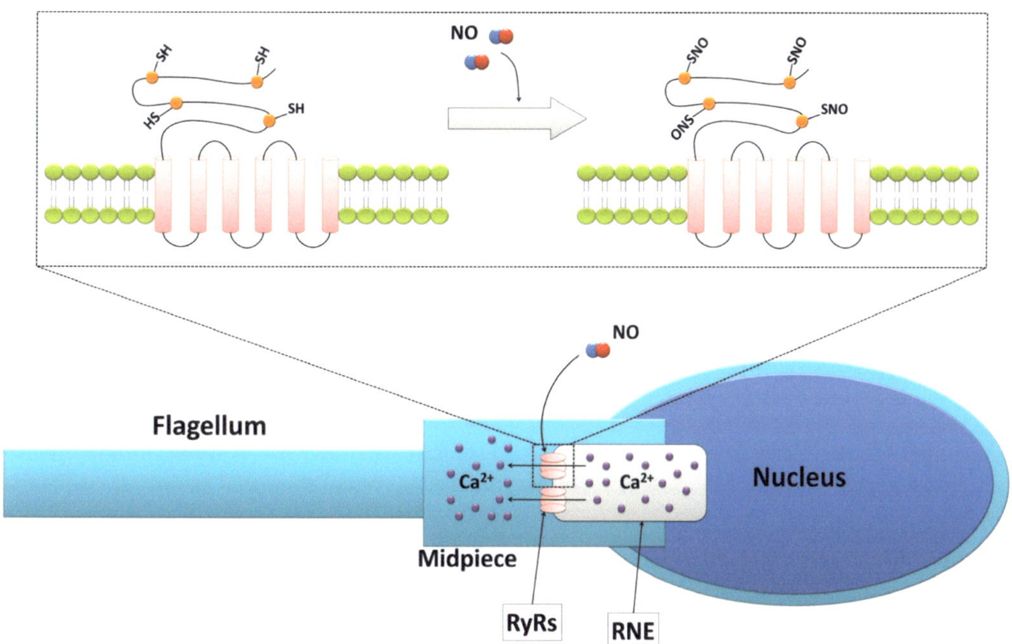

Figure 3. S-nitrosylation process. NO acts on the thiol groups (-SH) of the cysteines in proteins to form S-nitrosothiols (S-NO). At the sperm neck/midpiece, the S-nitrosylation occurs in ryanodine receptors (RyRs) allowing the release of calcium from the redundant nuclear envelope (RNE), which is required for sperm hyperactivation. Adapted and modified from López-Úbeda and Matás [13].

Other examples of proteins which can undergo S-nitrosylation in sperm and have a known biological significance are the A-kinase anchoring proteins (AKAPs) [64]. Both AKAP3 and AKAP4 are present in the fibrous sheath of the sperm flagellum, control PKA activity and undergo phosphorylation during the capacitation process [83–85]. AKAP complexes also modulate the motility of sperm. In fact, phosphodiesterase inhibitors were seen to significantly increase sperm motility [86], whereas PKA-anchoring inhibitor peptides arrested sperm motility [87]. Since the effects of NO on sperm motility are well established, the S-nitrosylation of AKAPs would be an interesting subject for additional studies.

A number of heat shock proteins (HSPs) may also be targets of S-nitrosylation in sperm [64], and some of them have been reported to act as important modulators of sperm capacitation. For instance, Asquith *et al.* [88] reported that heat shock protein 1 and endoplasmin undergo tyrosine phosphorylation during mouse sperm capacitation, whereas Nixon *et al.* [89] suggested that they form part of a zona pellucida complex, allowing successful sperm-egg interaction in the same species. Heat shock 70 kDa protein 8 and heat shock protein 90α also undergo tyrosine phosphorylation during human sperm capacitation [83], but whether they function in a zona receptor complex is still unknown [64]. Furthermore, HspA2 has been shown to be a marker of sperm maturity [90], its expression in infertile men with idiopathic oligoteratozoospermia being lower than in normozoospermic men [91].

7. Concluding remarks

In recent years, our knowledge of the involvement of the NOS/NO system in mammalian fertilization has grown, and there is clear evidence that NO acts as a significant modulator of the male and female gamete. However, many aspects regarding the NOS/NO duo, such as the presence of NOS-activating molecule(s) in the fertilization site or how the biological function of the S-nitrosylated proteins changes, remain to be discovered. Shedding light on these mechanisms will increase our understanding of the etiopathology of subfertility/infertility problems and how such problems can be overcome.

Acknowledgements

This work was supported by H2020 MSC-ITN-EJD 675526 REP-BIOTECH, the Spanish Ministry of Economy and Competitiveness (MINECO) and the European Regional Development Fund (FEDER), Grants AGL2015–66341-R.

Author details

Florentin-Daniel Staicu and Carmen Matas Parra*

*Address all correspondence to: cmatas@um.es

Department of Physiology, Faculty of Veterinary Science, University of Murcia, Spain

References

[1] Pacher P, Beckman JS, Liaudet L. Nitric oxide and peroxynitrite in health and disease. Physiol Rev. 2007;**87**(1):315–424. DOI: http://dx.doi.org/10.1152/physrev.00029.2006.

[2] Halliwell B, Gutteridge JMC. Free Radicals in Biology and Medicine. 3rd ed. Clarendon Press, Oxford:1999.

[3] Griffith OW, Stuehr DJ. Nitric oxide synthases: properties and catalytic mechanism. Annu Rev Physiol. 1995;**57**:707–736. DOI: http://dx.doi.org/10.1146/annurev.ph.57.030195.003423.

[4] Snyder SH. Nitric oxide. No endothelial NO. Nature. 1995;**377**(6546):196–197. DOI: http://dx.doi.org/10.1038/377196a0.

[5] Rosselli M, Keller PJ, Dubey RK. Role of nitric oxide in the biology, physiology and pathophysiology of reproduction. Hum Reprod Update. 1998;**4**(1):3–24. DOI: 10.1093/humupd/4.1.3.

[6] O'Flaherty C, Rodriguez P, Srivastava S. L-arginine promotes capacitation and acrosome reaction in cryopreserved bovine spermatozoa. Biochim Biophys Acta. 2004;**1674**(2):215–221. DOI: http://dx.doi.org/10.1016/j.bbagen.2004.06.020.

[7] Rosselli M, Dubey RK, Rosselli MA, Macas E, Fink D, Lauper U, Keller PJ, Imthurn B. Identification of nitric oxide synthase in human and bovine oviduct. Mol Hum Reprod. 1996;**2**(8):607–612. DOI: 10.1093/molehr/2.8.607.

[8] Lapointe J, Roy M, St-Pierre I, Kimmins S, Gauvreau D, MacLaren LA, Bilodeau JF. Hormonal and spatial regulation of nitric oxide synthases (NOS) (neuronal NOS, inducible NOS, and endothelial NOS) in the oviducts. Endocrinology. 2006;**147**(12):5600–5610. DOI: http://dx.doi.org/10.1210/en.2005–1548.

[9] Reyes R, Vázquez ML, Delgado NM. Detection and bioimaging of nitric oxide in bovine oocytes and sperm cells. Arch Androl. 2004;**50**(4):303–309. DOI: http://dx.doi.org/10.1080/01485010490448471.

[10] Tao Y, Fu Z, Zhang M, Xia G, Yang J, Xie H. Immunohistochemical localization of inducible and endothelial nitric oxide synthase in porcine ovaries and effects of NO on antrum formation and oocyte meiotic maturation. Mol Cell Endocrinol. 2004;**222**(1–2):93–103. DOI: http://dx.doi.org/10.1016/j.mce.2004.04.014.

[11] Ekerhovd E, Brännström M, Alexandersson M, Norström A. Evidence for nitric oxide mediation of contractile activity in isolated strips of the human fallopian tube. Hum Reprod. 1997;**12**(2):301–305. DOI: 10.1093/humrep/12.2.301.

[12] Bryant CE, Tomlinson A, Mitchell JA, Thiemermann C, Willoughby DA. Nitric oxide synthase in the rat fallopian tube is regulated during the oestrous cycle. J Endocrinol. 1995;**146**(1):149–157. DOI: 10.1677/joe.0.1460149.

[13] López-Úbeda R, Matás C. An approach to the factors related to sperm capacitation process. Andrology. 2015;**4**:1.

[14] Rosselli M, Imthurn B, Macas E, Keller PJ, Dubey RK. Endogenous nitric oxide modulates endothelin-1 induced contraction of bovine oviduct. Biochem Biophys Res Commun. 1994;**201**(1):143–148. DOI: http://dx.doi.org/10.1006/bbrc.1994.1680.

[15] Zhan X, Li D, Johns RA. Expression of endothelial nitric oxide synthase in ciliated epithelia of rats. J Histochem Cytochem. 2003;**51**(1):81–87. DOI: 10.1177/002215540305100110.

[16] Miraglia E, Rullo ML, Bosia A, Massobrio M, Revelli A, Ghigo D. Stimulation of the nitric oxide/cyclic guanosine monophosphate signaling pathway elicits human sperm chemotaxis *in vitro*. Fertil Steril. 2007;**87**(5):1059–1063. DOI: http://dx.doi.org/10.1016/j.fertnstert.2006.07.1540.

[17] Romero-Aguirregomezcorta J, Santa ÁP, García-Vázquez FA, Coy P, Matás C. Nitric oxide synthase (NOS) inhibition during porcine *in vitro* maturation modifies oocyte protein S-nitrosylation and *in vitro* fertilization. PLoS One. 2014;**9**(12):e115044. DOI: 10.1371/journal.pone.0115044.

[18] Goud AP, Goud PT, Diamond MP, Abu-Soud HM. Nitric oxide delays oocyte aging. Biochemistry. 2005;**44**(34):11361–11368. DOI: http://dx.doi.org/10.1021/bi050711f.

[19] Cameron IT, Campbell S. Nitric oxide in the endometrium. Hum Reprod Update. 1998;**4**(5):565–569. DOI: 10.1093/humupd/4.5.565.

[20] Herrero MB, Cebral E, Boquet M, Viggiano JM, Vitullo A, Gimeno MA. Effect of nitric oxide on mouse sperm hyperactivation. Acta Physiol Pharmacol Ther Latinoam. 1994;**44**(3):65–69.

[21] Creech MM, Arnold EV, Boyle B, Muzinich MC, Montville C, Bohle DS, Atherton RW. Sperm motility enhancement by nitric oxide produced by the oocytes of fathead minnows, *Pimephelas promelas*. J Androl. 1998;**19**(6):667–674. DOI: 10.1002/j.1939–4640.1998.tb02076.x.

[22] Yeoman RR, Jones WD, Rizk BM. Evidence for nitric oxide regulation of hamster sperm hyperactivation. J Androl. 1998;**19**(1):58–64. DOI: 10.1002/j.1939–4640.1998.tb02470.x.

[23] Hellstrom WJ, Bell M, Wang R, Sikka SC. Effect of sodium nitroprusside on sperm motility, viability, and lipid peroxidation. Fertil Steril. 1994;**61**(6):1117–1122.

[24] Rosselli M, Dubey RK, Imthurn B, Macas E, Keller PJ. Effects of nitric oxide on human spermatozoa: evidence that nitric oxide decreases sperm motility and induces sperm toxicity. Hum Reprod. 1995;**10**(7):1786–1790.

[25] Weinberg JB, Doty E, Bonaventura J, Haney AF. Nitric oxide inhibition of human sperm motility. Fertil Steril. 1995;**64**(2):408–413.

[26] Tomlinson MJ, East SJ, Barratt CL, Bolton AE, Cooke ID. Preliminary communication: possible role of reactive nitrogen intermediates in leucocyte-mediated sperm dysfunction. Am J Reprod Immunol. 1992;**27**(1–2):89–92. DOI: 10.1111/j.1600–0897.1992.tb00730.x.

[27] Herrero MB, Pérez Martínez S, Viggiano JM, Polak JM, de Gimeno MF. Localization by indirect immunofluorescence of nitric oxide synthase in mouse and human spermatozoa. Reprod Fertil Dev. 1996;**8**(5):931–934.

[28] O'Bryan MK, Zini A, Cheng CY, Schlegel PN. Human sperm endothelial nitric oxide synthase expression: correlation with sperm motility. Fertil Steril. 1998;**70**(6):1143–1147. DOI: http://dx.doi.org/10.1016/S0015–0282(98)00382–3.

[29] Herrero MB, Goin JC, Boquet M, Canteros MG, Franchi AM, Perez Martinez S, Polak JM, Viggiano JM, Gimeno MA. The nitric oxide synthase of mouse spermatozoa. FEBS Lett. 1997;**411**(1):39–42. DOI: 10.1016/S0014–5793(97)00570-X.

[30] Meiser H, Schulz R. Detection and localization of two constitutive NOS isoforms in bull spermatozoa. Anat Histol Embryol. 2003;**32**(6):321–325. DOI: 10.1111/j.1439–0264.2003.00459.x.

[31] Hou ML, Huang SY, Lai YK, Lee WC. Geldanamycin augments nitric oxide production and promotes capacitation in boar spermatozoa. Anim Reprod Sci. 2008;**104**(1):56–68. DOI: http://dx.doi.org/10.1016/j.anireprosci.2007.01.006.

[32] Aquila S, Giordano F, Guido C, Rago V, Carpino A. Nitric oxide involvement in the acrosome reaction triggered by leptin in pig sperm. Reprod Biol Endocrinol. 2011;**9**:133. DOI: http://dx.doi.org/10.1186/1477–7827-9–133.

[33] Ortega Ferrusola C, González Fernández L, Macías García B, Salazar-Sandoval C, Morillo Rodríguez A, Rodríguez Martinez H, Tapia JA, Peña FJ. Effect of cryopreservation on nitric oxide production by stallion spermatozoa. Biol Reprod. 2009;**81**(6):1106–1111. DOI: 10.1095/biolreprod.109.078220.

[34] Liman N, Alan E. Region-specific localization of NOS isoforms and NADPH-diaphorase activity in the intratesticular and excurrent duct systems of adult domestic cats (*Felis catus*). Microsc Res Tech. 2016;**79**(3):192–208. DOI: 10.1002/jemt.22619.

[35] Lewis SE, Donnelly ET, Sterling ES, Kennedy MS, Thompson W, Chakravarthy U. Nitric oxide synthase and nitrite production in human spermatozoa: evidence that endogenous nitric oxide is beneficial to sperm motility. Mol Hum Reprod. 1996;**2**(11):873–878. DOI: 10.1093/molehr/2.11.873.

[36] Donnelly ET, Lewis SE, Thompson W, Chakravarthy U. Sperm nitric oxide and motility: the effects of nitric oxide synthase stimulation and inhibition. Mol Hum Reprod. 1997;**3**(9):755–762. DOI: 10.1093/molehr/3.9.755.

[37] Sliwa L, Stochmal E. Effect of sodium nitroprusside on mouse sperm migration *in vitro*. Arch Androl. 2000;**45**(1):29–33.

[38] Herrero MB, de Lamirande E, Gagnon C. Nitric oxide regulates human sperm capacitation and protein-tyrosine phosphorylation *in vitro*. Biol Reprod. 1999;**61**(3):575–581. DOI: 10.1095/biolreprod61.3.575.

[39] Thundathil J, de Lamirande E, Gagnon C. Nitric oxide regulates the phosphorylation of the threonine-glutamine-tyrosine motif in proteins of human spermatozoa during capacitation. Biol Reprod. 2003;**68**(4):1291–1298. DOI: http://dx.doi.org/10.1095/biolreprod.102.008276.

[40] Sengoku K, Tamate K, Yoshida T, Takaoka Y, Miyamoto T, Ishikawa M. Effects of low concentrations of nitric oxide on the zona pellucida binding ability of human spermatozoa. Fertil Steril. 1998;**69**(3):522–527. DOI: http://dx.doi.org/10.1016/S0015–0282(97)00537–2.

[41] Revelli A, Soldati G, Costamagna C, Pellerey O, Aldieri E, Massobrio M, Bosia A, Ghigo D. Follicular fluid proteins stimulate nitric oxide (NO) synthesis in human sperm: a possible role for NO in acrosomal reaction. J Cell Physiol. 1999;**178**(1):85–92. DOI: http://dx.doi.org/10.1002/(SICI)1097–4652(199901)178:1%3C85::AID-JCP11%3E3.0.CO;2-Y.

[42] Revelli A, Costamagna C, Moffa F, Aldieri E, Ochetti S, Bosia A, Massobrio M, Lindblom B, Ghigo D. Signaling pathway of nitric oxide-induced acrosome reaction in human spermatozoa. Biol Reprod. 2001;**64**(6):1708–1712. DOI: 10.1095/biolreprod64.6.1708.

[43] Chian RC, Ao A, Clarke HJ, Tulandi T, Tan SL. Production of steroids from human cumulus cells treated with different concentrations of gonadotropins during culture *in vitro*. Fertil Steril. 1999;**71**(1):61–66. DOI: http://dx.doi.org/10.1016/S0015–0282(98)00416–6.

[44] Mingoti GZ, Garcia JM, Rosa-e-Silva AA. Steroidogenesis in cumulus cells of bovine cumulus-oocyte-complexes matured *in vitro* with BSA and different concentrations of steroids. Anim Reprod Sci. 2002;**69**(3–4):175–186. DOI: http://dx.doi.org/10.1016/S0378–4320(01)00187–7.

[45] Yamashita Y, Shimada M, Okazaki T, Maeda T, Terada T. Production of progesterone from de novo-synthesized cholesterol in cumulus cells and its physiological role during meiotic resumption of porcine oocytes. Biol Reprod. 2003;68(4):1193–1198. DOI: 10.1095/biolreprod.102.010934.

[46] Bedu-Addo K, Costello S, Harper C, Machado-Oliveira G, Lefievre L, Ford C, Barratt C, Publicover S. Mobilisation of stored calcium in the neck region of human sperm—a mechanism for regulation of flagellar activity. Int J Dev Biol. 2008;**52**(5–6):615–626. DOI: 10.1387/ijdb.072535kb.

[47] Machado-Oliveira G, Lefièvre L, Ford C, Herrero MB, Barratt C, Connolly TJ, Nash K, Morales-Garcia A, Kirkman-Brown J, Publicover S. Mobilisation of Ca^{2+} stores and flagellar regulation in human sperm by S-nitrosylation: a role for NO synthesised in the female reproductive tract. Development. 2008;**135**(22):3677–3686. DOI: 10.1242/dev.024521.

[48] Emiliozzi C, Fenichel P. Protein tyrosine phosphorylation is associated with capacitation of human sperm *in vitro* but is not sufficient for its completion. Biol Reprod. 1997;**56**(3):674–679. DOI: 10.1095/biolreprod56.3.674.

[49] Murad F. The nitric oxide-cyclic GMP signal transduction system for intracellular and intercellular communication. Recent Prog Horm Res. 1994;**49**:239–248.

[50] Wiesner B, Weiner J, Middendorff R, Hagen V, Kaupp UB, Weyand I. Cyclic nucleotide-gated channels on the flagellum control Ca^{2+} entry into sperm. J Cell Biol. 1998;**142**(2):473–484. DOI: 10.1083/jcb.142.2.473.

[51] Weyand I, Godde M, Frings S, Weiner J, Müller F, Altenhofen W, Hatt H, Kaupp UB. Cloning and functional expression of a cyclic-nucleotide-gated channel from mammalian sperm. Nature. 1994;**368**(6474):859–863. DOI: http://dx.doi.org/10.1038/368859a0.

[52] Cisneros-Mejorado A, Hernández-Soberanis L, Islas-Carbajal MC, Sánchez D. Capacitation and Ca(2+) influx in spermatozoa: role of CNG channels and protein kinase G. Andrology. 2014;**2**(1):145–154. DOI: 10.1111/j.2047–2927.2013.00169.x.

[53] Lohmann SM, Vaandrager AB, Smolenski A, Walter U, De Jonge HR. Distinct and specific functions of cGMP-dependent protein kinases. Trends Biochem Sci. 1997;**22**(8):307–312. DOI: http://dx.doi.org/10.1016/S0968–0004(97)01086–4.

[54] Pfeifer A, Ruth P, Dostmann W, Sausbier M, Klatt P, Hofmann F. Structure and function of cGMP-dependent protein kinases. Rev Physiol Biochem Pharmacol. 1999;**135**:105–149.

[55] Rahman MS, Kwon WS, Pang MG. Calcium influx and male fertility in the context of the sperm proteome: an update. Biomed Res Int. 2014;**2014**:841615. DOI: 10.1155/2014/841615.

[56] Miraglia E, De Angelis F, Gazzano E, Hassanpour H, Bertagna A, Aldieri E, Revelli A, Ghigo D. Nitric oxide stimulates human sperm motility via activation of the cyclic GMP/protein kinase G signaling pathway. Reproduction. 2011;**141**(1):47–54. DOI: 10.1530/REP-10–0151.

[57] Bender AT, Beavo JA. Cyclic nucleotide phosphodiesterases: molecular regulation to clinical use. Pharmacol Rev. 2006;**58**(3):488–520. DOI: http://dx.doi.org/10.1124/pr.58.3.5.

[58] Conti M, Beavo J. Biochemistry and physiology of cyclic nucleotide phosphodiesterases: essential components in cyclic nucleotide signaling. Annu Rev Biochem. 2007;**76**:481–511. DOI: http://dx.doi.org/10.1146/annurev.biochem.76.060305.150444.

[59] Kurtz A, Götz KH, Hamann M, Wagner C. Stimulation of renin secretion by nitric oxide is mediated by phosphodiesterase 3. Proc Natl Acad Sci U S A. 1998;**95**(8):4743–4747.

[60] Beavo JA. Cyclic nucleotide phosphodiesterases: functional implications of multiple isoforms. Physiol Rev. 1995;**75**(4):725–748.

[61] McVey M, Hill J, Howlett A, Klein C. Adenylyl cyclase, a coincidence detector for nitric oxide. J Biol Chem. 1999;**274**(27):18887–18892. DOI: 10.1074/jbc.274.27.18887.

[62] Belén Herrero M, Chatterjee S, Lefièvre L, de Lamirande E, Gagnon C. Nitric oxide interacts with the cAMP pathway to modulate capacitation of human spermatozoa. Free Radic Biol Med. 2000;**29**(6):522–536. DOI: http://dx.doi.org/10.1016/S0891–5849(00)00339–7.

[63] de Lamirande E, Gagnon C. The extracellular signal-regulated kinase (ERK) pathway is involved in human sperm function and modulated by the superoxide anion. Mol Hum Reprod. 2002;**8**(2):124–135. DOI: 10.1093/molehr/8.2.124.

[64] Lefièvre L, Chen Y, Conner SJ, Scott JL, Publicover SJ, Ford WC, Barratt CL. Human spermatozoa contain multiple targets for protein S-nitrosylation: an alternative mechanism of the modulation of sperm function by nitric oxide?. Proteomics. 2007;**7**(17):3066–3084. DOI: http://dx.doi.org/10.1002/pmic.200700254.

[65] Foster MW, Stamler JS. New insights into protein S-nitrosylation. Mitochondria as a model system. J Biol Chem. 2004;**279**(24):25891–25897. DOI: http://dx.doi.org/10.1074/jbc.M313853200.

[66] Hess DT, Matsumoto A, Kim SO, Marshall HE, Stamler JS. Protein S-nitrosylation: purview and parameters. Nat Rev Mol Cell Biol. 2005;**6**(2):150–166. DOI: http://dx.doi.org/10.1038/nrm1569.

[67] Martínez-Heredia J, Estanyol JM, Ballescà JL, Oliva R. Proteomic identification of human sperm proteins. Proteomics. 2006;**6**(15):4356–4369. DOI: http://dx.doi.org/10.1002/pmic.200600094.

[68] Lalancette C, Faure RL, Leclerc P. Identification of the proteins present in the bull sperm cytosolic fraction enriched in tyrosine kinase activity: a proteomic approach. Proteomics. 2006;**6**(16):4523–4540. DOI: http://dx.doi.org/10.1002/pmic.200500578.

[69] Stein KK, Go JC, Lane WS, Primakoff P, Myles DG. Proteomic analysis of sperm regions that mediate sperm-egg interactions. Proteomics. 2006;**6**(12):3533–3543. DOI: http://dx.doi.org/10.1002/pmic.200500845.

[70] Costello S, Michelangeli F, Nash K, Lefievre L, Morris J, Machado-Oliveira G, Barratt C, Kirkman-Brown J, Publicover S. Ca^{2+}-stores in sperm: their identities and functions. Reproduction. 2009;**138**(3):425–437. DOI: 10.1530/REP-09–0134.

[71] Kakizawa S, Yamazawa T, Iino M. Nitric oxide-induced calcium release: activation of type 1 ryanodine receptor by endogenous nitric oxide. Channels (Austin). 2013;**7**(1):1–5. DOI: 10.4161/chan.22555.

[72] Takeshima H, Nishimura S, Matsumoto T, Ishida H, Kangawa K, Minamino N, Matsuo H, Ueda M, Hanaoka M, Hirose T, et al. Primary structure and expression from complementary DNA of skeletal muscle ryanodine receptor. Nature. 1989;**339**(6224):439–445. DOI: http://dx.doi.org/10.1038/339439a0.

[73] Otsu K, Willard HF, Khanna VK, Zorzato F, Green NM, MacLennan DH. Molecular cloning of cDNA encoding the Ca^{2+} release channel (ryanodine receptor) of rabbit cardiac muscle sarcoplasmic reticulum. J Biol Chem. 1990;**265**(23):13472–13483.

[74] Stoyanovsky D, Murphy T, Anno PR, Kim YM, Salama G. Nitric oxide activates skeletal and cardiac ryanodine receptors. Cell Calcium. 1997;**21**(1):19–29. DOI: http://dx.doi.org/10.1016/S0143–4160(97)90093–2.

[75] Xu L, Eu JP, Meissner G, Stamler JS. Activation of the cardiac calcium release channel (ryanodine receptor) by poly-S-nitrosylation. Science. 1998;**279**(5348):234–237. DOI: 10.1126/science.279.5348.234.

[76] Hart JD, Dulhunty AF. Nitric oxide activates or inhibits skeletal muscle ryanodine receptors depending on its concentration, membrane potential and ligand binding. J Membr Biol. 2000;**173**(3):227–236.

[77] Li N, Zou AP, Ge ZD, Campbell WB, Li PL. Effect of nitric oxide on calcium-induced calcium release in coronary arterial smooth muscle. Gen Pharmacol. 2000;**35**(1):37–45. DOI: http://dx.doi.org/10.1016/S0306–3623(01)00089–1.

[78] Heunks LM, Machiels HA, Dekhuijzen PN, Prakash YS, Sieck GC. Nitric oxide affects sarcoplasmic calcium release in skeletal myotubes. J Appl Physiol. 1985;**91**(5):2117–2124.

[79] Zima AV, Blatter LA. Redox regulation of cardiac calcium channels and transporters. Cardiovasc Res. 2006;**71**(2):310–321. DOI: http://dx.doi.org/10.1016/j.cardiores.2006.02.019.

[80] Zhang Y, Hogg N. The mechanism of transmembrane S-nitrosothiol transport. Proc Natl Acad Sci U S A. 2004;**101**(21):7891–7896. DOI: http://dx.doi.org/10.1073/pnas.0401167101.

[81] Kakizawa S. Nitric oxide-induced calcium release: activation of type 1 ryanodine receptor, a calcium release channel, through non-enzymatic post-translational modification by nitric oxide. Front Endocrinol (Lausanne). 2013;**4**:142. DOI: http://dx.doi.org/10.3389%2Ffendo.2013.00142.

[82] Zahradníková A, Minarovic I, Venema RC, Mészáros LG. Inactivation of the cardiac ryanodine receptor calcium release channel by nitric oxide. Cell Calcium. 1997;**22**(6):447–454.

[83] Ficarro S, Chertihin O, Westbrook VA, White F, Jayes F, Kalab P, Marto JA, Shabanowitz J, Herr JC, Hunt DF, Visconti PE. Phosphoproteome analysis of capacitated human sperm. Evidence of tyrosine phosphorylation of a kinase-anchoring protein 3 and valosin-containing protein/p97 during capacitation. J Biol Chem. 2003;**278**(13):11579–11589. DOI: http://dx.doi.org/10.1074/jbc.M202325200.

[84] Visconti PE, Johnson LR, Oyaski M, Fornés M, Moss SB, Gerton GL, Kopf GS. Regulation, localization, and anchoring of protein kinase A subunits during mouse sperm capacitation. Dev Biol. 1997;**192**(2):351–363. DOI: http://dx.doi.org/10.1006/dbio.1997.8768.

[85] Bajpai M, Fiedler SE, Huang Z, Vijayaraghavan S, Olson GE, Livera G, Conti M, Carr DW. AKAP3 selectively binds PDE4A isoforms in bovine spermatozoa. Biol Reprod. 2006;**74**(1):109–118. DOI: http://dx.doi.org/10.1095/biolreprod.105.043588.

[86] Fisch JD, Behr B, Conti M. Enhancement of motility and acrosome reaction in human spermatozoa: differential activation by type-specific phosphodiesterase inhibitors. Hum Reprod. 1998;**13**(5):1248–1254. DOI: 10.1093/humrep/13.5.1248.

[87] Vijayaraghavan S, Goueli SA, Davey MP, Carr DW. Protein kinase A-anchoring inhibitor peptides arrest mammalian sperm motility. J Biol Chem. 1997;**272**(8):4747–4752. DOI: 10.1074/jbc.272.8.4747.

[88] Asquith KL, Baleato RM, McLaughlin EA, Nixon B, Aitken RJ. Tyrosine phosphorylation activates surface chaperones facilitating sperm-zona recognition. J Cell Sci. 2004;**117**(Pt 16):3645–3657. DOI: http://dx.doi.org/10.1242/jcs.01214.

[89] Nixon B, Asquith KL, John Aitken R. The role of molecular chaperones in mouse sperm-egg interactions. Mol Cell Endocrinol. 2005;**240**(1–2):1–10. DOI: http://dx.doi.org/10.1016/j.mce.2005.06.004.

[90] Ergur AR, Dokras A, Giraldo JL, Habana A, Kovanci E, Huszar G. Sperm maturity and treatment choice of *in vitro* fertilization (IVF) or intracytoplasmic sperm injection: diminished sperm HspA2 chaperone levels predict IVF failure. Fertil Steril. 2002;**77**(5):910–918. DOI: http://dx.doi.org/10.1016/S0015–0282(02)03073-X.

[91] Cedenho AP, Lima SB, Cenedeze MA, Spaine DM, Ortiz V, Oehninger S. Oligozoospermia and heat-shock protein expression in ejaculated spermatozoa. Hum Reprod. 2006;**21**(7): 1791–1794. DOI: http://dx.doi.org/10.1093/humrep/del055.

[92] Guevara I, Iwanejko J, Dembińska-Kieć A, Pankiewicz J, Wanat A, Anna P, Gołabek I, Bartuś S, Malczewska-Malec M, Szczudlik A. Determination of nitrite/nitrate in human biological material by the simple Griess reaction. Clin Chim Acta. 1998;**274**(2):177–188. DOI: http://dx.doi.org/10.1016/S0009–8981(98)00060–6.

Chapter 9

From Nitric Oxide Toward S-Nitrosylation: Expanding Roles in Gametes and Embryos

Ješeta Michal, Marketa Sedmikova and
Jean-François Bodart

Additional information is available at the end of the chapter

http://dx.doi.org/10.5772/67270

Abstract

Nitric oxide (NO) is a gasotransmitter involved in various aspects of reproduction. The observational data from different species, such as sea urchin, ascidians, amphibians, rodents, porcine, bovine, and human, suggest that NO might have a significant role in reproduction through several mechanisms. This proposed role might appear preserved through evolution; however, the effects of NO also depend on the species or stages considered. There has been debate over the physiological relevance of NO, though the benefits of its use in assisted reproduction are now widely recognized. Over the past years, S-nitrosylation has provided a new angle to decipher the mechanisms through which NO exerts its actions. This chapter summarizes, in a nonexhaustive manner, research that explores the role of NO in gametes and embryos.

Keywords: nitric oxide, gamete, meiosis, oocyte, spermatozoa, nitrosylation, cell cycle, embryo, reproduction

1. Introduction

Nitric oxide (NO) is a gaseous free radical that plays a key role both in intra- and extracellular signaling pathways in a wide variety of organisms. The role of NO has been emphasized in many physiological processes including reproduction. NO is generated by nitric oxide synthases (NOS), whose isoforms have been detected in a variety of mammalian reproductive tissues such as ovary, uterus, testis, or epididymis. Nitric oxide has been involved in the regulation of follicle growth and ovulation in mice, spermatogenesis in humans, embryo implantation in rats, and meiosis in pigs and in mice. Data collected from the various abovementioned species suggest that NO might have a significant role in reproduction through mechanisms preserved through evolution; however, one cannot discard that the effects of NO may also be dependent on the species or stages considered. Therefore, nitric oxide could be considered as a gasotransmitter ruling out several aspects of reproduction, from gametes to early embryogenesis, though there have been a debate over the physiological relevance of NO. This chapter summarizes, in a nonexhaustive manner, the research that explores the role of NO in gametes and embryos.

2. NO in sea urchin

Nitric oxide was first reported to trigger parthenogenetic activation in sea urchin oocytes and was suggested as a potential physiological regulator for egg activation. Twenty years ago, an increase in NO levels at fertilization was reported [1, 2], and NO was hypothesized as the primary activator for sea urchin egg activation [3]. Enthusiasm has been lately shaded by a report of NO increases occurring lately during fertilization in comparison to the rise of intracellular calcium level [4, 5]. From this observation, it has been suggested that the role of NO could be limited to sustaining the duration of the calcium transient [4]. NO increases are likely correlated to the calcium increases but not by being its primary activator [4, 5]. Nevertheless, NO increases may play a role in the hardening of the fertilization envelope surrounding the fertilized egg, thereby protecting the embryo from severe environmental conditions during its early development [5].

3. NO in ascidians oocytes, eggs, and embryos

In contrast to the observations performed in sea urchins, Hyslop et al. [6] reported that NO was not likely to be involved in the physiological process of fertilization of *Ascidiella aspersa* eggs. This report was in contrast to a previous study examining the inward current induced by NO, sharing similarities with the ascidians sperm current, which suggested that fertilization in ascidians could use NO as a second messenger [7]. Though nitric oxide donors induce increases in free calcium in ascidians and sea urchin eggs at the intracellular level, NOS inhibitor L-NAME (N(G)-nitro-L-arginine methyl ester) did not prevent sperm-induced fertilization, and not so much as a discrete increase in NO preceded the calcium wave [6].

However, NO pathways and production were related to metamorphosis, notochord, and tail regression in *Ciona intestinalis*, which are correlated to caspase-dependent apoptosis [8]. The nitric oxide synthase spatial pattern expression was highly dynamic during this larval development [8]. Experimental increases and decreases in NO levels can drive delays or accelerations in tail regression, respectively, with NO changes being related to the modulation of Caspase 3-like activity [8]. In most of the considered marine larvae of ascidians, NO acted as a repressor of the initiation of metamorphosis [9, 10]. Further works had been undertaken in *C. intestinalis* to unravel the role of NO during larval development. Mitogen-activated protein kinase (MAPK) and extracellular-regulated kinase (ERK) phosphorylations levels appeared closely related to NO levels [11].

In conclusion, the fine tuning of NO pathways and levels are physiologically involved in ascidian larvae development and metamorphosis. However, NO did not seem to be required at very early steps of development.

4. NO in amphibian oocytes, eggs, and embryos

Amphibian oocytes offer a typical playground for deciphering the biochemical mechanisms underlying cell cycle transition. Meiosis or the M-phase–promoting factor (MPF) was discovered in amphibian oocytes [12] and characterized in this model to be made up of at least two subunits: Cdk1 (catalytic) and Cyclin B (regulatory) [13]. In *Xenopus* oocytes, NO-scavenging did not appear to impair the progression of M-phase entry and meiotic maturation because CPTIO (nitric oxide scavenger)-treated oocytes resumed and completed meiosis after hormonal stimulation by progesterone [14]. If NO is not required for meiotic progression, an excess in NO provided by a donor, such as S-nitroso-N-acetyl penicillamine (SNAP), leads to meiotic maturation inhibition. Under these conditions, SNAP lead heterogeneous response at the biochemical level with respect to the two pathways involved in meiotic resumption (MPF and MAPK). SNAP hindered the all-or-none response of MPF and MAPK pathways, especially in *Xenopus* oocytes; most oocytes exhibited partially active MPF and MAPK in the absence of any external signs of meiotic resumption [15]. Noticeably, SNAP altered meiotic spindle formation, therefore impairing proper genomic transmission [15]. This observation corroborates reports of mitosis inhibition through Tyrosine residue nitrosylation in plant models, and Alteration of cross walls orientation, presumably by impairing microtubules organization [16, 17]. Nevertheless, the mechanisms hindered by NO in vertebrate oocyte spindle formations remain undetermined.

Studies carried out in amphibians reported parthenogenetic activation of *Xenopus* eggs with nitric oxide donor SNAP. This parthenogenetic activation induced M-phase exit through an atypical mechanism involving calcium-dependent pathway and MAPK inactivation, while MPF activity was maintained [14] (**Figure 1A**).

During early development, the *Xenopus* nitric oxide synthase 1 (XNOS1) accounts for most of the NOS detected, being expressed in oocytes and eggs during segmentation [18]. At later stages, XNOS1 expression is restricted to the notochord, eyes, and developing neural system [18]. Exploration of NO function in *Xenopus* oocytes indicates that NO increased cell proliferation but impaired cell movements at gastrulation. Inhibition of NOS affected cell movements both in neural and mesodermal extension, through cGMPS-independent pathways involving dishevelled (Dsh) and the central components of the planary cell polarity (PCP) pathway [18]. Cell division during early development has been proposed according to this model to be impacted by NO through the ROCK-cGMP pathway (**Figure 1B**).

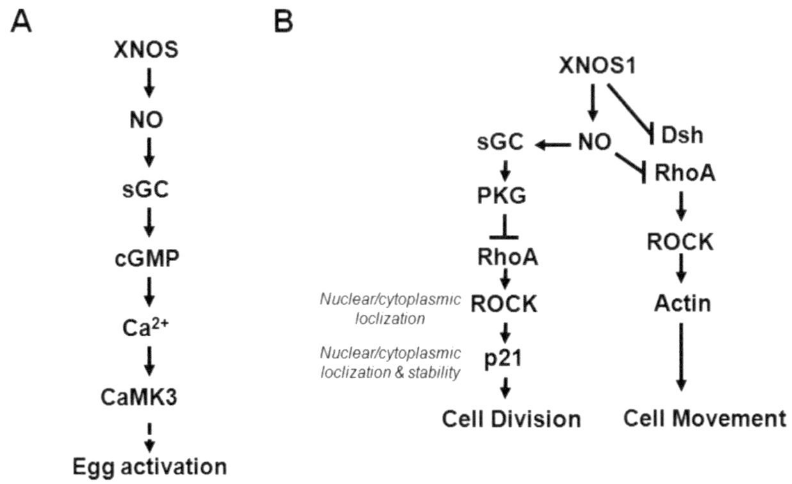

Figure 1. (A) Proposed mechanisms for NO action for parthenogenetic activation in *Xenopus* eggs. (B) Schematic representation of the pathways mediating the action of nitric oxide during early development in *Xenopus* embryos. Adapted from reference [18]. NO, nitric oxide; XNOS, *Xenopus* nitric oxide synthase; sGC, soluble guanylyl cyclase; cGMP, cyclic guanosine monophosphate; CaMK3, calmodulin kinase 3; Dsh, disheveled; PKG, protein kinase G.

5. NO in rodent oocytes, eggs, and embryos

Observations gathered in *Xenopus* oocytes may be to put in perspective with earlier reports endothelial NOS (eNOs) knock-out mice, in which meiotic abnormalities suggested that eNOS-derived NO is a modulator of oocyte meiotic maturation [19]. Indeed, the ovulation rates decreased in eNOS nullizygous mice, and oocytes often exhibited blocks during metaphase I or indicated various meiotic abnormalities with degenerative/atypical morphology of meiotic stages [19]. One should also note that in such conditions, oocytes from eNos nullizygous mice exhibit a higher rate of cell death than those in control ones. The importance of NOS for rodent oocyte meiotic maturation was confirmed by the expression of NOS isoforms in mouse or rat oocytes [20–22]. If NO was acknowledged to be important for meiotic maturation, high concentrations of NO can damage mouse oocyte and impair their further development [23]. Positive effects of low doses of NO donor SNAP during *in vitro* maturation was observed in mouse [24] and rat [25] oocytes. NO has been reported to play a dual role in oocyte meiotic maturation in mice, depending on its concentrations; however, the mechanism by which it influences oocyte maturation has not been fully clarified. In addition to these results, reports have shown that NO was most likely not essential for mouse oocyte fertilization [6].

Though NO is not a primary stimulus for oocyte activation through calcium mobilization, as observed in sea urchin and ascidians, it has been reported to play several potential roles during embryogenesis. The NO donor SNP brought about the arrest of embryonic development at early stages in mice; only half of the treated embryos reached blastocysts stages, with concentrations ranging from 10 nM to 1 mM [26]. Inhibition of NO production also suggested that NO plays a role in preimplantation embryos, which can be achieved through oxygen consumption limitation and mitochondrial activity or apoptosis modulation [27, 28]. Recently, it has been suggested that NO, through the use of the NO donor SNP and NOS inhibitor L-NAME, may regulate blastocyst hatching, which is a crucial step for embryo survival and implantation [29].

6. NO in porcine oocytes, eggs, and embryos

The production of low concentrations of NO concentrations has been reported to be necessary for meiotic maturation in porcine oocytes, whereas high concentrations of NO damage oocytes integrity [30]. Inhibition of NO synthase suppressed *in vitro* maturation of porcine oocytes, as attested by the absence of GVBD or meiosis I to meiosis II transitions [31]. NO synthases were detected in porcine oocytes [32, 33], but the involvement of NO/NOS in oocyte development has not been fully elucidated. Previously, it was assumed that NO acted *via* the cGMP cascade, similar to mechanisms observed for muscle contractions [19], or *via* the cascade-activating kinases, which are necessary for the resumption of meiotic maturation [34] (**Figure 2**). Nitric oxide had been reported to act on NO-sensitive guanyl cyclases, but new articles suggest that NO could also exert its functions by non-cGMP-mediated pathways by protein S-nitrosylation. It was reported that NOS inhibition decreases the amount of S-nitrosylated proteins in porcine oocytes [35].

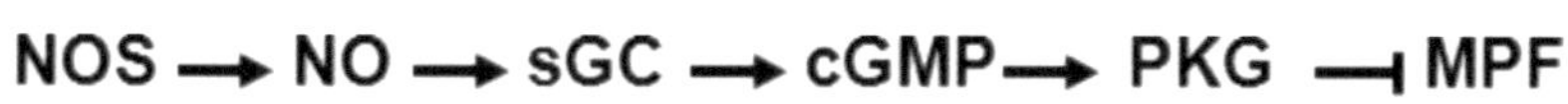

Figure 2. Scheme of NO/GMPc/PKG pathway in porcine oocytes and likely impact on basic regulation factor of meiosis, MPF. Nitric oxide synthase (NOS) increase enhances the concentration of nitric oxide (NO) in oocyte. NO stimulates the activity of soluble guanylyl cyclase (sGC), which catalyzes the production of cyclic guanosine monophosphate (cGMP). cGMP is necessary for protein kinase G (PKG) activation, which in turn suppresses MPF activation, and therefore, impairs meiosis resumption and maturation.

In porcine oocytes, NO donors are potent to induce parthenogenetic activation, resulting in early embryonic development [36]. NO production is also important for embryonic development since the NOS inhibitor L-NAME strongly decreases the proportion of porcine embryo blastocysts after 6 days of *in vitro* cultivation [37].

7. NO in bovine oocytes, eggs, and embryos

As well as in porcine oocytes, nitric oxide synthases (NOS) were detected in bovine oocytes [38]. Oocyte maturation was noticeably inhibited by the use of NOS inhibitors [39], which strengthens the hypothesis that NO plays a role during oocyte meiosis in this model. Regarding the molecular mechanisms involved, Bilodeau-Goeseels was first to suggest that the inhibitory effect of NO on bovine oocyte meiotic resumption did not appear to be mediated by the cGMP/PKG pathway; this is in contrast to previous observations gathered in mice [40].

Inhibition of NOS during maturation of bovine oocytes affected the quality of resulting bovine embryos by increasing the number of apoptotic blastomeres [41]. The importance of NO for correct embryonic development was observed, mainly for transition from morulae to the blastocyst stage. Treatment of embryos with the NOS inhibitor L-NAME reduced blastocysts numbers and hatching rates [42]. Contrarily, exposing bovine mature oocytes to a nitric oxide donor short term did not induce stress tolerance and had no positive effect on the *in vitro* embryo production of bovine embryos, as was expected [43].

8. A role for NO in human follicles?

Only a limited number of studies have addressed NO in human reproduction. Anteby et al. [44] observed an increase in NO concentrations in follicular fluid from bigger follicles, which positively correlated follicular volume and oestradiol concentrations. NO most likely acts as an important endocrinological regulator of ovulation. NO may be involved in an autocrine/paracrine regulation of the developing follicle and have a direct effect on granulosa cells, theca cells, and the developing oocyte. In a recent study, a possible association between idiopathic recurrent spontaneous abortion and variations in the gene encoding endothelial nitric oxide synthase was proposed [45].

9. NO roles in spermatozoa

In spermatozoa, NO was described to be involved in the regulation of viability, motility, capacitation, hyperactivation, acrosome reaction, and fusion with oocytes. Thus nitric oxide appears to be crucial for the processes driving successful fertilization (reviewed in [46, 47]). NOS isoforms were found in sperm of different mammal species such as mice [48], bulls [49], boars [50], and humans [51]. The abovementioned and well-known duality of NO's effects, depending upon concentrations, was also described in spermatozoa. Low levels of NO stimulate hyperactivation and increase motility of cryopreserved sperm after thawing [52]; conversely, high concentrations of NO decrease sperm motility [46, 53, 54] and inhibit sperm-oocyte fusion [55].

NO was reported to have a positive impact on sperm motility [55], whereas NO donor increased sperm motility [56, 57], inhibition of nitric oxide synthases by L-NAME negatively affected this motility [58]. Miraglia and colleagues [53] observed the existence of NO signaling pathways in human spermatozoa. NO stimulates sperm motility *via* the activation of soluble guanylate cyclase (sGC), the subsequent synthesis of cGMP, and the activation of cGMP-dependent protein kinase. The level of cGMP is modulated by cGMP-dependent phosphodiesterase (PDE). These observations concur with a former report that PDE inhibitor sidenafil citrate increased sperm motility [59]. NO is, on the other hand, considered a major free radical involved in sperm damage at sperm motility level. Nitrosative stress produced by high levels of reactive nitrogen species decreases progressive and total motility, as well as several sperm kinetic parameters, meanwhile, sperm viability is not affected [60, 61].

Sperm capacitation and acrosome reaction of mammal spermatozoa are essential for the fertilization process to occur. Both of them are NO-dependent. The final step in spermatozoa maturation, capacitation, involves a cascade of events such as the removal of cholesterol from plasma membrane, an influx of Ca^{2+} followed by an increase in intracellular cAMP levels, change of pH, and hyperactivation of sperm [62] (**Figure 3**). Acrosome reaction is a precondition for sperm fusion with the oocyte. It encompasses the release of proteolytic enzymes from the acrosome cap of the spermatozoa and an influx of Ca^{2+} and phosphorylation of tyrosine residues at the molecular level [63]. It has been reported that NO donors support acrosome reaction and accelerate capacitation and hyperactivation, whereas NOS inhibitors such as L-NAME significantly decrease or block this processes [46, 56, 58]. L-Arginin has similar effects as NO donors. Supplementation by L-arginin increases intracellular NO levels and supports sperm capacitation and acrosomal reaction without decreasing sperm viability [64]. Herrero et al. [65] also reported that capacitation was regulated by NO *via* cAMP levels and protein tyrosine phosphorylation. Moreover, it was proven that exogenous NO induces acrosomal reactions in human spermatozoa, and the process was mediated by the stimulation of a NO-sensitive sGC, cGMP synthesis, and the activation of PKC. However, the presence of extracellular Ca^{2+} was required for PKC activation in such conditions [66].

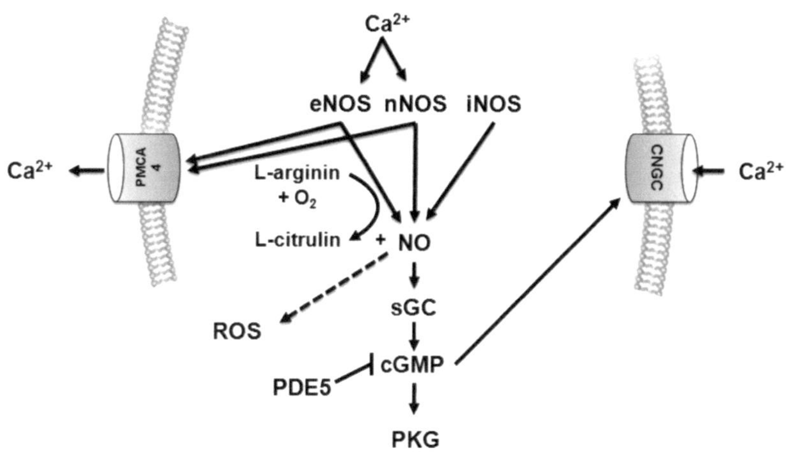

Figure 3. Scheme of NO/NOS role in sperm capacitation and acrosome reaction; eNOS—endothelial nitric oxide synthase; nNOS—neuronal nitric oxide synthase; iNOS—inducible nitric oxide synthase; sCG—soluble guanylyl cyclase; cGMP—cyclic guanosine monophosphate; PKG—protein kinase G; PDE5—phosphodiesterase 5; CNGC—cyclic nucleotide-gated channels; PMCA4—plasma membrane calcium ATPase 4; ROS—reactive oxygen species.

NO has the ability to improve the quality of freshly ejaculated sperm as well as thawed sperm. The NO donor sodium nitroprusside (SNP) was found to increase motility and viability of sperm after thawing and reduced membrane lipid peroxidation levels [57, 67].

10. NO effects on oocyte aging

Mammal oocytes are normally fertilized soon after the completion of meiotic maturation during the MII stage. If ovulated *in vitro*, matured oocytes are not fertilized, they undergo a process called aging, which is characterized by numerous changes. Oocyte aging rapidly decreases their quality and capacity to undergo embryonic development after fertilization. Functional and morphological changes associated with oocyte aging include decreased fertilization rates, polyspermy, parthenogenetic activation, apoptosis, chromosomal abnormalities, cortical granules exocytosis, ooplasmic microtubule dynamics, zona pellucida hardening, decreases in MPF and MAPK activities, epigenetic changes, and abnormal or delayed embryo development [24, 68–71]. Pathological conditions of oocyte aging impose limits for assisted reproduction technologies in animals as well as in humans [72].

It has been described that nitric oxide plays a part in oocyte aging, but it appears to do so by mobilizing more than one signaling pathway. NO can act either as a decelerating factor in oocyte aging [24] or, conversely, as an important cause of unwholesome aging-associated changes, which are caused by high ROS production [73].

Although high levels of NO are related mostly to pathological conditions, e.g., poor oocyte quality, increased protein nitration, and resistance to IVM in women with endometriosis [74], supplementing culture medium with low doses of the NO donor S-nitroso acetyl penicillamine (SNAP) delays manifestation of oocyte aging, *i.e.*, decrease of spontaneous cortical granule exocytosis, zona pellucida hardening, and the rate of spindle abnormalities [24]. The significance of NO in sustaining oocyte quality was demonstrated by Goud et al. [75]. Exposure of aged oocytes to L-NAME resulted in a significant disruption of fertilization and apoptosis during early embryonic development.

Different NOS isoforms could play different roles in aging. Lower NO levels produced by eNOS and nNOS could participate in delaying oocyte aging through the activation of sGC, which leads to an increased production of cyclic guanosine monophosphate [76]. However, high NO levels generated by iNOS were associated with higher $O_2^{\bullet-}$ production and promoted oocyte fragmentation and apoptosis [75]. Contrarily, Tripathi and colleagues [73] reported that the generation of NO through iNOS-mediated pathways was associated with the maintenance of meiotic arrest in diplotene-arrested oocytes and the sustained reduction of iNOS expression. Furthermore, they reported that intracellular NO level may induce apoptosis in aged rat oocytes cultured *in vitro*. Similarly, Nevoral et al. [77] described the suppression of apoptosis and lysis after prolonged cultivation of porcine oocytes in media supplemented by the NOS nonspecific inhibitor L-NAME. The decrease of intracellular levels of NO interrupts intracellular signal transduction pathways, especially Ca^{2+}-mediated pathways [75]. Premkumar and Chaube [78] reported that NO increases levels of cytosolic free Ca^{2+}, cGMP, and Wee 1 through an iNOS-mediated pathway. High levels of these signaling molecules trigger parthenogenetic activation of aged oocytes *via* the accumulation of phosphorylated Cdk1 (pThr-14/Tyr-15), a catalytic subunit of MPF. These findings indicate that NO can influence changes associated with oocyte aging in various manners.

11. S-nitrosylation as a posttranslational modification potentially regulating cell cycle

Though cGMP pathway has been reported to be the main road for NO's involvement, evidences have been raised for c-GMP independent pathway, through protein S-nitrosylation (i.e., in porcine oocytes [35]). S-nitrosylation is an established posttranslation modification, whose potential spectra of involvements in cancer cell lines, oocytes, and embryos ranges from cell cycle regulation to embryo implantation [18, 79–81]. The effects of NO may differ among the cellular models considered. For instance, whereas low concentrations of NO donors DETA-NO promote cell proliferation in promyelotic HL-60 cells [82], nitric oxide synthase inhibition drives cell proliferation at blastula stages in *Xenopus* embryos [18]. Such a duality in the effects certainly rely on the diversity of S-nitrosylated proteins.

S-nitrosylation has been reported to modify several regulators of cell cycle progression (**Figure 4**), such as cyclin-dependent kinases (CDKs). CDKs' S-nitrosylation was observed for CDK2, CDK5, and CDK6. While CDK2-nitrosylation enhances its activity independently of any effects on protein levels expression [82], the effect of S-nitrosylation on CDK5 and CDK6 remains elusive. Though G2/M arrests might be observed associated with NO release [83, 84], no S-nitrosylation has been so far reported for Cdk1, the catalytic subunit of MPF. S-nitrosylation of cyclin B was seek in HL-60 cells, but not observed [82]. As well, no S-nitrosylations have yet been reported for polo-like kinases (PLKs), anaphase promoting factor/cyclosome (APC/C), Wee1, and Myt1, which are MPF regulators. Nevertheless, the dual specificity phosphatase Cdc25, which is the main activator of MPF, is clearly impacted by its S-nitrosylation because it annihilates its phosphatase activity [82, 87]. Recently, an

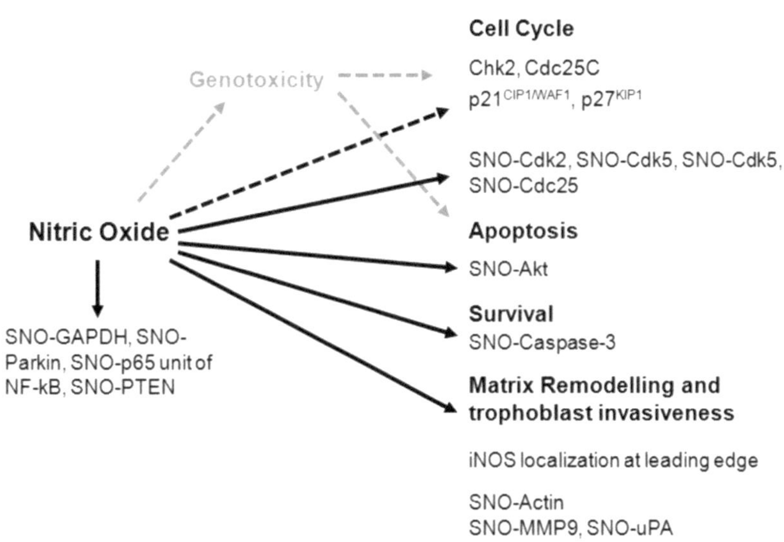

Figure 4. Comprehensive scheme of nitrosylation pivotal role in cell cycle, apoptosis, survival, matrix remodelling, and trophoblasts invasiveness.

alternate mechanism than direct S-nitrosylation of Cdc25C for its inhibition was proposed since NOAD, a nitric oxide-releasing derivative of oleanolic acid, induced activation of Chk2, resulting in an increase of the inhibitory phosphorylation of Cdc25C on its residue Serine 216 [84]. However, genotoxicity of nitric oxide might account for the activation of Chk2, the latter being involved in DNA damage response machinery. In the same study, the arrest in G2/M was associated with the upregulation of Cdk inhibitors $p21^{WAF1/CIP1}$ and $P27^{KIP1}$, without providing evidence for direct S-nitrosylation of these proteins. $p21^{WAF1/CIP1}$ downregulation by NO was also reported in *Xenopus* embryos, but through nitric oxide modulation of the RhoA-ROCK pathway [18] (**Figure 1B**). Thus, though there are converging evidences for role of NO in cell cycle regulation [85], the exact mechanisms remain to be fully deciphered.

12. S-nitrosylation plays pivotal role in modulating trophoblast motility and survival

As mentioned above, S-nitrosylation has been also called to play a role in preimplantation embryos and implantation (**Figure 4**). Microenvironnemental presence of NO was reported to contribute to the pathologic effects of endometriose on the development potential of embryos. In this context, NO effects on embryos survival could either rely upon S-nitrosylation, NO/GC/cGMP, or peroxynitrite formation. Lee and collaborators [28] suggested that the apoptotic effects of NO excess on mice embryos could be related to S-nitrosylation, in exclusion to other any mechanisms. The latter effects were closely associated with lipid-rich organelles (mitochondria and endoplasmic reticulum) [28, 86]. On the other hand, trophoblasts might also be protected from apoptosis *via* S-nitrosylation of caspase 3 [87].

Moreover, NO was shown to influence trophoblasts motility [88, 89]: it was further proposed that NO effects trophoblasts migration and invasion, which are critical processes for the successful embryonic development. In human trophoblasts, NO was required for outgrowth since L-NAME prevented this phenomenon in a dose-dependent manner [90]. Nevertheless, one shall keep in mind that high concentrations of NO in the environment have deleterious effects on trophoblasts outgrowth [90]. Effects of NO on trophoblasts motility has been proposed to be mediated by nitrosylation of the matrix metalloprotease MMP9 [79], based on (1) iNOS and MMP9 colocalization in front migration in trophoblast (leading edge with lamellipodium) and (2) observations of MMP9 being S-nitrosylated [91]. The dynamics of iNOS and S-nitrosylated proteins accumulation at the leading edge in trophoblasts led the authors to conclude that iNOS was not likely to be passively piling up [79] (**Figure 5**). In these conditions, iNOS also accumulates in aggregates in the cytoplasm. Taken together, colocalization of Actin, MMP9, and iNOS at the leading edge suggested indeed an active role for S-nitrosylation in cell migration. Invasiveness of cytotrophoblast-derived cell lines induced by adrenomedullin was also associated with an increase in S-nitrosylated protein rate [92]. S-nitrosylation of urokinase plasminogen activator (uPA), whose involvement in extracellular matrix is acknowledged, was reported to increase in these conditions [92]; the mechanisms through which S-nitrosylation increased this enzyme activity remain to be clarified.

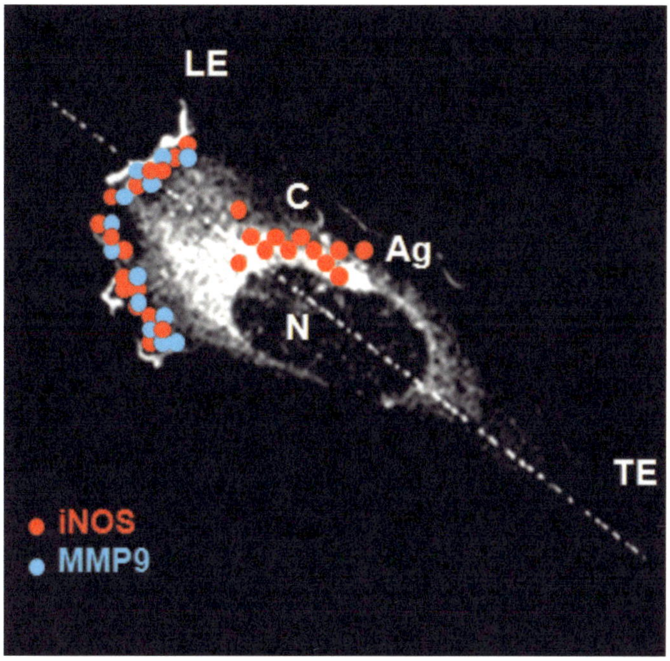

Figure 5. Schematic view of iNOS and MMP9 localization in trophoblasts. Adapted from reference [79]. LE, leading edge; TE, trailing edge; N, nucleus; C, cytoplasm; Ag, aggresome (iNOS particle). Actin polymerisation sites were also detected in LE. Noteworthy, Harris and colleagues noted a colocalization of eNOS and actin in trophoblasts.

13. A yin-yang relationship for S-nitrosylation and S-sulfhydration?

Along with NO, hydrogen sulfide (H_2S) is another gasotransmitter involved in regulating various aspects of cellular life. Many protein sites have been reported to undergo both S-nitrosylation and S-sulfhydration (brought by H_2S), such as Actin [93, 94], GAPDH [95], Parkin [96], PTEN [97], and the p65 subunit of NF-κB [98]. If S-sulfhydration and nitrosylation may occur on the same residue [99], reactive site cysteine, they generally promote different and opposing effects. Indeed, S-nitrosylation typically reduces cysteine thiols reactivity, while S-sulfhydration increases cysteine thiols reactivity, thereby making them more nucleophilic. If one wants to compare S-sulhydration and nitrosylation, it has mainly to outline that (1) proteins are rather S-sulhydrated than S-nitrosylated, and (2) nitrosylation rather inhibits and impairs protein functions. In this regard, the case of tensin (PTEN) provides an example of yin-yang relationship for S-nitrosylation and S-sulfhydration (**Figure 6**). Gene suppressor PTEN acts as an inhibitor of the PI-3 kinase/Akt signaling pathway, attenuating cellular growth and survival. S-nitrosylation enabled Akt hyperactivity, and thereby is associated with observed neuroprotective effects [100]. Oxidation of the active site cysteine is acknowledged as a common mechanism for regulating protein tyrosine phosphatases [101]. In addition, a NO-mediated PTEN degradation mechanism has been suggested to be common in neurodegenerative conditions where NO exerts a critical physiopathological role [102]. Finally, S-sulfhydration was

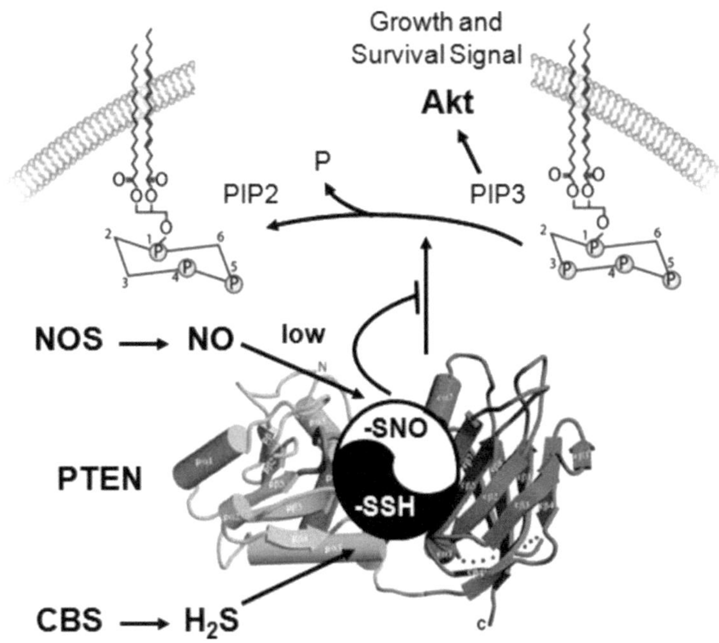

Figure 6. PTEN (left: phosphatase domain; right: C2 domain) is either S-nitrosylated (SNO) or S-sulfhydrated (SSH). When S-sulfhydrated, PTEN exerts its activity of phosphatase on PIP3 (phosphatidylinositol—3,4,5-triphosphate), in order to generate PIP2 (phosphatidylinositol-4,5-bisphosphate). In the absence of hydrogen sulfide, low concentrations of NO drive S-nitrosylation of PTEN, leading to its inactivation. In these conditions, Akt activity is maintained. If low concentration of NO drives SNO-PTEN, higher concentrations lead to SNO-Akt and abrogation of survival signal.

reported to maintain the activity of this lipid tyrosine-phosphatase, thereby preventing its S-nitrosylation [97]. PTEN structure reports disparate sites for S-nitrosylation (Cys 83 [100]), S-sulfhydration, and hydrogen peroxide-induced disulfide bond formation (Cys 71 and Cys 124 [103]). One should keep in mind that in this particular example, S-nitrosylation targets a different site than the Cys 124, mandatory in the phosphatase activity of PTEN.

It is tantalizing to hypothesize that sequence of S-nitrosylation and S-sulfhydration could provide a way for fine tuning of signaling pathways and cellular functions regulation. Because protein S-nitrosylation can foster intramolecular disulfide bond formation, a protein S-nitrosylation event might promote the formation of a more enduring S-sulfhydration reaction.

14. Conclusion

While NO is not necessary for meiotic resumption *per se* in amphibians, NO clearly influences meiotic processes in rodents, porcine, and bovine oocytes. During fertilization, the role of nitric oxide evolved from the hypothesis of being a primary activator to being solely correlated to fertilization, and its role was limited to a particular function such as hardening of the fertilization envelope in sea urchins. Therefore, the physiological relevance of NO has

http://dx.doi.org/10.5772/67270

been debated. Nevertheless, NO, together with its effects on spermatozoa (viability, motility, capacitation acrosome reaction, and fusion with oocytes) appeared as a modulator of oocyte aging. The benefits of nitric oxide use in assisted reproduction are now well-considered.

Though NO was not reported to be involved in early developmental processes in ascidians, NO positively affects cell proliferation in early *Xenopus* embryos and impairs cell movement during gastrulation in this model. The involvement of NO during segmentation is emphasized in mammalian models, where NO seemed to be requested for segmentation and blastocysts survival (rodents, porcine, and bovine) and for implantation through blastocysts hatching (rodents and bovine) and trophoblasts motility (humans).

One of the main difficulties when considering the effects of NO remains in the existence of the multiple pathways that can be activated by this gasotransmitter: cGMP-dependent pathway, calcium-related pathways, and reactive oxygen species production. Over the past few years, S-nitrosylation has offered a new angle to decipher NO's actions since S-nitrosylation modulates the activities of many key regulators such as members of the RhoA-ROCK pathway, Cdk2, Cc25, or PTEN. PTEN regulation by nitrosylation offers a new paradigm since sulfhydration and nitrosylation, both provided by gasotransmitters, appeared to play reciprocally in a yin-yang manner.

Acknowledgements

JFB is affiliated with the Site de Recherche Intégrée en Cancérologie (SIRIC ONCOLILLE). We would like to thank the personnel of the BICeL-Lille1-HB Facility for access to the microscopy systems and technical advices. We are indebted to the Research Federation FRABio for providing the scientific and technical environment to achieving our work [14, 15]. MJ work in Department of Obstetrics and Gynaecology, Center of Assisted Reproduction, University Hospital Brno, and Masaryk University. MJ's work was supported by grant MH CZ—DRO (FNBr, 65269705) and funds from the Faculty of Medicine, Masaryk University Brno, Czech Republic. We thank Brian Kavalir for helpful discussions and comments.

Author details

Ješeta Michal[1], Marketa Sedmikova[2], and Jean-François Bodart[3]*

*Address all correspondence to: jean-francois.bodart@univ-lille1.fr

1 Department of Obstetrics and Gynaecology, Center of Assisted Reproduction, University Hospital Brno and Masaryk University, Brno, Czech Republic

2 Department of Veterinary Sciences, Faculty of Agrobiology, Food and Natural Resources, Czech University of Life Sciences in Prague, Prague-Suchdol, Czech Republic

3 University of Lille, CNRS, UMR 8576 - UGSF - Unité de Glycobiologie Structurale et Fonctionnelle, Lille, France

References

[1] Sethi JK, Empson RM, Galione A. Nicotinamide inhibits cyclic ADP-ribose-mediated calcium signalling in sea urchin eggs. Biochem J. 1996;319(Pt 2):613–617.

[2] Willmott N, Sethi JK, Walseth TF, Lee HC, White AM, Galione A. Nitric oxide-induced mobilization of intracellular calcium via the cyclic ADP-ribose signaling pathway. J Biol Chem. 1996;271(7):3699–3705.

[3] Kuo RC, Baxter GT, Thompson SH, Stricker SA, Patton C, Bonaventura J, Epel D. NO is necessary and sufficient for egg activation at fertilization. Nature. 2000;406(6796):633–636.

[4] Leckie C, Empson R, Becchetti A, Thomas J, Galione A, Whitaker M. The NO pathway acts late during the fertilization response in sea urchin eggs. J Biol Chem. 2003;278(14):12247–12254.

[5] Mohri T, Sokabe M, Kyozuka K. Nitric oxide (NO) increase at fertilization in sea urchin eggs upregulates fertilization envelope hardening. Dev Biol. 2008;322(2):251–262.

[6] Hyslop LA, Carroll M, Nixon VL, McDougall A, Jones KT. Simultaneous measurement of intracellular nitric oxide and free calcium levels in chordate eggs demonstrates that nitric oxide has no role at fertilization. Dev Biol. 2001;234(1):216–230.

[7] Grumetto L, Wilding M, De Simone ML, Tosti E, Galione A, Dale B. Nitric oxide gates fertilization channels in ascidian oocytes through nicotinamide nucleotide metabolism. Biochem Biophys Res Commun. 1997;239(3):723–728.

[8] Comes S, Locascio A, Silvestre F, d'Ischia M, Russo GL, Tosti E, Branno M, Palumbo A. Regulatory roles of nitric oxide during larval development and metamorphosis in *Ciona intestinalis*. Dev Biol. 2007;306(2):772–784.

[9] Bishop CD, Brandhorst BP. On nitric oxide signaling, metamorphosis, and the evolution of biphasic life cycles. Evol Dev. 2003;5(5):542–550.

[10] Ueda N, Degnan SM. Nitric oxide acts as a positive regulator to induce metamorphosis of the ascidian *Herdmania momus*. PLoS One. 2013;8(9):e72797.

[11] Castellano I, Ercolesi E, Palumbo A. Nitric oxide affects ERK signaling through down-regulation of MAP kinase phosphatase levels during larval development of the ascidian *Ciona intestinalis*. PLoS One. 2014;9(7):e102907.

[12] Masui Y, Markert CL. Cytoplasmic control of nuclear behavior during meiotic maturation of frog oocytes. J Exp Zool. 1971;177(2):129–145.

[13] Lohka MJ, Hayes MK, Maller JL. Purification of maturation-promoting factor, an intracellular regulator of early mitotic events. Proc Natl Acad Sci U S A. 1988;85(9):3009–3013.

[14] Jeseta M, Marin M, Tichovska H, Melicharova P, Cailliau-Maggio K, Martoriati A, Lescuyer-Rousseau A, Beaujois R, Petr J, Sedmikova M, Bodart JF. Nitric oxide-donor SNAP induces Xenopus eggs activation. PLoS One. 2012;7(7):e41509. doi: 10.1371/journal.pone.0041509.

[15] Gelaude A, Marin M, Cailliau K, Jeseta M, Lescuyer-Rousseau A, Vandame P, Nevoral J, Sedmikova M, Martoriati A, Bodart JF. Nitric oxide donor S-nitroso-N-acetyl penicillamine (SNAP) alters meiotic spindle morphogenesis in Xenopus oocytes. J Cell Biochem. 2015;116(11):2445–2454.

[16] Jovanović AM, Durst S, Nick P. Plant cell division is specifically affected by nitrotyrosine. J Exp Bot. 2010;61(3):901–909. doi: 10.1093/jxb/erp369.

[17] Blume YB, Krasylenko YA, Demchuk OM, Yemets AI. Tubulin tyrosine nitration regulates microtubule organization in plant cells. Front Plant Sci. 2013;4:530.

[18] Peunova N, Scheinker V, Ravi K, Enikolopov G. Nitric oxide coordinates cell proliferation and cell movements during early development of Xenopus. Cell Cycle. 2007;6(24):3132–3144.

[19] Jablonka-Shariff A, Olson LM The role of nitric oxide in oocyte meiotic maturation and ovulation: meiotic abnormalities of endothelial nitric oxide synthase knock-out mouse oocytes. Endocrinology. 1998;139:2944–2954.

[20] Zackrisson U, Mikuni M, Wallin A, Delbro D, Hedin L, Brannstrom M Cell-specific localization of nitric oxide synthases (NOS) in the rat ovary during follicular development, ovulation and luteal formation. Hum Reprod. 1996;11:2667–2673.

[21] Jablonka-Shariff A, Ravi S, Beltsos AN, Murphy LL, Olson LM Abnormal estrous cyclicity after disruption of endothelial and inducible nitric oxide synthase in mice. Biol Reprod. 1999;61:171–177.

[22] Nishikimi A, Matsukawa T, Hoshino K, Ikeda S, Kira Y, Sato EF, Inoue M, Yamada M Localization of nitric oxide synthase activity in unfertilized oocytes and fertilized embryos during preimplantation development in mice. Reproduction. 2001;122:957–963.

[23] Bu S, Xia G, Tao Y, Lei L, Zhou B.Dual effects of nitric oxide on meiotic maturation of mouse cumulus cell-enclosed oocytes in vitro. Mol Cell Endocrinol. 2003;207:21–30.

[24] Goud AP, Goud PT, Diamond MP, Abu-Soud HM Nitric oxide delays oocyte aging. Biochemistry. 2005;44(34):11361–11368.

[25] Nakamura Y, Yamagata Y, Sugino N, Takayama H, Kato H. Nitric oxide inhibits oocyte meiotic maturation. Biol Reprod. 2002;67:1588–1592.

[26] Wu TP, Huang BM, Tsai HC, Lui MC, Liu MY. Effects of nitric oxide on human spermatozoa activity, fertilization and mouse embryonic development. Arch Androl. 2004;50(3):173–179.

[27] Manser RC, Leese HJ, Houghton FD. Effect of inhibiting nitric oxide production on mouse preimplantation embryo development and metabolism. Biol Reprod. 2004;71(2):528–533.

[28] Lee TH, Lee MS, Huang CC, Tsao HM, Lin PM, Ho HN, Shew JY, Yang YS. Nitric oxide modulates mitochondrial activity and apoptosis through protein S-nitrosylation for preimplantation embryo development. J Assist Reprod Genet. 2013;30(8):1063–1072.

[29] Pan X, Wang X, Wang X, Sun Z, Zhang X, Liang X, Li Z, Dou Z. Nitric oxide regulates blastocyst hatching in mice. Int J Clin Exp Med. 2015;8(5):6994–7001.

[30] Tichovska H, Petr J, Chmelikova E, Sedmikova M, Tumova L, Krejcova M, Dorflerova A, Rajmon R. Nitric oxide and meiotic competence of porcine oocytes. Animal. 2011;5: 1398–1405.

[31] Chmelikova E, Ješeta M, Sedmikova M, Petr J, Tumova L, Kott T, Lipovova P, Jilek F Nitric oxide synthase isoforms and the effect of their inhibition on meiotic maturation of porcine oocytes. Zygote. 2010;18:235–244.

[32] Hattori MA, Takesue K, Kato Y, Fujihara N Expression of endothelial nitric oxide synthase in the porcine oocyte and its possible function. Mol Cell Biochem. 2001;219:121–126.

[33] Chmelikova E, Sedmikova M, Petr J, Kott T, Lanska V, Tumova L, Tichovska H, Jeseta M. Expression and localization of nitric oxide synthase isoforms during porcine oocyte growth and acquisition of meiotic competence. Czech J Anim Sci. 2009;54:137–149.

[34] Sela-Abramovich S, Galiani D, Nevo N, Dekel N. Inhibition of rat oocyte maturation and ovulation by nitric oxide: mechanism of action. Biol Reprod. 2008;78:1111–1118.

[35] Romero-Aguirregomezcorta J, Santa ÁP, García-Vázquez FA, Coy P, Matás C. Nitric oxide synthase (NOS) inhibition during porcine in vitro maturation modifies oocyte protein S-nitrosylation and in vitro fertilization. PLoS One. 2014 Dec 26;9(12)

[36] Petr J, Rajmon R, Rozinek J, Sedmikova M, Jeseta M, Activation of pig oocytes using nitric oxide donors. Mol Reprod Dev. 2005;71:115–122.

[37] Redel BK, Tessane KJ, Spate LD, Murphy CN, Prather RS. Arginine increases development of in vitro-produced porcine embryos and affects the protein arginine methyltransferase-dimethylarginine dimethylaminohydrolase-nitric oxide axis. Reprod Fertil Dev. 27(4):655–666, 2015

[38] Pires PR, Santos NP, Adona PR, Natori MM, Schwarz KR, de Bem TH, Leal CL. Endothelial and inducible nitric oxide synthases in oocytes of cattle. Anim Reprod Sci. 2009 Dec;116(3–4):233–243.

[39] Matta SGC, Caldas-Bussiere MC, Viana KS, Faes MR, Paes de Carvalho CS, Dias BL, Quirino CR. Effect of inhibition of synthesis of inducible nitric oxide synthase-derived nitric oxide by aminoguanidine on the in vitro maturation of oocyte–cumulus complexes of cattle. Anim Reprod Sci. 2009;111(2–4):189–201.

[40] Bilodeau-Goeseels S. Effects of manipulating the nitric oxide/cyclic GMP pathway on bovine oocyte meiotic resumption in vitro. Theriogenology. 2007;68:693–701.

[41] Schwarz KR, Pires PR, de Bem TH, Adona PR, Leal CL. Consequences of nitric oxide synthase inhibition during bovine oocyte maturation on meiosis and embryo development. Reprod Domest Anim. 2010;45(1):75–80.

[42] Santana PP, da Silva BB, Silva TV, Costa NN, Cordeiro MS, Santos SS, Ohashi OM, Miranda MS. Addition of L-arginine to the fertilization medium enhances subsequent bovine embryo development rates. Theriogenology. 2016;85(6):1132–1138.

[43] Cheuquemán C, Loren P, Arias M, Risopatrón J, Felmer R, Álvarez J, Mogas T, Sánchez R. Effects of short-term exposure of mature oocytes to sodium nitroprusside on in vitro embryo production and gene expression in bovine. Theriogenology. 2015;84(8):1431–1437.

[44] Anteby EY, Hurwitz A, Korach O, Revel A, Simon A, Finci-Yeheskel Z, Mayer M, Laufer N. Human follicular nitric oxide pathway: relationship to follicular size, oestradiol concentrations and ovarian blood flow. Hum Reprod. 1996;11(9):1947–1951.

[45] Pereza N, Peterlin B, Volk M, Kapovic M, Ostojic S. A critical update on endothelial nitric oxide synthase gene variations in women with idiopathic recurrent spontaneous abortion: genetic association study, systematic review and meta-analyses. Mol Hum Reprod. 2015;21(5):466–478.

[46] Herrero, MB, Lamirande, E, Gagnon, C Nitric oxide is a signaling molecule in spermatozoa. Curr Pharm Des. 2003;9(5):419–425.

[47] Buzadzic, B, Vucetic, M, Jankovic, A, Stancic, A, Korac, A, Korac, B, Otasevic, V New insights into male (in)fertility: the importance of NO. Br J Pharmacol. 2014;172:1455–1467.

[48] Herrero, MB, Goin, JC, Canteros, MG, Franchi, AM, Perez Martinez, S, Polak, JM, Viggiano, JM, Gimeno MA The nitric oxide synthase of mouse spermatozoa. FEBS Lett. 1997;411:39–42.

[49] Meisner, H, Schulz, R. Detection and localization of two constitutive NOS isoforms in bull spermatozoa. Anat Histol Embryol. 2003;32:321–325.

[50] Aquila, S, Giordano, F, Guido, C, Rago, V, Carpino, A Nitric oxide involvement in the acrosome reaction triggered by leptin in pig sperm. Reprod Biol Endocrinol. 2011;9(133).

[51] O'Bryan, MK, Zini, A, Cheng, CY, Chlengel, PN Human sperm endothelial nitric oxide synthase expression: correlation with sperm motility. Fertil Steril. 1998;70(6):1143–1147.

[52] Hellstrom, WJGH, Bell, M, Wang, R, Sikka, SC. Effect of sodium nitroprusside on sperm motility, viability, and lipid peroxidation. Fertil Steril. 1994;61(6):1117–1122.

[53] Miraglia E, De Angelis F, Gazzano E, Hassanpour H, Bertagna A, Aldieri E, Revelli A, Ghigo D Nitric oxide stimulates human sperm motility via activation of the cyclic GMP/protein kinase G signaling pathway. Reproduction, 2011;141(1):47–54.

[54] Hassapour H, Mirshokrail P, Amnian A, Effect of nitric oxide on rat sperm motility *in vitro*. Pak J Biol Sci. 2007; 10 (14): 2374–8.

[55] Lewis SEM, Donnelly ET, Sterling ESL, Kennedy MS, Thompson W, Chakravarthy U Nitric oxide synthase and nitrite production in human spermatozoa: evidence that endogenous nitric oxide is beneficial to sperm motility. Mol Hum Reprod. 1996;2(11):873 878.

[56] Wang J, He Q, Yan X, Cai Y, Chen J Effect of exogenous nitric oxide on sperm motility in vitro. Biol Res. 2014;47(44).

[57] Khodaei, H Chamani, M, Mahdavi, B, Akhondi, AA Effects of adding sodium nitroprusside to semen diluents on motility, viability and lipid peroxidation of sperm in Holstein Bulls. Int J Fertil Steril. 2016. 9(4):521–526.

[58] Kameshwari DB, Siva AB, Shivaji S Inhibition of in vitro capacitation of hamster spermatozoa by nitric oxide synthase inhibitors. Cell Mol Biol. 2003;49(3):421–428.

[59] Dimitriadis, F, Giannakis, D, Pardalidis, N, Zikopoulos, K, Paraskevaidis, E, Giotitsas, N, Kalaboki, V, Tsounapi, P, Baltogiannis, D, Georgiou, I, Saito, M, Watanabe, T, Miyagawa, I, Sofikitis, N. Effects of phosphodiesterase 5 inhibitors on sperm parameters and fertilizing capacity. Asian J Androl. 2008;10(1):115–133.

[60] Uribe P, Boguen R, Treulen F, Sánchez R, and Villegas JV. Peroxynitrite-mediated nitrosative stress decreases motility and mitochondrial membrane potential in human spermatozoa. Mol Hum Reprod. 2015;21(3):237–243.

[61] Moran JM, Madejon L, Ferrusola CO, Pena FJ, Nitric oxide induces caspase activity in boar spermatozoa. Theriogenology. 2008;70(1):91–96.

[62] Visconti, PE, Galantino-Homer, H, Moore, GD, Bailey, JL, Ning, XP, Fornes, M, Kopf, GS The molecular basis of sperm capacitation. J Androl. 1998;19(2):242–248.

[63] Abou-Haila A, Tulsiani DRP Mammalian sperm acrosome: formation, contents, and function. Arch Biochem Biophys. 2000;379(2):173–182.

[64] Funahashi H Induction of capacitation and the acrosome reaction of boar spermatozoa by L-arginine and nitric oxide synthesis associated with the anion transport systém. Reproduction. 2002;124(6):857–864.

[65] Herrero, MB, Chatterjee, S, Lefievre, L, de Lamirande, E, Gagnon, C Nitric oxide interacts with the cAMP pathway to modulate capacitation of human spermatozoa. Free Radic Biol Med. 2000;29(6):522–536.

[66] Revelli, A, Costamagna, C, Moffa, F, Aldieri, E, Ochetti, S, Bosia, A, Massobrio, M, Lindblom, B, Chigo, D Signaling pathway of nitric oxide-induced acrosome reaction in human spermatozoa. Biol Reprod. 2001. 64(6):1708–1712.

[67] Sharma, RK, Agarwal, A Artificial stimulation of cryopreserved human spermatozoa by sodium nitroprusside, 2-chloroadenosine, and 2-deoxyadenosine. Eur Urol. 1997;32(3):344–352.

[68] Liang XW, Ma JY, Schatten H, Sun QY. Epigenetic changes associated with oocyte aging Sci China Life Sci. 2012;55(8):670–676.

[69] Miao YL, Kikuchi K, Sun QY, Schatten H. Oocyte aging: cellular and molecular changes, developmental potential and reversal possibility. Hum Reprod Update. 2009;15(5): 573–585.

[70] Petrova I, Sedmikova M, Petr J, Vodkova Z, Pytloun P. The role of c-Jun N-terminal kinase (JNK) and p38 Mitogen-activated protein kinase (p38 MAPK) in aged pig oocytes. J Reprod Dev. 2009;55:75–82.

[71] Jiang GJ, Wang K, Miao DQ, Guo L, Hou Y, et al. Protein profile changes during porcine oocyte aging and effects of caffeine on protein expression patterns. PLoS One. 2011;6(12):e28996.

[72] Fissore RA, Kurokawa M, Knott J, Zhang M, Smyth J. Mechanisms underlying oocyte activation and postovulatory ageing. Reproduction. 2002;124:745–774.

[73] Tripathi A, Khatun S, Pandey AN, Mishra SK, Chaube R, Shrivastav TG, Chaube SK. Intracellular levels of hydrogen peroxide and nitric oxide in oocytes at various stages of meiotic cell cycle and apoptosis. Free Radic Res. 2009;43(3):287–294.

[74] Goud PT, Goud AP, Diamond MP et al, "Chronological age enhances oocyte post ovulatory aging, protein nitration and nitric oxide insufficiency in oocytes and their microenvironment". Fertil Steril. 2014;3(102):e329.

[75] Goud AP, Goud PT, Diamond MP, Gonik B Abu-Soud HM Reactive oxygen species and oocyte aging: role of superoxide, hydrogen peroxide, and hypochlorous acid. Free Radic Biol Med. 2008;44(7):1295–1304.

[76] Goud PT, Goud AP, Diamond MP, Gonik B Abu-Soud HM. Nitric oxide extends the oocyte temporal window for optimal fertilization. Free Radic Biol Med. 2008;45(4):453–459

[77] Nevoral, J, Krejcova, T, Petr, J, Melicharova, P, Vyskocilova, A, Dvorakova, M, Weingartova, I, Chmelikova, E, Tumova, L, Hoskova, K, Kucerova-Chrpova, V, Sedmikova, M The role of nitric oxide synthase isoforms in aged porcine oocytes. Czech J Anim Sci. 2013;58(10): 453–459.

[78] Premkumar KV, Chaube SK. Nitric oxide signals postovulatory aging-induced abortive spontaneous egg activation in rats. Redox Rep. 2015;20(4):184–192.

[79] Harris LK, McCormick J, Cartwright JE, Whitley GS, Dash PR. S-nitrosylation of proteins at the leading edge of migrating trophoblasts by inducible nitric oxide synthase promotes trophoblast invasion. Exp Cell Res. 2008;314(8):1765–1776.

[80] Foster MW, Forrester MT, Stamler JS. A protein microarray-based analysis of S-nitrosylation. Proc Natl Acad Sci. 2009;106(45):18948–18953.

[81] Ben-Lulu S, Ziv T, Admon A, Weisman-Shomer P, Benhar M. A substrate trapping approach identifies proteins regulated by reversible S-nitrosylation. Mol Cell Proteom. 2014;13(10):2573–2583.

[82] Kumar S, Barthwal MK, Dikshit M. Cdk2 nitrosylation and loss of mitochondrial potential mediate NO-dependent biphasic effect on HL-60 cell cycle. Free Radic Biol Med. 2010;48(6):851–861.

[83] Gao L, Williams JL. Nitric oxide-donating aspirin induces G2/M phase cell cycle arrest in human cancer cells by regulating phase transition proteins. Int J Oncol. 2012;41(1):325–330.

[84] Liu L, Wang D, Wang J, Ji H, Zhang Y. NOAD, a novel nitric oxide donor, induces G2/M phase arrest and apoptosis in human hepatocellular carcinoma Bel-7402 cells. Toxicol In Vitro. 2015;29(7):1289–1297.

[85] Majumdar U, Biswas P, Subhra Sarkar T, Maiti D, Ghosh S. Regulation of cell cycle and stress responses under nitrosative stress in *Schizosaccharomyces pombe*. Free Radic Biol Med. 2012;52(11–12):2186–2200.

[86] Guo W, Kan JT, Cheng ZY, Chen JF, Shen YQ, Xu J, Wu D, Zhu YZ. Hydrogen sulfide as an endogenous modulator in mitochondria and mitochondria dysfunction. Oxid Med Cell Longev. 2012;2012:878052.

[87] Dash PR, Cartwright JE, Baker PN, Johnstone AP, Whitley GS. Nitric oxide protects human extravillous trophoblast cells from apoptosis by a cyclic GMP-dependent mechanism and independently of caspase 3 nitrosylation. Exp Cell Res. 2003;287(2):314–324.

[88] Cartwright JE, Holden DP, Whitley GS. Hepatocyte growth factor regulates human trophoblast motility and invasion: a role for nitric oxide. Br J Pharmacol. 1999;128(1):181–189.

[89] Cartwright JE, Tse WK, Whitley GS. Hepatocyte growth factor induced human trophoblast motility involves phosphatidylinositol-3-kinase, mitogen-activated protein kinase, and inducible nitric oxide synthase. Exp Cell Res. 2002;279(2):219–226.

[90] Tsui KH, Li HY, Cheng JT, Sung YJ, Yen MS, Hsieh SL, Wang PH. The role of nitric oxide in the outgrowth of trophoblast cells on human umbilical vein endothelial cells. Taiwan J Obstet Gynecol. 2015;54(3):227–231.

[91] Gu Z, Kaul M, Yan B, Kridel SJ, Cui J, Strongin A, Smith JW, Liddington RC, Lipton SA. S-nitrosylation of matrix metalloproteinases: signaling pathway to neuronal cell death. Science. 2002;297(5584):1186–1190.

[92] Wong BS, Lam KK, Lee CL, Wong VH, Lam MP, Chu IK, Yeung WS, Chiu PC. Adrenomedullin enhances invasion of human extravillous cytotrophoblast-derived cell lines by regulation of urokinase plasminogen activator expression and s-nitrosylation. Biol Reprod. 2013;88(2):34.

[93] Thom SR, Bhopale VM, Mancini DJ, Milovanova TN. Actin S-nitrosylation inhibits neutrophil beta2 integrin function. J Biol Chem. 2008;283(16):10822–10834.

[94] Mustafa AK, Gadalla MM, Sen N, Kim S, Mu W, Gazi SK, Barrow RK, Yang G, Wang R, Snyder SH. H2S signals through protein S-sulfhydration. Sci Signal. 2009;2(96):ra72.

[95] Hao G, Derakhshan B, Shi L, Campagne F, Gross SS. SNOSID, a proteomic method for identification of cysteine S-nitrosylation sites in complex protein mixtures. Proc Natl Acad Sci U S A. 2006;103(4):1012–1017.

[96] Vandiver MS, Paul BD, Xu R, Karuppagounder S, Rao F, Snowman AM, Ko HS, Lee YI, Dawson VL, Dawson TM, Sen N, Snyder SH. Sulfhydration mediates neuroprotective actions of parkin. Nat Commun. 2013;4:1626.

[97] Ohno K, Okuda K, Uehara T. Endogenous S-sulfhydration of PTEN helps protect against modification by nitric oxide. Biochem Biophys Res Commun. 2015 ;456(1):245–249.

[98] Sen N, Paul BD, Gadalla MM, Mustafa AK, Sen T, Xu R, Kim S, Snyder SH. Hydrogen sulfide-linked sulfhydration of NF-κB mediates its antiapoptotic actions. Mol Cell. 2012;45(1):13–24.

[99] Lu C, Kavalier A, Lukyanov E, Gross SS. S-Sulfhydration/desulfhydration and S-nitrosylation/denitrosylation: a common paradigm for gasotransmitter signaling by H_2S and NO. Methods. 2013;62(2):177–181.

[100] Numajiri N, Takasawa K, Nishiya T, Tanaka H, Ohno K, Hayakawa W, Asada M, Matsuda H, Azumi K, Kamata H, Nakamura T, Hara H, Minami M, Lipton SA, Uehara T. On-off system for PI3-kinase-Akt signaling through S-nitrosylation of phosphatase with sequence homology to tensin (PTEN). Proc Natl Acad Sci U S A. 2011;108(25):10349–10354.

[101] Heneberg P Reactive nitrogen species and hydrogen sulfide as regulators of protein tyrosine phosphatase activity. Antioxid Redox Signal. 2014;20(14):2191–2209.

[102] Kwak YD, Ma T, Diao S, Zhang X, Chen Y, Hsu J, Lipton SA, Masliah E, Xu H, Liao FF. NO signaling and S-nitrosylation regulate PTEN inhibition in neurodegeneration. Mol Neurodegener. 2010;5:49.

[103] Lee SR, Yang KS, Kwon J, Lee C, Jeong W, Rhee SG. Reversible inactivation of the tumor suppressor PTEN by H_2O_2. J Biol Chem. 2002;277(23):20336–20342.

Section 4

Miscellaneous Roles of Nitric Oxide Synthase

Chapter 10

Role of Endothelial Nitric Oxide Synthase in Breast Cancer

Tupurani Mohini Aiyengar, Padala Chiranjeevi and
Hanumanth Surekha Rani

Additional information is available at the end of the chapter

http://dx.doi.org/10.5772/67493

Abstract

Breast cancer (BC) is the most common form of carcinoma and a primary cause of morbidity and mortality globally. Oxidative stress represents as an important factor in carcinogenesis and may play a role in initiation and progression of tumors. Oxidative stress–induced NO• damage to DNA includes a multitude of lesions, many of which are mutagenic and have multiple roles in cancer and aging. It is caused by an unfavorable balance between reactive oxygen species/reactive nitrogen species (ROS/RNS) and antioxidant defenses. ROS/RNS are generated during normal cellular metabolism, as a result of the influence of various environmental factors, as well as during pathological processes. Nitric oxide (NO•) is a ubiquitous, short-lived free radical produced from L-arginine by nitric oxide synthases (NOSs), and isoforms of NOS exist, depending on the site of origin: endothelial (eNOS), neuronal (nNOS), mitochondrial (mtNOS), and inducible (iNOS). eNOS is responsible for the endothelial synthesis of NO• and has shown to modulate cancer-related events such as inflammation, angiogenesis, apoptosis, cell cycle, invasion, and metastasis. Genetic studies also showed that eNOS gene polymorphisms are associated with the development of breast cancer. Therefore, selective targeting of eNOS may prove a potential strategy for prevention and treatment of breast cancer.

Keywords: breast cancer, oxidative stress, nitric oxide, endothelial nitric oxide synthase, therapeutics

1. Introduction

Cancer, a multifaceted disorder, represents one of the most important health problems worldwide, with approximately 14 million new cases and 8.2 million cancer-related deaths every year [1]. The transformation from a normal cell into a tumor cell is a multistage process, typically a progression from a precancerous lesion to malignant tumors.

Cancers originate from a single abnormal cell (clonal origin) with an altered DNA sequence (mutation). Successive rounds of mutation and selective expansion of these cells result in the formation of a tumor mass and leads to tumor growth and progression, which eventually breaks through the basal membrane barrier of surrounding tissues and spreads to other parts of the body (metastasis) (**Figure 1**).

Cancers may be classified by their primary site of origin such as brain cancer, oral cancer, lung cancer, prostate cancer, liver cancer, renal cell carcinoma (kidney cancer), breast cancer, etc.

Breast cancer is defined as a malignant growth that begins in the epithelium of the breast. It is estimated as one of the most commonly diagnosed cancers worldwide (11.9%) [2]. In India, breast cancer has overtaken cervical cancer and has become the most prevailing cancer among women [3]. The most frequent type of breast cancer is ductal carcinoma *in situ* (DCIS), which affects the cells of the milk ducts. Cancer that starts in lobes or lobules is called lobular carcinoma *in situ* (LCIS); it is the second most common type of breast cancer. While rare breast cancer types include tubular, medullary, metaplastic, mucinous carcinoma, and Paget's disease.

Breast cancer is a clinically heterogeneous disease. Breast cancer cells generally overexpress estrogen receptor (ER)/progesterone receptor (PR), and/or human epidermal growth factor-2 (HER-2) receptor and lead to the tumor formation and progression. Thus, breast cancer can be classified into two subgroups on the basis of receptor status namely: (i) ER/PR positive (luminal A/B) and (ii) triple negative (ER, PR, and HER-2 negative) (basal-like) to know the prognosis and clinical outcome of breast cancer (**Figure 2**). Early breast cancer usually does not cause symptoms. As the cancer grows, symptoms may include: lump in the armpit, change in the size, shape, or feel of the breast or nipples, nipple discharge, etc. Symptoms of advanced breast cancer may include: breast pain or discomfort, bone pain, skin ulcers, and weight loss.

Breast cancer etiology is complex and multifactorial where there is a strong interplay between genetic and environmental factors. The strongest nonmodifiable determinants of breast cancer risk are female gender and age. Other risk factors associated with breast cancer can be grouped into three broad determinants: family history (hereditary) factors, hormonal and reproductive factors, and environmental (including lifestyle) factors (**Figure 3**).

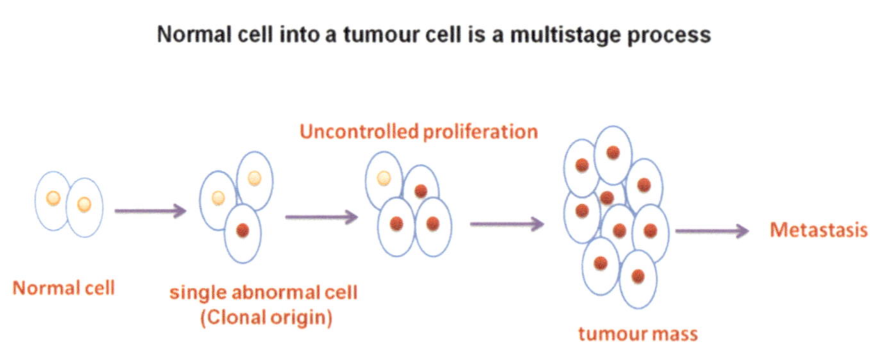

Figure 1. Carcinogenesis, a multistep process.

http://dx.doi.org/10.5772/67493

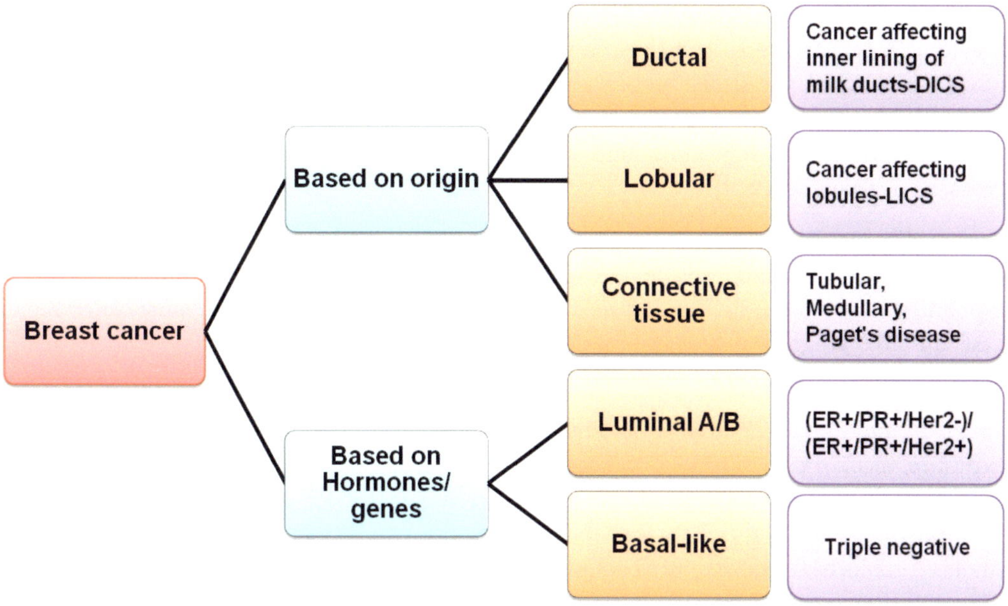

Figure 2. Classification of breast cancer.

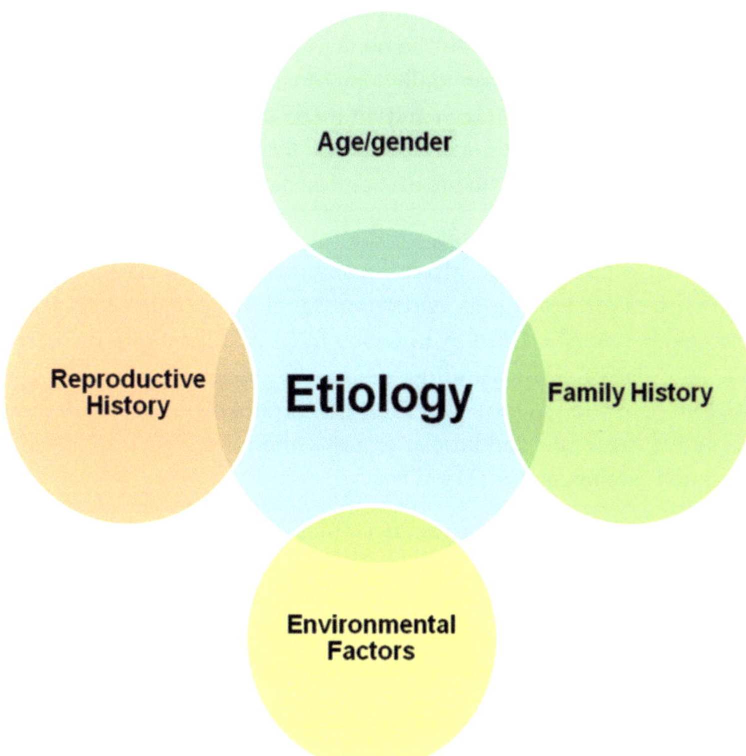

Figure 3. Breast cancer etiology.

Family history (hereditary) factors are associated primarily with early-onset of breast cancer. Previous genetic analyses of breast cancer–prone families have identified the BRCA1 and BRCA2 genes (breast cancer type 1 and 2 susceptibility protein). Women with mutations in either BRCA1 or BRCA2 (over 250 mutations have been identified) are at a significantly elevated lifetime risk (55–85% compared with 12% for the general population) for developing breast or ovarian cancer [4].

Reproductive and hormonal factors may increase the time and/level of steroid hormone exposure, consequently stimulating cell growth, and have been associated with breast cancer susceptibility. Major reproductive factors such as early menarche, late menopause, later age at first full-term pregnancy, and nulliparity are known to be associated with a higher risk of breast cancer [5].

Environmental (including lifestyle) factors such as the use of exogenous estrogens, radiation exposure, alcohol consumption, and socioeconomic status are also well-known risk factors for the disease [6]. Physical inactivity is a major risk factor for several types of cancer [7–9]. Exposure to ionizing radiation such as X-rays at a young age, alcohol consumption, and smoking habits have consistently been shown to increase breast cancer risk.

Radiations are classified into two major types namely: ionizing radiation and nonionizing radiation. The ionizing radiation transmits energy via X-rays and alpha particles disrupt chemical bonds, which results in chemically reactive free radical formation, a phenomenon known as ionization. These ionizing radiations may directly pass through the cell and may cause DNA damage. Such damage if unrepaired can result in nonlethal DNA modifications or cell death. The nonlethal DNA modifications thus eventually may cause malignant transformations. Ionizing radiation is, therefore, a known mutagen and an established breast carcinogen. Nonionizing radiation (e.g., microwaves and extremely low-frequency electric and magnetic fields (ELF-EMF)) does not have enough energy to break chemical bonds and produce ionization [10].

Several epidemiological studies have shown that alcohol consumption has been associated with breast cancer susceptibility. The alcohol metabolism occurs via multiple stages which may increase the risk of carcinogenesis. Alcohol metabolized into an acetaldehyde and other products subsequently damages DNA by inducing DNA modifications. Apart from this, acetaldehyde alone may also cause breast tumorigenesis by interfering with DNA repair mechanisms. Free radicals generated in the second stage of alcohol metabolism are thought to cause DNA damage, strand breakage, and base alterations and have been implicated for their role in alcohol-associated carcinogenesis [11].

There are over 60 carcinogens in cigarette smoke, which have been evaluated by the International Agency for Research on Cancer [12]. Cigarette smoke is rich in carcinogens such as nitrosamines and polycyclic aromatic hydrocarbons and induces oxidative damage. The gas phase of freshly generated cigarette smoke has large amounts of nitric oxide and other unstable oxidants. The presence of such free radicals and oxidants can lead to oxidative DNA damage [13]. The environmental risk factors that alter the levels of free radicals (reactive species) generated in the body are known to react with DNA to cause mutations in critical genes and consequently promote carcinogenesis.

http://dx.doi.org/10.5772/67493

2. Free radicals

Free radicals are molecules with high instability and reactivity due to the presence of an odd number of electrons in the outermost orbit of their atoms; their action derives from their attempts to attain "balance" by binding with electrons of neighboring atoms, giving rise to chain reactions (**Figure 4**) [14].

Free radical can be classified as reactive oxygen species (ROS) and reactive nitrogen species (RNS). ROS include the superoxide anion (O_2^-), hydroxyl radical (OH^-), hydrogen peroxide (H_2O_2), while RNS include nitric oxide ($NO\bullet^-$) and peroxynitrite ($ONOO^-$) the radical nitrogen dioxide (NO_2^-), and nitrite (NO_2^-) [15].

Free radicals are key players in the initiation and progression of tumor cells and enhance their metastatic potential. They are now considered a hallmark of cancer.

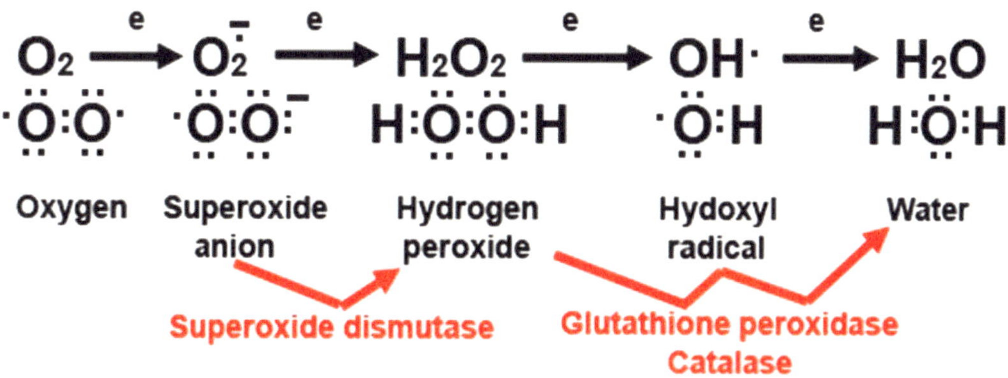

Figure 4. Formation and elimination of ROS.

3. Oxidative stress

The cell generates free radicals and also degrades, which is strictly necessary to avoid the damage derived from free radicals. However, various intrinsic and extrinsic circumstances and the biochemical activity of the cell can make it lose control over the formation and management of free radicals and results in "oxidative stress." It results from a disturbance in balance between the formation of ROS/RNS and the defense provided by cell antioxidants. This imbalance may cause damage related to various human diseases (**Figure 5**) [16].

During endogenous metabolic reactions, aerobic cells produce ROS such as superoxide anion (O_2^-), hydrogen peroxide (H_2O_2), hydroxyl radical ($OH\bullet$), and organic peroxides as normal products of the biological reduction of molecular oxygen [17]. Under hypoxic conditions, the mitochondrial respiratory chain also produces nitric oxide ($NO\bullet^-$), which can generate other reactive nitrogen species (RNS) [18].

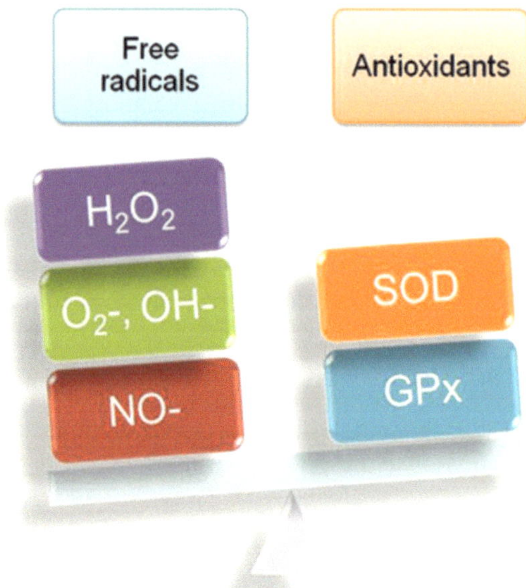

Figure 5. Oxidative stress.

4. Antioxidants

Cells have natural defense systems against ROS that consist of antioxidant enzymes, vitamins, etc., some of these antioxidants are produced inside the human body, mostly falling into the enzymatic category, as they are predominantly protein in nature. These proteins include the superoxide dismutase (SOD) enzymes (which have differential subcellular localization and superoxide dismutation to H_2O_2), glutathione peroxidase (GPx), and catalase (both of which clear peroxide), thioredoxins (Trxs; reduce oxidized proteins), and glutathione synthetase (GSS; synthesizes glutathione [GSH], an important antioxidant), among others. Vitamins are mostly obtained from nutritional sources and include ascorbic acid (vitamin A), tocopherol. Therefore, a fine balance exists between the levels of ROS and antioxidants within the cell [19]. Increased ROS/RNS can result in a greater number of mutations, oxidation of critical proteins, and other alterations, finally culminating in cell death. Identification of potentially modifiable factors that affect oxidative stress in breast cancer patients is an increasingly important task.

5. Nitric oxide

Nitric oxide (NO•) is a ubiquitous, short-lived, water soluble and endogenously produced gas that exerts a wide range of biological effects. It is a pleiotropic regulator, critical to numerous biological processes, including vasodilation, neurotransmission, and macrophage-mediated immunity.

http://dx.doi.org/10.5772/67493

The general function of NO• protects against the effects of free radicals but at excessive concentrations, NO• or its derivatives may lead to DNA damage and impair the tumor suppressor function of p53, which may cause cancer development [20].

5.1. Intracellular mechanisms of NO•⁻

When NO• is synthesized, it has a half-life of only a few seconds. Its bioavailability is reduced due to the high affinity binding of superoxide anion (high reactivity in both molecules is due to the unpaired electrons). NO• has an ability to tightly bind to the heme moiety of both hemoglobin (Hb) and guanylyl cyclase (GC) and found mostly in the vascular smooth muscle cells and other cells. Thus, NO• when produced in endothelium is quickly diffused into the blood circulation, binds to hemoglobin in blood and forms nitrates. NO• may also activate guanylyl cyclase (GC), an enzyme that catalyzes the dephosphorylation of guanosine triphosphate (GTP) to 3′,5′-cyclic guanosine monophosphate (cGMP) and serves as a second messenger for many important cellular functions, particularly for signaling smooth muscle relaxation [21].

5.2. NO•⁻ biological functions

Nitric oxide (NO•⁻) is known to play important functional roles in a variety of physiological systems [22].

1. NO•⁻ induces vasodilation—NO initiates and maintains vasodilation through a cascade of biological events that culminate in the relaxation of smooth muscle cells that line arteries, veins, lymphatics. NO also inhibits the aggregation of platelets and thus keeps inappropriate clotting from interfering with blood flow.
2. NO•⁻ regulates programmed cell death (apoptosis)—NO has been shown to both induce and inhibit apoptosis. The activity of the caspases can be directly affected by antiapoptotic effects of NO through nitrosylation of the active site, leading to inhibition of protein function. NO can also affect the expression of many members of the Bcl-2 protein family, including both proapoptotic proteins.
3. NO•⁻ alters kidney function—The release of NO around the glomeruli of the kidneys increases blood flow through them thus increasing the rate of filtration and urine formation.
4. NO•⁻ induces smooth muscle cell contractility—The relaxing effect of NO on the smooth muscle allows wavelike motions in the gastrointestinal tract.

While, NO also inhibits the contractility of the smooth muscle wall of the uterus. During child birth, as the moment of birth approaches the production of NO decreases.

5. NO•⁻ affects secretion from several endocrine glands—NO regulates the release of Gonadotropin-releasing hormone (GnRH) from the hypothalamus and adrenaline from the adrenal medulla.

In mammals, under normoxic conditions, NO• is generated endogenously "on demand" when the guanidino nitrogen of L-arginine undergoes a five-electron oxidation to yield the gaseous free radical, nitric oxide, and citrulline in equimolar concentrations (**Figure 6**) [23].

The constitutive forms of NOS are endothelial NOS (eNOS; type III) and inducible NOS (iNOS; type II). Cofactors for NOS include oxygen, NADPH, tetrahydrobiopterin, and flavin adenine nucleotides.

The inducible (calcium-independent) isoform (iNOS) produces much larger amounts of NO and is only expressed during inflammation. Whereas iNOS can produce injurious amounts of RNS (check), eNOS and nNOS produce beneficial amounts under physiological conditions. The constitutive (calcium-dependent) isoform and endothelial NOS (eNOS) produce small amounts of NO, which act as a vasodilator. The third form, neural NOS (nNOS; type I) serves as a transmitter in the brain and in different nerves of the peripheral nervous system to produce vasodilation. mtNOS constitutively expressed and membrane-bound nNOS isoform alpha, precluding a novel alternative spliced product. The mitochondrial production of nitric oxide is catalyzed nitric-oxide synthase (mtNOS). This enzyme has the same cofactor and substrate requirements as other constitutive nitric-oxide synthases (**Figure 6**).

NOS1 is the neural (or brain) isoform, also known as *nNOS*. It plays an important role in neural communication via synaptic transmission from nerve to nerve across synapses, and from peripheral nerves to the brain.

NOS2 is also known as *iNOS*. It generates extremely elevated concentrations of NO, to participate in a host defense mechanism. It also takes several hours to be activated in response to an injury or infection. Unlike nNOS, which takes part in normal neural communication, an abnormal stimulus (a wound, tissue damage, hypoxia, bacterial infection, etc.) may induce iNOS.

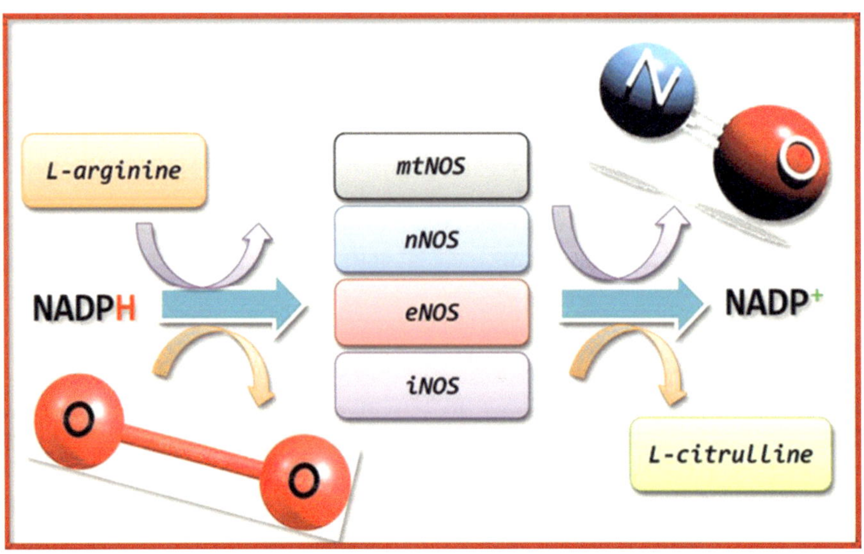

Figure 6. Mechanism of NO production from NOS.

The third isoform is *eNOS* (or NOS3) which stands for "endothelial cell" NOS. This isoform is active at all times and is found in endothelial cells which are the cells that line the inner surface of all blood vessels and lymph ducts. The eNOS is activated by the pulsatile flow of blood through vessels and maintains the diameter of the blood vessels at an optimal level. In addition, it also promotes angiogenesis, a process of new blood vessels formation.

A fourth type, mitochondrial NOS (*mtNOS*), differentiated by its subcellular localization in the mitochondrial inner membrane, regulates various functions of mitochondria in liver, heart, kidney, breast, etc. The mitochondrial utilization of NO involves superoxide anion and H_2O_2, a species freely diffusible outside the mitochondria that participate in the modulation of cell proliferation and apoptosis and in cell transformation leading to cancer [24, 25].

6. Endothelial nitric oxide synthase (eNOS)

Endothelial nitric oxide synthase enzyme, also known as nitric oxide synthase-3 (NOS-3) or constitutive NOS (cNOS), has been shown to be a critical regulator of carcinogenesis. It is a dimeric structure in their active form containing two identical monomers of 134 kD represented by *a reductase domain* containing the binding sites for nicotinamide adenine dinucleotide phosphate (NADPH), flavin mononucleotide (FMN), and flavin adenine dinucleotide (FAD). While *an oxidase domain* comprises the binding sites for the heme group, zinc, the cofactor tetrahydrobiopterin (BH4), and the substrate L-arginine [26]. The reductase domain is attached to the oxidase domain via calmodulin-binding sequence [27].

The eNOS expression is regulated by a range of transcriptional and posttranscriptional mechanisms, generates nitric oxide (NO•) in response to a number of stimuli. Constitutively expressed eNOS oxidizes L-arginine to generate L-citrulline and NO•. The essential cofactors for catalysis of this reaction include as calmodulin (CaM), flavin mononucleotide, flavin adenine dinucleotide, tetrahydrobiopterin (H4B) and NADPH.

Typically, the eNOS isoforms can be activated as a result of calmodulin (CaM) binding following a rise in intracellular calcium. They may also be activated and/or inhibited by phosphorylation via various protein kinases. Oxygen levels also regulate NOS levels in a cell type and isoform-specific manner by altering enzyme expression and by limiting the availability of oxygen, a key substrate for NO• synthesis.

Experimental and epidemiological evidence has shown the contributory role of eNOS induced NO• in tumor progression, suggesting a possible implication of endothelium expressed eNOS in tumors. The role of NO• in cancer biology is widespread, including its involvement in cellular transformation, malignant lesions, initiation, progression of the metastatic process, and induction of genotoxicity [28]. NO•-mediated DNA damage may be due to direct modification of DNA, or inhibition of DNA repair mechanisms [29]. Similarly, most of the RNS can cause DNA strand breaks and result in multiple mutations in DNA [30].

Breast carcinogenesis involves the transformation of a normal cell into a tumor cell mediated by a sequence of cellular and molecular events. These events consist of the attainment of precise characteristics, such as uncontrolled proliferation, avoidance of mitotic control, resistance to apoptosis, replicative immortality, escape from immune surveillance, progression by stimulating invasion and metastasis, angiogenesis, genomic instability, and deregulated metabolism [31].

Tumor-derived eNOS has shown to modulate cancer-related events (inflammation, apoptosis, cell cycle, angiogenesis, invasion, and metastasis) (**Figure 7**) and genetic studies showed that eNOS gene polymorphisms are associated with the development of multiple cancers [32].

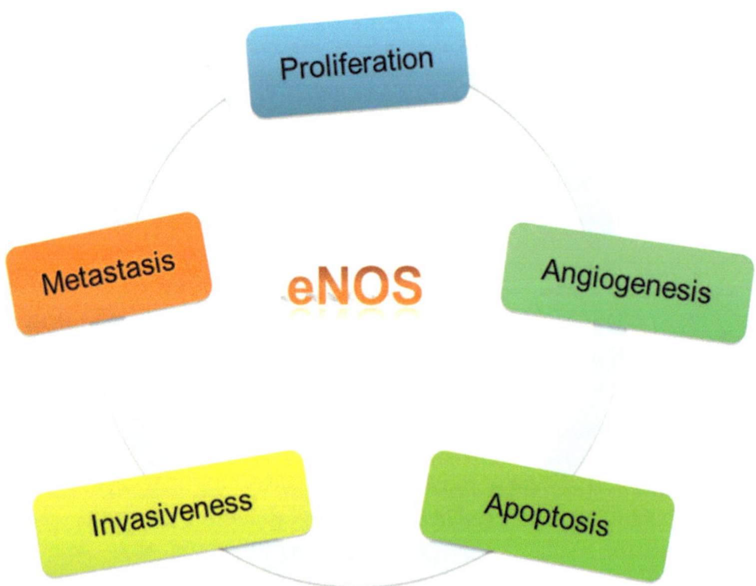

Figure 7. Multiple roles of eNOS in tumor development.

6.1. Inflammation

Inflammation a localized protective response elicited by injury is known to cause DNA damage [33]. Chronic inflammation due to infection or injury is estimated to contribute to 25% of all cancers in the world [34]. A growing body of laboratory research has shown that inflammation is a key mediator in the promotion of malignant transformation, where pro-inflammatory cytokines can facilitate tumor growth and metastasis by altering tumor cell biology and activating stromal cells in the tumor microenvironment, such as vascular endothelial cells, tumor-associated macrophages, and fibroblasts.

NO• is closely related to inflammatory status and regarded as a critical inflammation mediator. Pro-inflammatory cytokines can modulate the expression of eNOS and can accelerate the growth and development of cancer. eNOS, for example, can regulate the expression of the pro-inflammatory molecules nuclear factor-κB (NF-κB) and cyclooxygenase-2 [32, 35, 36].

6.2. Apoptosis

Apoptosis is an ordered and orchestrated cellular process that occurs in physiological and pathological conditions. Impaired apoptosis has been associated with initiation and development of cancer. The mechanism of apoptosis is regulated by an array of factors triggering and activating signaling cascade, subsequently leading to cellular death [33]. eNOS may be a molecular node in growth factor–mediated inhibition of apoptosis [37]. The antiapoptotic mechanism is understood on the basis of gene transcription of protective proteins and direct inhibition of the apoptotic executive effectors (caspase family protease).

The mechanisms of action that lead to the proneoplastic activity of NO• are via apoptosis inhibition by S-nitrosylation-inactivation of caspases- 1, 2, 4, 8 and 3, 6, 7 and disruption of the apoptotic protease activating factor 1/caspase-9 complex (Apaf-1/caspase-9 apoptosome is an essential initiator of caspase activation that initiates an apoptotic protease cascade) [38]. Other antiapoptotic effects of NO• depends on the interaction of NO•/cGMP, which inhibits the release of cytochrome C, stimulation of heat-shock protein (Hsp) 70 and Hsp 32, elevated Bcl-2 expression, repression of ceramide generation [39], and induces cyclooxygenase-2 activity [40]. In the animal model study, the lipopolysaccharide (LPS)-induced hepatic apoptosis was increased by the administration of NOS inhibitors [41]. Thus at high concentration, NO• has a potential role in cancer, i.e., inhibition of eNOS activity specific to tumor cells may be a viable option for the stimulation of apoptosis and treatment of cancer alone or in combination with chemotherapeutic agents.

6.3. Angiogenesis

Angiogenesis in mammary tumors can be stimulated by inflammation, which induces proliferation and morphogenesis of vascular endothelial cells in response to a large number of cytokines or angiogenic molecules produced by tumor and host cells [42]. Angiogenesis is essential for tumor growth and metastasis and has been considered the most important prognostic indicator for predicting overall survival [43]. eNOS strongly affects tumor growth by promoting angiogenesis [44]. Tumor growth–enhancing effects of vascular endothelial growth factor (VEGF) are associated with increased NOS activity and inhibition of apoptosis in human breast carcinoma xenografts [45].

Endogenous NO• promotes tumor blood flow via dilatation of arteriolar vessels. It decreases leukocyte-endothelial adhesive interactions and increases vascular permeability [46]. Several cancer treatment methods influence eNOS expression and activity. Low-dose irradiation-induced angiogenesis is believed to be mediated by NO• from eNOS [47]. Studies have shown that VEGF released as a purified protein or produced by tumor cells requires a functional NO•/cGMP pathway within the end compartment to promote neovascular growth. NO• also has an invasion-stimulating effect, which is mediated by upregulation of MMP-2 and MMP-9 (matrix metalloproteinases) and downregulation of TIMP-2 and possibly TIMP-3 (tissue inhibitors of MMP) [48].

6.4. Tumor progression/invasion

NO• has been investigated regarding its possible involvement in the promotion of breast carcinoma. Increased amounts of NO• have been observed in the blood circulation of advanced grade breast cancer patients [49] where the increased levels of NO• have shown to promote tumor angiogenesis. Nitrotyrosine, a marker derived from NO•, was correlated with expression of VEGF-C and has been associated with lymph node metastasis in breast carcinoma patients, implicating the role of NO• in the development and progression of breast cancer [50]. In relation to the link of eNOS with cell proliferation, the eNOS expression has been detected in tumor cells specifically in breast cancer [51].

Several studies have observed NO• released by eNOS, can stimulate cancer cell cycle progression and proliferation. More specific to eNOS, studies have shown that the eNOS/NO• pathway plays a role in cancer cell DNA/RNA synthesis and proliferation apart from promoting angiogenesis [52]. The eNOS gene plays an essential role in endothelial cell proliferation in cell culture models and is a central mediator of several endothelium growth stimulators, such as vascular endothelial growth factor (VEGF) and prostaglandin E2. In human breast cancer, eNOS appears to be expressed in tumor epithelial cells, and its presence is correlated with histological grade and lymph node status. Higher NOS activity has been found in invasive breast tumors when compared with benign or normal breast tissue carcinoma [53].

6.5. Metastasis

Metastasis is a complex process by which the malignant cancer cells from the breast expand into other regions of the body. Lymphatic metastasis is a critical determinant of cancer prognosis. Recent findings indicate that eNOS mediates VEGF-C-induced lymph-angiogenesis and, consequently, plays a critical role in lymphatic metastasis [54]. Investigational studies on tumors have provided substantial evidence of the contributory role of NO• in tumor development, where series of tests were performed on tumor-bearing mice using NOS inhibitors showed a delayed tumor growth through eNOS inhibition and barred metastasis, signifying the potential role of endothelium-derived eNOS in metastasis [55].

7. Regulation OF *NOS3* gene expression

The most fundamental level of NOS regulation is reflected in the tissue-specific expression of the different isoforms. The amount of NO produced results from the expression level and activity of eNOS. It is regulated by several interlinking mechanism such as transcriptional, post-transcriptional, and posttranslational modifications. The activity of NOS3 is also controlled by avid binding transcription factors namely Ets-1, Elf-1, Sp1, Sp3, and YY1 to the NOS3 promoter region. Posttranscriptionally, NOS3 activity is controlled by modifications of the primary transcript, stability of mRNA, its subcellular localization, and nucleocytoplasmic transport. Posttranslational modifications of NOS3 consist of fatty acid acylation, substrate, and cofactor availability, protein-protein interactions, and amount of phosphorylation. Another significant epigenetic mechanism for NOS3 gene expression is differential promoter methylation [47].

The gene-encoding eNOS located on chromosome 7q35-36 and is composed of 26 exons (coding sequences) and introns (sequences between exons) with an entire length of 21 kb and has more than 168 polymorphisms [56]. Over the last few years, polymorphisms of the gene have been identified, and their association with various diseases has been explored. Genetic comparison studies on healthy people and cancer patients have shown that polymorphisms in eNOS are associated with the development of cancers.

A single nucleotide polymorphism (SNP), T-786C, was identified in the 5′ flanking region involving a substitution of thymine (T) to cytosine (C) at a locus 786 base pairs upstream [57]. Another common variant of eNOS with a G to T transversion at nucleotide position 894 (G894T) leading to a change in amino acid at 298 (Glu298Asp) has been reported [58], and a 27-bp variable number of tandem repeats (VNTR) polymorphism in intron 4 (intron 4b/4a) [59] and high numbers of CA, which have been repeated in intron 13 of eNOS gene, are also known to be associated with complex disorders. These polymorphisms seem to be functional and have been widely investigated for their associations with cancer risk [60, 61]. Molecular studies of eNOS −786T > C, intron 4b/4a, and 894G > T polymorphisms (**Figure 8**) if performed in large and unbiased can provide valuable insights into the association between the eNOS gene and breast cancer risk.

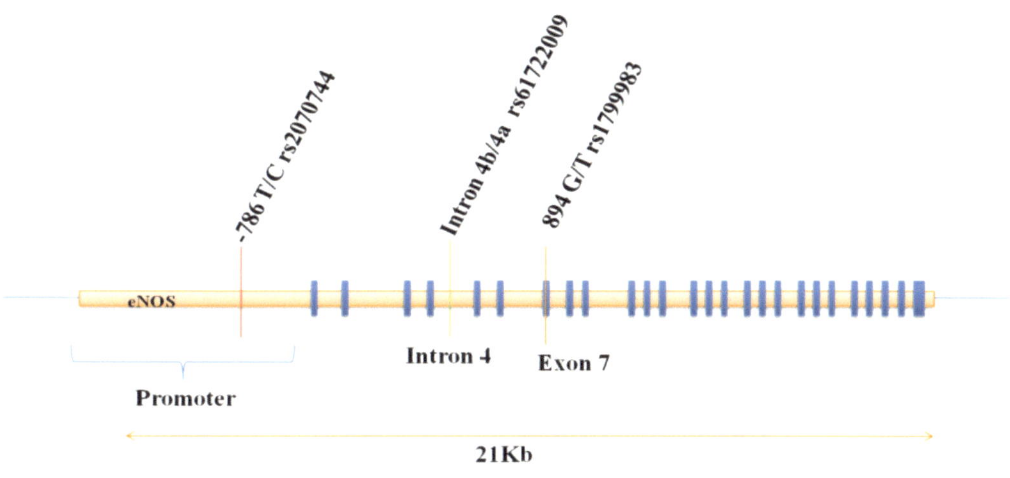

Figure 8. Organization of eNOS gene.

8. Conclusion

All these findings suggest that the expression of eNOS in breast cancer may be a critical event in carcinogenesis. Understanding different actions of NO• induced by eNOS in breast cancer at the molecular level can help in providing diagnostic or prognostic markers and also in devising potential strategies for prevention of breast cancer. The ability of many tumors to exploit eNOS/NO for a survival, proliferative, and metastatic advantage suggests that pharmacological use of eNOS inhibitors might attenuate these effects. Therefore, selective targeting of eNOS may prove a useful therapeutic or chemopreventive measure. However, further careful studies are needed to confirm the potential therapeutic role of eNOS as a novel target for breast cancer therapy.

Author details

Tupurani Mohini Aiyengar, Padala Chiranjeevi and Hanumanth Surekha Rani*

*Address all correspondence to: surekharanih@gmail.com

Department of Genetics, University College of Science, Osmania University, Hyderabad, Telangana State, India

References

[1] Torre LA, Bray F, Siegel RL, Ferlay J, Lortet-Tieulent J, Jemal A. Global cancer statistics, 2012. CA: A Cancer Journal for Clinicians. 2015;**65**:87–108. doi: 10.3322/caac.21262.

[2] Abdelgawad IA, El-Mously RH, Saber MM, Mansour OA, Shouman SA. Significance of serum levels of vitamin D and some related minerals in breast cancer patients. International Journal of Clinical and Experimental Pathology. 2015;**8**:4074–4082.

[3] Gangane N, Anshu, Manvatkar S, Ng N, Hurtig AK, San Sebastián M. Prevalence and risk factors for patient delay among women with breast cancer in rural India. Asia Pacific Journal of Public Health. 2016;**28**:72–82. doi: 10.1177/1010539515620630.

[4] Bennet LB, Taurog JD, Bowcock AM. Hereditary breast cancer genes. In: Bowcock AM, editor. Breast Cancer: Molecular Genetics, Pathogenesis, and Therapeutics. Totowa, NJ: Humana Press, 1999;199–224.

[5] Colditz GA, Baer HJ, Tamimi RM. Breast cancer. In: Schottenfeld D, Fraumeni JF, editors. Cancer Epidemiology and Prevention. 3rd ed. New York: Oxford University Press, 2006.

[6] Anand P, Kunnumakara AB, Sundaram C, et al. Cancer is a preventable disease that requires major lifestyle changes. Pharmaceutical Research. 2008;**25**:2097–2116. doi: 10.1007/s11095-008-9661-9.

[7] Vainio H, Kaaks R, Bianchini F. Weight control and physical activity in cancer prevention: international evaluation of the evidence. European Journal of Cancer Prevention. 2002;**11**:S94–S100.

[8] Friedenreich CM. Physical activity and cancer prevention: from observational to intervention research. Cancer Epidemiology Biomarkers & Prevention. 2001;**10**:287–301.

[9] Gammon MD, John EM, Britton JA. Recreational and occupational physical activities and risk of breast cancer. Journal of the National Cancer Institute. 1998;**90**:100–117.

[10] Wakeford R. The cancer epidemiology of radiation. Oncogene. 2004;**23**:6404–6428.

[11] Coronado GD, Beasley J, Livaudais J. Alcohol consumption and the risk of breast cancer. Salud pública de México. 2011;**53**:440–447.

[12] Hoffmann D, Hoffmann I, El-Bayoumy K. The less harmful cigarette: a controversial issue. A tribute to Ernst L. Wynder. Chemical Research in Toxicology. 2001;**14**:767–790.

[13] Pfeifer GP, Denissenko MF, Olivier M, Tretyakova N, Hecht SS, Hainaut P. Tobacco smoke carcinogens, DNA damage and p53 mutations in smoking-associated cancers. Oncogene. 2002;**21**:7435–7451.

[14] Griendling KK, Sorescu D, Lasse`gue B, Ushio-Fukai M. Modulation of protein kinase activity and gene expression by reactive oxygen species and their role in vascular physiology and pathophysiology. Arteriosclerosis, Thrombosis, and Vascular Biology. 2000;**20**:2175–2183. doi: 10.1161/01.ATV.20.10.2175

[15] Choudhari SK, Chaudhary M, Bagde S, Gadbail AR, Joshi V. Nitric oxide and cancer: a review. World Journal of Surgical Oncology. 2013;**11**:118. doi: 10.1186/1477-7819-11-118

[16] Ríos-Arrabal S, Artacho-Cordón F, León J, et al. Involvement of free radicals in breast cancer. SpringerPlus. 2013;**2**:404. doi: 10.1186/2193-1801-2-404

[17] Fridovich I. The biology of oxygen radicals. Science. 1978;**201**:875–880.

[18] Poyton RO, Ball KA, Castello PR. Mitochondrial generation of free radicals and hypoxic signaling. Trends Endocrinology Metabolism. 2009;**20**:332–340.

[19] Droge W. Free radicals in the physiological control of cell function. Physiological Reviews. 2002;**82**:47–95.

[20] Valko M, Leibfritz D, Moncol J, Cronin MT, Mazur M, Telser J. Free radicals and antioxidants in normal physiological functions and human disease. International Journal of Biochemistry & Cell Biology. 2007;**39**:44–84. doi: 10.1016/j.biocel.2006.07.001

[21] Moncada S, Higgs A. The L-arginine-nitric oxide pathway. The New England Journal of Medicine. 1993;**329**:2002–2012.

[22] Cohen RA, Weisbrod RM, Gericke M, Yaghoubi M, Bierl C, Bolotina VM. Mechanism of nitric oxide-induced vasodilatation. Circulation Research. 1999;**84**:210–219.

[23] Antosova M, Plevkova J, Strapkova A. Buday T. Nitric oxide—Important messenger in human body. Open Journal of Molecular and Integrative Physiology. 2012;**2**:9(Art No. 21845):98–106. doi: 10.4236/ojmip.2012.23014.

[24] Haynes V, Elfering SL, Squires RJ, Traaseth N, Solien J, Ettl A, Giulivi C. Mitochondrial nitric-oxide synthase: role in pathophysiology. IUBMB Life. 2003;**55**:599–603.

[25] Sen S, Kawahara B, Chaudhuri G. Mitochondrial-associated nitric oxide synthase activity inhibits cytochrome c oxidase: implications for breast cancer. Free Radical Biology and Medicine. 2013;**57**:210–220. doi: 10.1016/j.freeradbiomed.2012.10.545.

[26] Qian J, Fulton D. Post-translational regulation of endothelial nitric oxide synthase in vascular endothelium. Frontiers in Physiology. 2013;**4**:347. doi: 10.3389/fphys.2013. 00347

[27] Alderton WK, Cooper CE, Knowles RG. Nitric oxide synthases: structure, function and inhibition. The Biochemical Journal. 2001;**357**(Pt 3):593–615.

[28] Gal A, Wogan GN. Mutagenesis associated with nitric oxide production in transgenic SJL mice. Proceedings of the National Academy of Sciences of the U S A. 1996;**93**:15102–15107. doi: 10.1073/pnas.93.26.15102

[29] Wink DA, Vodovotz Y, Laval J, Laval F, Dewhirst MW, Mitchell JB. The multifaceted roles of nitric oxide in cancer. Carcinogenesis. 1998;**19**:711–721. doi: 10.1093/carcin/19.5.711

[30] Sun Y. Free radicals, antioxidant enzymes, and carcinogenesis. Free Radical Biology and Medicine. 1990;**8**:583–599. doi: 10.1016/0891-5849(90)90156-D

[31] Hanahan D, Weinberg RA. Hallmarks of cancer: the next generation. Cell. 2011;**144**: 646–674.

[32] Grumbach IM, Chen W, Mertens SA, Harrison DG. A negative feedback mechanism involving nitric oxide and nuclear factor κ-B modulates endothelial nitric oxide synthase transcription. Journal of Molecular and Cellular Cardiology. 2005;**39**:595–603. doi: http://dx.doi.org/10.1016/j.yjmcc.2005.06.012

[33] Cooke MS, Evans MD, Dizdaroglu M, Lunec J. Oxidative DNA damage: mechanisms, mutation, and disease. FASEB Journal. 2003;**17**:1195–1214.

[34] Coussens LM, Werb Z. Inflammation and cancer. Nature. 2002;**420**:860–867.

[35] Blais V, Rivest S. Inhibitory action of nitric oxide on circulating tumor necrosis factor-induced NF-κB activity and COX-2 transcription in the endothelium of the brain capillaries. Journal of Neuropathology & Experimental Neurology. 2001;**60**:893–905. doi: http://dx.doi.org/10.1093/jnen/60.9.893

[36] Conelly L, Jacobs AT, Palacios-Callender M, Moncada S, Hobbs AJ. Macrophage endothelial nitric-oxide synthase autoregulates cellular activation and pro-inflammatory protein expression. Journal of Biological Chemistry. 2003;**278**:26480–26487. doi:10.1074/jbc.M302238200

[37] Murphy PR, Limoges M, Dodd F Boudreau RT, Too CK. Fibroblast growth factor-2 stimulates endothelial nitric oxide synthase expression and inhibits apoptosis by a nitric oxide-dependent pathway in Nb2 lymphoma cells. Endocrinology. 2001;**142**:81–88.

[38] Korde Choudhari S, Chaudhary M, Bagde S, Gadbail AR, Joshi V. Nitric oxide and cancer: a review. World Journal of Surgical Oncology. 2013;**11**:118. doi: 10.1186/1477-7819-11-118

[39] Khalkhali-Ellis Z, Hendrix MJ. Nitric oxide regulation of maspin expression in normal mammary epithelial and breast cancer cells. American Journal of Pathology. 2003;**162**:1411–1417.

[40] Choi BM, Pae HO, Jang SI, Kim YM, Chung HT. Nitric oxide as a pro-apoptotic as well as anti-apoptotic modulator. Journal of Biochemistry and Molecular Biology. 2002;**35:**116–126. doi: 10.5483/BMBRep.2002.35.1.116

[41] Von KA, Brune B. Cyclooxygenase-2: an essential regulator of NO mediated apoptosis. FASEBJ 1997;**11:**887–895.

[42] Pervin S, Chaudhuri G, Singh R. NO to breast: when, why and why not? Current Pharmaceutical Design. 2010;**16:**451–462.

[43] Wang L, Shi GG, Yao JC, Gong W, Wei D, Wu TT, Ajani JA, Huang S, Xie K. Expression of endothelial nitric oxide synthase correlates with the angiogenic phenotype of and predicts poor prognosis in human gastric cancer. Gastric Cancer. 2005;**8**:18–28.

[44] Babaei S, Teichert-Kuliszewska K, Zhang Q, Jones N, Dumont DJ, Stewart DJ. Angiogenic actions of angiopoietin-1 require endothelium-derived nitric oxide. American Journal of Pathology. 2003;**162**:1927–1936.

[45] Harris SR, Schoeffner DJ, Yoshiji H, Thorgeirsson UP. Tumor growth enhancing effects of vascular endothelial growth factor are associated with increased nitric oxide synthase activity and inhibition of apoptosis in human breast carcinoma xenografts. Cancer Letter. 2002;**179**:95–101.

[46] Ziche M1, Morbidelli L. Nitric oxide and angiogenesis. Journal of Neurooncology. 2000;**50**:139–148. doi: 10.1023/A:1006431309841

[47] Sonveaux P, Brouet A, Havaux X, et al. Irradiation induced angiogenesis through the up-regulation of the nitric oxide pathway: implications for tumor radiotherapy. Cancer Research. 2003;**63**:1012–1019.

[48] Lala PK, Orucevic A. Role of nitric oxide in tumor progression: lessons from experimental tumors. Cancer Metastasis Review. 1998;**17**:91–106.

[49] Thomsen LL, Miles DW, Happerfield L, Bobrow LG, Knowles RG, Moncada S. Nitric oxide synthase activity in human breast cancer. British Journal of Cancer 1995;**72**:41–44.

[50] Nakamura Y, Yasuoka H, Tsujimoto M, Yoshidome K, Nakashara M, Nakao K, Nakamura M, Kakudo K: NO in breast cancer: induction of vascular endothelial growth factor-C and correlation with metastasis and poor prognosis. Clinical Cancer Research. 2006;**12**:1201–1207. doi: 10.1158/1078-0432.CCR-05-1269

[51] Martin JH, Begum S, Alalami O, Harrison A, Scott KW. Endothelial nitric oxide synthase: correlation with histologic grade, lymph node status and estrogen receptor expression in human breast cancer. Tumour Biology. 2000;**21**:90–97.

[52] Shang ZJ, Li ZB, Li JR. In vitro effects of nitric oxide synthase inhibitor L-NAME on oral squamous cell carcinoma: a preliminary study. International Journal of Oral & Maxillofacial Surgery. 2006;**35**:539–543.

[53] Loibl S, von Minckwitz G, Weber S, Sinn HP, Schini-Kerth VB, Lobysheva I, Nepveu F, Wolf G, Strebhardt K, Kaufmann M. Expression of endothelial and inducible nitric oxide synthase in benign and malignant lesions of the breast and measurement of nitric oxide using electron paramagnetic resonance spectroscopy. Cancer. 2002;**95**:1191–1198. doi: 10.1002/cncr.10817

[54] Lahdenranta J, Hagendoorn J, Padera TP, Hoshida T, Nelson G, Kashiwagi S, Jain RK, Fukumura D. Endothelial nitric oxide synthase mediates lymphangiogenesis and lymphatic metastasis. Cancer Research. 2009;**69**:2801–2808.

[55] Jadeski LC, Lala PK. Nitric oxide synthase inhibition by N(G)-nitro-L-arginine methyl ester inhibits tumor-induced angiogenesis in mammary tumors. American Journal of Pathology. 1999;**155**:1381–1390.

[56] Safarinejad MR, Safarinejad S, Shafiei N, et al. Effects of the T-786C, G894T, and Intron 4 VNTR (4a/b) polymorphisms of the endothelial nitric oxide synthase gene on the risk of prostate cancer. Urologic Oncology. 2013;**31**:1132–1140. doi: 10.1016/j.urolonc.2012.01.002

[57] Nakayama M, Yasue H, Yoshimura M, Shimasaki Y, Kugiyama K, Ogawa H, Motoyama T, Saito Y, Ogawa Y, Miyamoto Y, Nakao K. T-786→C mutation in the 5′-flanking region of the endothelial nitric oxide synthase gene is associated with coronary spasm. Circulation. 1999;**99**:2864–2870.

[58] Yoshimura M, Yasue H, Nakayama M, Shimasaki Y, Sumida H, Sugiyama S, Kugiyama K, Ogawa H, Ogawa Y, Saito Y, Miyamoto Y, Nakao K. A missense Glu298Asp variant in the endothelial nitric oxide synthase gene is associated with coronary spasm in the Japanese. Human Genetics. 1998;**103**:65–69.

[59] Wang XL, Sim AS, Badenhop RF, McCredie RM, Wilcken DE. A smoking-dependent risk of coronary artery disease associated with a polymorphism of the endothelial nitric oxide synthase gene. Nature Medicine.1996;**2**:41–45.

[60] Jang MJ, Jeon YJ, Kim JW, et al. Association of eNOS polymorphisms (−786T > C, 4a4b, 894G > T) with colorectal cancer susceptibility in the Korean population. Gene. 2013;**512**:275–281. doi: 10.1016/j.gene.2012.10.032

[61] Oztürk E, Dikensoy E, Balat O, et al. Association of endothelial nitric oxide synthase gene polymorphisms with endometrial carcinoma: a preliminary study. Journal of the Turkish-German Gynecological Association. 2011;**12**:229–233. doi: 10.5152/jtgga.2011.47

Chapter 11

Nitric Oxide Synthase and Nitric Oxide Involvement in Different Toxicities

Emine Atakisi and Oguz Merhan

Additional information is available at the end of the chapter

http://dx.doi.org/10.5772/intechopen.68494

Abstract

Nitric oxide (NO) is known to have a very short half-life, and it is oxidized to nitrate (NO_3^-) and nitrite (NO_2^-). The activity and/or expression of nitric oxide synthases (NOSs) can change in response to toxins or therapeutic medications. For example, in recent studies in our laboratory and others, it has been reported that the amount of NO was increased in the serum of N-nitroso compounds-treated animals. N-nitroso compounds, which are found in different types of foodstuffs, including meat, salted fish, alcoholic beverages, agricultural drugs, insecticides, cigarettes, and several vegetables, are known to have carcinogenic effects. In addition, it is experimentally used to induce liver carcinoma to study the mechanisms of liver cytotoxic injury. Uncontrolled, prolonged, and/or massive production of NO by inducible NOS may cause liver damage, inflammation, and even tumor development during N-nitroso compound toxicity. In this chapter, we explain the roles of NOS and NO in various toxicity conditions, such as toxicity in environment pollutant or food additive, and present the evaluation of the toxicity and the importance of NOSs in human health.

Keywords: nitric oxide synthase, nitrate, nitrite, nitric oxide, N-nitroso compounds, toxicity

1. Introduction

Nitric oxide (NO) is synthesized by nitric oxide synthase (NOS) (EC: 1.14.13.39) through oxidation of L-arginine to L-citrulline [1–5]. NO is a biologically significant molecule for many species from bacteria to mammals. Mechanisms for NO synthesis in an organism are extremely limited. Nitric oxide synthase enzyme is the only source of endogenous NO, except NO formed by metabolism of the nitro compounds entering the organism [6]. Three different types of NOS isoforms have been isolated from different tissues, such as vascular endothelium, brain, macrophage, and urinary system, of mammals, including (a) neuronal NOS (nNOS), (b) inducible NOS (iNOS), and (c) endothelial NOS (eNOS) [5, 7]. Neuronal NOS (nNOS, NOS1) is a Ca^{+2}-dependent, ~160 kDa enzyme that is found in the central and peripheral nervous system cells and striated muscle [4, 8]. Inducible NOS (iNOS, NOS2) is a calcium-insensitive, ~130 kDa enzyme that was first isolated from activated macrophages and that can be activated by some cytokines (IL-1, TNF, IF-γ) or bacterial endotoxins [4, 9, 10]. Endothelial NOS (eNOS, NOS3) is a Ca^{+2}/calmodulin-dependent, ~135 kDa enzyme that is localized in vascular endothelial cells, hippocampal neural cells, pulmonary and renal epithelial cells, and cardiac myocytes [4, 9, 11].

Level of NO can be determined indirectly by measuring the concentration of nitrate (NO_3^-) and nitrite (NO_2^-) using an acidic Griess reaction. In recent studies in our laboratory and others, it has been reported that the amount of NO can change in response to various toxicity conditions [12–18] which are closely associated with animal and human disease conditions. In this chapter, we mention NO_3^- and NO_2^-, the molecules which are naturally found in foods as NO sources, agricultural activities such as the use of artificial fertilizers, polluted water, curing process to give a natural smell and taste to meat. The conversion of NO_3^- and NO_2^- into NO in the gastrointestinal tract, the effect of NOS, and possible mechanisms for how it is converted back into NO_3^- and NO_2^- in the bloodstream will also be covered. Studies on N-nitrosamines, which are formed by reaction of NO_3^- and NO_2^- with amines, and which can be seen in cured meats, cigarette smoke, and rubber industry, reveal the carcinogenic effects of these molecules (**Figure 1**).

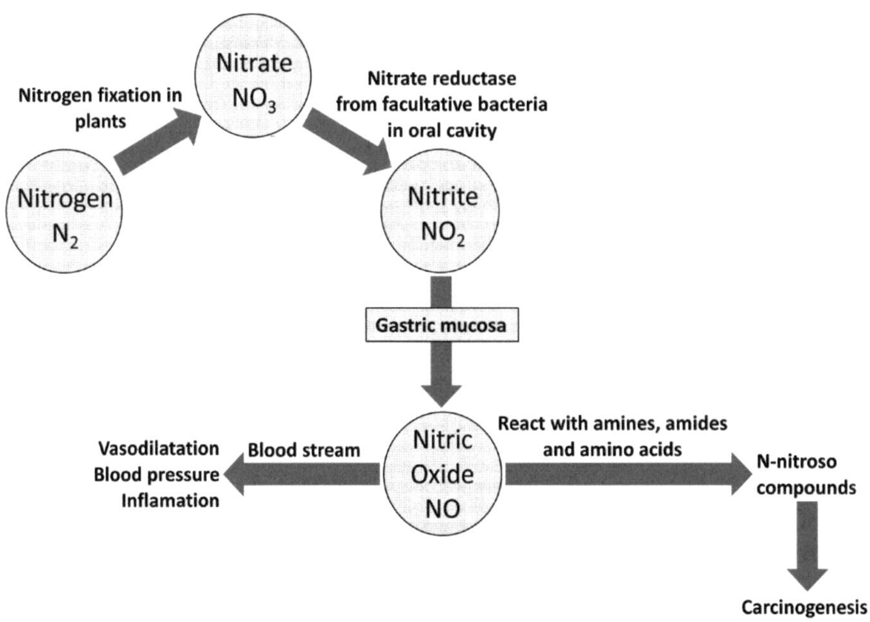

Figure 1. Formation of N-nitroso compounds from NO_3^-, NO_2^-, NO, and their effects on human health.

2. NO_3^- and NO_2^- as environment pollutants and food additives

Nitrogen is the basic element for essential micromolecules, such as amino acids, proteins, and nucleic acids. Nitrogen is the most abundant gas in the atmosphere; however, it has to be fixated before it is taken by plants and animals. Fixation is an important part of the nitrogen cycle. In this cycle, N_2 is converted into ammonium and various nitrogen oxides. These higher nitrogen oxides are eventually gradually reduced. The nitrogen is then freed into the atmosphere and the cycle is completed. Bacteria play an important role in the cycle as they can catalyze each step, including the interconversion of different nitrogen oxides. NO_3^-, NO_2^-, and NO are all necessary intermediate products in the denitrification process and are catalyzed by NO_3^-, NO_2^-, and NO reductases, respectively [19]. Bacteria use these molecules as terminal electron acceptors in the absence of oxygen. The production and metabolism of nitrogen oxides also occur in mammals. NO_3^- is easily converted into NO_2^- in mammals by the activity of enzymes in both bacteria and mammals. The NO_2^- then later can react with different molecules, such as amines, amides, and to form N-nitroso compounds which can be carcinogenic [20]. Potentially carcinogenic or inert oxidizing molecules, such as NO_3^- and NO_2^-, occur as a result of endogenous NO metabolism during the food chain (**Figure 1**) [21].

Although nitrogen is found naturally in surface waters, its amount increases in many parts of the world. The reason for this is the pollution caused by commonly used inorganic fertilizers, soil drainage, or contamination of water resources by sewage [22]. The main causes of water pollution are pollution from industrial and agricultural activities. Chemical fertilizers used in agricultural production have an important role. NO_3^- is applied in increasing amounts in the fertilizers for agricultural production, and they accumulate in the soil. This accumulated NO_3^-, in varying amounts depending on the conditions, moves toward the deeper parts of the ground especially with rainwater, and some of it reaches underground and some to surface waters. High NO_3^- concentrations in water resources pose a potential risk to human health, because sunlight and some bacteria can easily convert NO_3^- into NO_2^- [23]. NO_3^- and NO_2^- can also occur spontaneously in vegetables and fruits consumed by humans and especially in animal feed [24]. Vegetables and fruits usually receive NO_3^- and NO_2^- from the soil [25]. As a result of nitrogenous fertilizers being used in excess to increase appearance and yield in the plants, plants store NO_3^- in excess of their need. When the amount of received NO_3^- is high, the reduction to ammonia is limited and NO_2^- accumulates as an intermediate metabolism product [26, 27].

Excess NO_3^- and NO_2^- can also be utilized to cure meats. In order to improve the taste, protection, appearance, and quality of the meat, NO_3^- and NO_2^- are used for curing purposes. NO_2^- can also be used as a preservative against the proliferation of microorganisms, especially *Clostridium botulinum*. It also inhibits lipid peroxidation and prevents putrefaction [28].

Dietary NO_3^- and NO_2^-, which are taken by the organisms, could cause various physiological and pathological outcomes.

3. Conversion of NO_3^- and NO_2^- into NO in the gastrointestinal system: possible role of NOSs

In an organism, NO_2^- is converted into NO in three ways:

(a) It is enzymatically (via NOSs) reduced to NO in the circulation and tissues.

(b) It is non-enzymatically reduced to NO in acidic stomach environment, capillary beds, and at low pH and hypoxic conditions that occur during intense exercise.

(c) NO can be produced by acidification of NO_2^- in the oral cavity.

In this section, the conversion of NO_3^- and NO_2^-, taken from foods in the gastrointestinal tract, into NO will be mentioned. At the end of this pathway, NO_3^- and NO_2^- are synthesized again from NO. This synthesis in the tissues is catalyzed by NOSs. With the identification of the NO_3^--NO_2^--NO pathway, the importance of the diet, rather than the biological significance of systemic NO_3^- and NO_2^-, in the physiological regulation of NO has arisen. NO_3^-, rich in green leafy vegetables such as beetroot, is reduced to NO_2^- by bacterial NO_3^- reductase in the commensal anaerobic microflora in the oral cavity by saliva secretion and is reduced to NO in the stomach [29–31]. The highest NO concentration is obtained from an acidic stomach pH after a NO_3^--rich meal (**Figure 1**) [29].

The rapid postprandial increase in gastric NO is directly proportional to many actions, such as mucus production in the gastrointestinal tract, increased vascular tone, antimicrobial effect, and immunomodulation. It has also been shown that this increased NO is related to many physiological mechanisms, such as the prevention of ischemia-reperfusion injury and increased cerebral blood flow [21, 29].

How endogenous NO_3^- and NO_2^- can be synthesized in the body if NO_3^- and NO_2^- are not taken into the body with nutrients? The inorganic NO_3^- and NO_2^-, which cannot be taken up with nutrients in starving mammals, are mainly derived from NOSs. These enzymes form NO by using L-arginine and oxygen, and then this NO is rapidly degraded to NO_3^- and NO_2^-. Endogenous NO_3^- and NO_2^- are synthesized in this way [30]. Endogenous NO can also be produced by NOSs using NO_3^- and NO_2^- in our daily nutrients [30, 32].

NO is a highly diffusible free radical that participates in various in vivo signal pathways and is involved in critical physiological events, such as regulation of vascular tone and immune response [31]. NO also exhibits antimicrobial activity [33] other than its regulatory role in vascular tone. Numerous intracellular pathogenic parasites [34] and bacteria [35] are susceptible to NO.

NO_2^- in saliva is converted non-enzymatically into NO and some other nitrogen oxide species when it enters into the stomach with low pH [30]. Increasing number of studies on cardiovascular, inflammatory, and gastrointestinal diseases reported that the NO-related effects of dietary NO_3^- and NO_2^- are protective and preventive. Recent data suggest that these anions are beneficial to gastrointestinal cancer [36] and cardiovascular diseases rather than having harmful effects. Dietary NO_3^- and salivary NO_2^- have been shown to protect gastric mucus from experimentally induced gastric damage by increasing gastric mucus thickness and mucosal blood flow [30].

NO plays an important role for the intestine. It is produced from arginine by eNOS and iNOS in a reaction catalyzed in the intestine. eNOS is structurally expressed in low levels in intestinal microcapillaries and is responsible for the initial levels of NO. Low NO levels regulate vascular tone and mucosal blood flow in cyclic guanosine monophosphate and neuron-dependent manner and are also crucial for mucosal homeostasis. Additionally, NO can protect from oxidative stress by diminishing oxygen radicals. eNOS-derived NO facilitates leucocyte uptake by supporting endothelial adhesion of leukocytes. iNOS is upregulated during inflammation and increases NO synthesis. NO also allows dilatation of capillary vessels. Excess NO secreted during inflammation has harmful effects on the intestinal barrier [37, 38].

NO reacts with the superoxide ion to form a reactive oxygen and nitrogen type, peroxynitrite, which can be harmful for epithelial cells. It can induce enterocyte cell apoptosis and inhibit proliferation. iNOS is expressed in intestinal smooth muscle cells, endothelial and epithelial cells [38]. During inflammatory conditions in the intestine, such as necrotizing enterocolitis [39], ulcer [29], and colon cancer [40], expression of iNOS mRNA increases.

Numerous studies have shown that NO_3^- and NO_2^- obtained from pharmacologic supplements or diet have obvious effects on gastrointestinal function [29, 39, 40]. However, it is still unclear whether endogenous NO_3^- and NO_2^- derived from NOSs in the endothelium and elsewhere affect gastric function. This situation has been tried to be illuminated in germ-free and starved animals [30, 41].

The gastric NO levels are very low in germ-free animals lacking microflora even after dietary NO_3^- load. A significant amount of NO_2^- is produced in the saliva even in the case of fasting, indicating NO_3^- production due to endogenous NOS production [30]. Petersson et al. [30] reported that three doses of NO_2^- given to germ-free rats not only increased stomach mucus thickness by more than fourfold but also unexpectedly had an effect in the non-NO_2^- group. This suggests that endogenous NO_3^- from NOSs also plays a role in the regulation of gastric physiology. In a similar study in humans, individuals were given a low NO_3^- diet with an antibacterial mouthwash containing chlorhexidine to lower the reduction of oral NO_3^-, and it was determined that the levels of circulating NO_2^- lowered, and this then increased the blood pressure [42]. These studies show that NO_3^- and NO_2^- have a NO_3^--NO_2^--NO pathway that starts from the mouth and ends in the mouth through digestive and circulatory system. It is possible that NO_3^- and NO_2^- in the circulation and saliva may originate from the endogenous NOS pathway.

In addition to NO_3^-, NO_2^-, NO, sodium nitrite ($NaNO_2$) is an inorganic compound taken by endogenous sources. $NaNO_2$ may have some beneficial and undesirable effects. $NaNO_2$ is a preservative used in processed meats, such as salami and bacon. $NaNO_2$ is synthesized by several chemical reactions, including the reduction of sodium nitrate. $NaNO_2$ is used as an additive in foods. There are some suspects about the safety of use in foods, but it is still being used, and, on the contrary, there is information that $NaNO_2$ may actually be healthy [43]. There are studies on the effects of $NaNO_2$ on human health from 1945 [44] to present date [45].

NO_3^- salts are used as a cheap nitrogen source in fertilizers. Therefore, with the widespread use of nitrogenous fertilizers in agriculture and inappropriate disposal of nitrogenous wastes, humans are exposed to high NO_3^-/NO_2^- levels at an alarming rate, especially through contaminated food and water [46]. Infants and individuals with deficiency of glucose-6-phosphate dehydrogenase are particularly sensitive to high levels of NO_3^-/NO_2^- [45]. The digestive system of newborn babies convert NO_3^- to NO_2^-, and NO_2^- reacts with hemoglobin and prevents oxygen transport to the tissues. As a result, "methemoglobinemia" known as "blue baby syndrome" occurs in infants [47].

Prolonged non-lethal exposure to high levels of NO_3^-/NO_2^- may cause respiratory failure, growth failure, diabetes, neurological disorders, and cancer [48]. $NaNO_2$ causes oxidative stress in human erythrocytes in vitro by increasing lipid and protein oxidation, osmotic fragility, and membrane damage [49].

Despite the fact that $NaNO_2$ has not been reliable in the past and has the potential to cause many cancers, it has recently been reported that it can prevent myocardial ischemia-reperfusion injury in diabetic rats by regulating eNOS and iNOS expression and inhibiting lipid peroxidation in the heart [50]. It also prevents hypertension and increases endothelium-dependent relaxation and total NO by regulating eNOS activity [51].

Serum malondialdehyde, NO, arginase, and glutathione S-transferase activities were increased, and glutathione and catalase activities were decreased in $NaNO_2$-treated rats [52]. In another study, decrease in GSH and catalase activity was reported in $NaNO_2$-intoxicated rats [53]. In their study investigating the histopathological, biochemical, and genotoxic effects of low dose $NaNO_2$ administration for 8 months, Ozen et al. [54] reported that the liver and kidney NO levels were decreased in rats. The reduction in NO levels may be explained by the rapid and/or efficient removal of this molecule from these tissues, resulting in an increase in serum levels due to reduced NO by metabolic depletion. Therefore, the investigators reported that the increased serum NO level did not contradict with the decreasing NO level in the tissues [54]. In addition, these and other investigators indicated that chronic administration of $NaNO_2$ increased iNOS activity in experimental animals [54, 55].

Peroxynitrite can interact with tyrosine residues to form nitrotyrosine. Ozen et al. [54] showed that the expression of iNOS and nitrotyrosine was increased in the liver and kidney tissues of $NaNO_2$-treated mice, and it caused tissue degeneration in both organs. Peroxynitrite can be decomposed to form NO_3^- and NO_2^- which can cause DNA damage, as well [56, 57].

The mechanisms of $NaNO_2$ are still not fully understood; this suggests that further work needs to be performed in the future.

4. N-nitroso compounds

4.1. The chemical structure and sources of N-nitroso compounds

NO_2^- is the precursor of N-nitroso compounds that have carcinogenic effect [58, 59]. NO_2^- is converted into nitrous acid in acidic environment, and nitrous acids react with secondary amines to form nitrosamine compounds (**Figure 2**) [60].

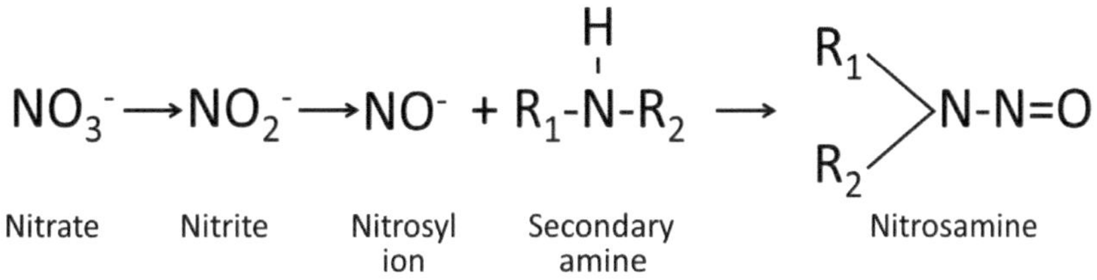

Figure 2. Formation of nitrosamine from NO_3^-.

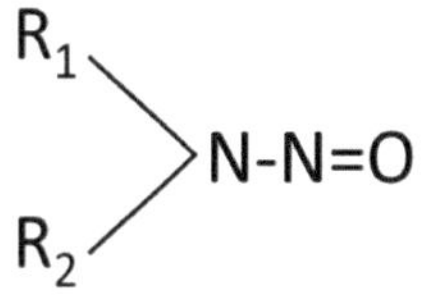

Figure 3. Chemical formula for nitrosamine.

Nitrosamines are chemical compounds with general formulas as shown in (**Figure 3**).

Nitrosamines are used in the manufacture of some cosmetics, pesticides, and most rubber products [61, 62]. Nitrosamines are found in latex products, cereal, tea, many foods, cigarettes, and cigarette smoke [60]. They are also formed by the reduction of NO_3^-, which is abundant in nature, into NO_2^- by bacteria [27].

The most commonly used N-nitroso compounds for the purpose of toxicity are N-nitrosodimethylamine, N-nitrosodiethylamine, N-nitrosopyrrolidine, and N-nitrosopiperidine [63, 64].

Dimethylnitrosamine (also called N-nitrosodimethylamine, DMN, DMNA, NDMA—$C_2H_6N_2O$) is found in wheat flour, cheese, smoked meat, fish, and other food products (**Figure 4**) [65]. It can be formed by reaction of dimethylamine with NO_2^-. In addition, it can be formed by nitrozation and decarboxylation of amino acids, such as glycine and alanine [66].

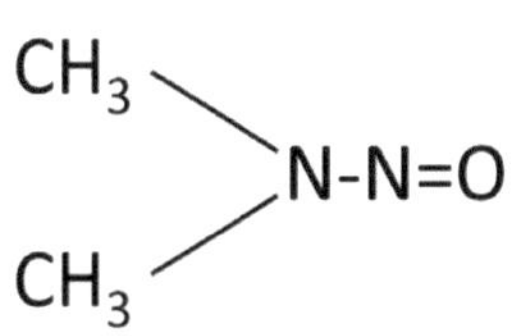

Figure 4. Chemical formula for dimethylnitrosamine.

Diethylnitrosamine (also called N-nitrosodiethylamine, DEN, DENA, NDEA—$C_4H_{10}N_2O$) is found in chemicals used in agriculture and rubber industry, cigarette smoke, alcoholic beverages, and processed meat products (**Figure 5**) [67]. It can be formed by reaction of diethylamine with NO_2^-. Additionally, it can be formed by nitrozation and decarboxylation of amino acids, such as glycine alanine [66].

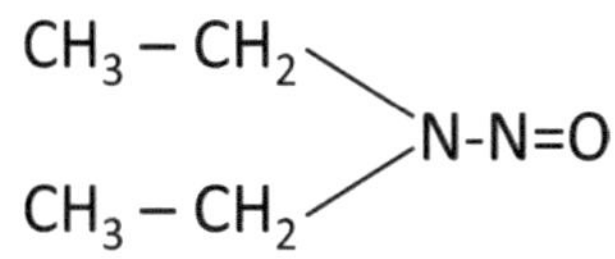

Figure 5. Chemical formula for diethylnitrosamine.

N-nitrosopyrrolidine (also called NPYR—$C_4H_8N_2O$) is found in cigarette smoke, meat and fish products (**Figure 6**) [68]. In meat products, this compound is formed by the nitrozation and decarboxylation of L-proline [69].

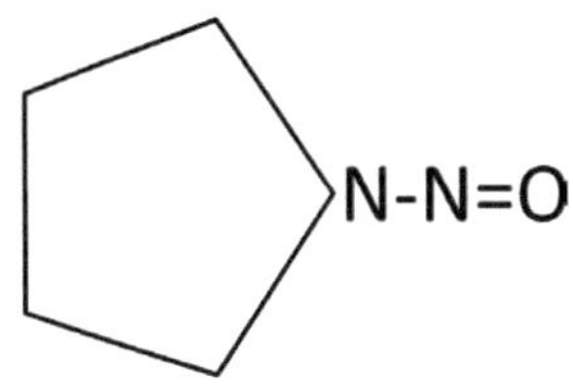

Figure 6. Chemical formula for N-nitrosopyrrolidine.

N-nitrosopiperidine (also called NPIP—$C_5H_{10}N_2O$) formation follows these steps: decarboxylation of lysine results in cadaverine; cadaverine is converted into piperidine by heat and the reaction of the resulting piperidine with NO_2^- (**Figure 7**) [69, 70].

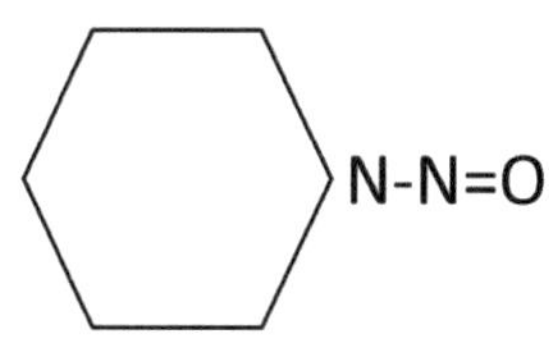

Figure 7. Chemical formula for N-nitrosopiperidine.

N-nitrosamines can be found in food products and might cause serious health problems for humans.

4.2. N-nitrosamines in meat and dairy products

N-nitrosamines taken with food and found in the environment have been found to cause serious health risks above a certain level, even though they have both food processing and protection functions as additives.

Nitrosamines occur mostly in meat and dairy products [71]. Since meat, an important nutrient, is easily decomposed by different factors, it is necessary to protect it with various methods and to increase its durability. For this purpose, some ingredients are added to meat and meat products. Cured meat products, unlike fresh meat or salted meat with only table salt, have a pleasant smell, flavor, and a natural looking but a heat-resistant color. Today, this process is applied to most of the meat products consumed [72, 73]. It has been reported that the degradation products of NO_3^- and NO_2^- result in the formation of carcinogenic nitrosamines by combining with amino acids, such as putrescine, thiamine, piperidine, pyrrolidine, histamine, cadaverine, trimethylamine, β-phenylethylamine, n-propylamine, and isopropylamine [74–77].

The most important sources of nitrosamines in dairy products, such as cheese and butter, are NO_3^-, NO_2^-, and amine compounds [78]. The metabolic activities of some microorganisms in milk and dairy products result in the formation of histamine and tyramines, and nitrosamines are formed by the reaction of the secondary amines, which are the degradation products of these biogenic amines in various ways, with NO_2^- [65, 69].

The first formation of these nitrosamines in meat and dairy products is seen in the oral cavity [79]. The salivary secretion contains abundant NO_3^- and this NO_3^- is reduced to NO_2^- by nitrate reductase enzyme [72]. This NO_2^- causes the formation of nitrosamines [80]. These compounds can be taken into the stomach in various ways, such as ingestion or smoking, or can also be formed by the reaction of NO_2^- and amines in acidic conditions [81]. Some bacteria in the stomach and intestines increase the formation of nitrosamines by facilitating the conversion of NO_3^- into NO_2^-. NO_2^- transforms into nitrous acid in the acidic environment of the stomach, and nitrous acid reacts with amines in the environment to form nitrosamines [60, 82]. Nitrosamines are usually excreted through urine [83, 84].

5. The roles of NOS isoforms and *N*-nitrosamine compounds in liver toxicity or carcinogenesis

NO_3^- can easily be formed in mammalian systems through bacterial and mammalian enzymes. The resulting NO_3^- can then react with amines, amides, and amino acids to form N-nitroso compounds. While NO_3^- has relatively low toxicity, NO_2^- and N-nitroso compounds have higher toxicity in mammals. For this reason, there are many studies investigating the toxicity of these two molecules, as well as studies investigating ways to decrease the detrimental effects of these two molecules [20]. It has been suggested in long-term experimental animal studies that nitrosamines cause cancer in many tissues, but the role of nitrosamines in the formation of cancers is still being investigated.

5.1. NOS isoforms in carcinogenesis

Various studies have indicated that three NOS isoforms both trigger and prevent cancer etiology. Nitric oxide synthase activity has been detected in a variety of tumor cells, and it has been shown to be closely related to tumor grade, proliferation rate, and cancer development. High NOS expression can be cytotoxic for cancer cells. On the other hand, low NOS expression may have an adverse effect and may increase tumor development [85, 86]. For this reason, NOS can be both genotoxic and angiogenic. High NO production leads to angiogenesis by increasing the VEGF gene, especially in p53 mutant cells. In addition, NO can alter the expressions of DNA repair proteins, such as poly (ADP-ribose) polymerase and DNA-protein kinase in tumor cells. NO may exhibit carcinogenic effect by the production of different NO metabolites. For example, NO may rapidly react with intracellular environment to form N-nitroso compounds. These metabolites, for example, can cause genotoxic effects by creating DNA damage [86]. In some other studies, N-nitroso compounds have been reported to alter the activity of creatine kinase, lactate dehydrogenase [87], pyruvate kinase [88], and Na/K-ATPase [89] enzymes.

5.2. *N*-nitrosamine compounds in liver toxicity and cancer

Certain levels of nitrosamines taken in the body with any food ingredient are less likely to cause cancer in the human body alone. However, different types of nitrosamines, which are continuously taken from different sources, such as air and cigarettes, increase the risk of developing cancer [65, 90]. Although NO_3^- and NO_2^- create a toxicological problem, the main problem is that they turn into nitrosamines, which are carcinogenic. Nitrosamines are known to exhibit carcinogenic effect through binding to proteins and nucleic acids [91]. Nitrosamines also have mutagenic and teratogenic effects [61]. Because many organ-specific nitrosamines are metabolized in the same way in human and animal tissues, humans are very sensitive to the carcinogenic properties of nitrosamines [67]. N-nitroso compounds are potent alkylating agents that can form endogenously and can cause cancer in surrounding animals [64]. Bacterial decarboxylation of amino acids in NO_3^- taken with nutrients results in amines and amides [58]. There is a relationship between the formation of N-nitroso compounds by bacterial catalysis and increased risk for liver, stomach, esophagus, nasopharynx, chronic urinary tract infections, and bladder squamous cell carcinoma [77, 92].

Metabolic activations of the nitrosamines occur primarily in the liver, and this transformation can occur in all cells. Dimethylnitrosamine is a potent carcinogen that can induce malignant tumors in various animal species in various tissues, including the liver, lungs, and stomach [79]. Various studies on different species of mice have shown that adenomas and adenocarcinomas in the lungs and hepatocarcinoma in the liver are formed by dimethylnitrosamine. The target organs of dimethylnitrosamine are the liver, lungs, and kidneys [65, 79].

It is suggested that diethylnitrosamine metabolism is catalyzed by the enzymes of the multifunctional cytochrome P-450 monooxygenase system and toxic effects are initiated by its metabolic activation, and that the resulting reactive intermediate products have little affinity for the catalytic domains of the binding enzymes, so that instead of being excreted with urine, they stimulate the onset of mutation, cancer, and necrosis by forming covalent bonds with important cellular components [93, 94]. Low concentrations of diethylnitrosamine cause mutations and cancer which was shown by the Ames assay [94]. Many studies suggest that the harmful effects of diethylnitrosamine may be reduced by various antioxidant molecules. The administration of molecules such as α-lipoic acid [95], omega-3 [96], blueberry [97], and beta-carotene [98] were stated to reduce the carcinogenic effect of nitrosamines in experimental animals.

N-nitrosopiperidine and N-nitrosopyrrolidine are structurally cyclic nitrosamines with different carcinogenic activities. Comparative carcinogenicity studies of these two nitrosamines in rats revealed that N-nitrosopiperidine caused esophagus, liver, and stomach tumors, and N-nitrosopyrrolidine caused tumors mainly in the liver [63].

6. Conclusion

NO-mediated responses are cell specific, and they depend on the existence of different NOS isoforms at different concentrations, and their regulations at pre- and post-transcriptional levels are quite complex. The latest developments on strategies for treating or preventing pathological events in association with the stimulation or inhibition of excessive production of NO and N-nitroso compounds present a crucial importance in medicine.

Author details

Emine Atakisi* and Oguz Merhan

*Address all correspondence to: et_tasci@hotmail.com

Department of Biochemistry, Faculty of Veterinary, Kafkas University, Kars, Turkey

References

[1] Rosen GM, Tsai P, Pou S. Mechanism of free-radical generation by nitric oxide synthase. Chemical Reviews. 2002;**102**:1191–1199

[2] Porasuphatana S, Tsai P, Rosen GM. The generation of free radicals by nitric oxide synthase. Comparative Biochemistry and Physiology C Toxicology & Pharmacology. 2003;**134**:281–289

[3] Förstermann U, Sessa WC. Nitric oxide synthases: Regulation and function. European Heart Journal. 2012;**33**:829–837. DOI: 10.1093/eurheartj/ehr304

[4] Kopincova J, Puzserova A, Bernatova I. Biochemical aspects of nitric oxide synthase feedback regulation by nitric oxide. Interdisciplinary Toxicology. 2011;**4**:63–68. DOI: 10.2478/v10102-011-0012-z

[5] Smith BC, Fernhoff NB, Marletta MA. Mechanism and kinetics of inducible nitric oxide synthase auto-s-nitrosation and inactivation. Biochemistry. 2012;**51**:1028–1040. DOI: 10.1021/bi201818c

[6] Caplin B, Leiper J. Endogenous nitric oxide synthase inhibitors in the biology of disease: Markers, mediators and regulators. Arteriosclerosis Thrombosis and Vascular Biology. 2012;**32**:1343–1353. DOI: 10.1161/ATVBAHA.112.247726

[7] Daff S. NO synthase: Structures and mechanisms. Nitric Oxide. 2010;**23**:1–11. DOI: 10.1016/j.niox.2010.03.001

[8] Zhang YH, Jin CZ, Jang JH, Wang Y. Molecular mechanisms of neuronal nitric oxide synthase in cardiac function and pathophysiology. The Journal of Physiology. 2014;**15**:3189–3200. DOI: 10.1113/jphysiol.2013.270306

[9] Millatt LJ, Abdel-Rahman EM, Siragy HM. Angiotensin II and nitric oxide: A question of balance. Regulatory Peptides. 1999;**81**:1–10

[10] Choudhari SK, Chaudhary M, Bagde S, Gadbail AR, Joshi V. Nitric oxide and cancer: A review. World Journal of Surgical Oncology. 2013;**11**:118. DOI: 10.1186/1477-7819-11-118

[11] Förstermann U, Boissel JP, Kleinert H. Expressional control of the constitutive isoforms of nitric oxide synthase (NOS I and NOS III). FASEB Journal. 1998;**12**:773–790

[12] Atakisi E, Atakisi O, Topcu B, Uzun M. Effects of therapeutic dose of ivermectin on plasma nitric oxide and total antioxidant capacity in rabbits. European Review for Medical and Pharmacological Sciences. 2009;**13**:425–429

[13] Atakisi O, Erdogan HM, Atakisi E, Citil M, Kanici A, Merhan O, et al. Effects of reduced glutathione on nitric oxide level, total antioxidant and oxidant capacity and adenosine deaminase activity. European Review for Medical and Pharmacological Sciences. 2010;**14**:19–23

[14] Atakisi O, Oral H, Atakisi E, Merhan O, Pancarci SM, Ozcan A, et al. Subclinical mastitis causes alterations in nitric oxide, total oxidant and antioxidant capacity in cow milk. Research in Veterinary Science. 2010;**89**:10–13. DOI: 10.1016/j.rvsc.2010.01.008

[15] Atakisi E, Bozukluhan K, Atakisi O, Gokce HI. Total oxidant and antioxidant capacities and nitric oxide levels in cattle with traumatic reticuloperitonitis. Veterinary Record. 2010;**167**:908–909. DOI: 10.1136/vr.c3664

[16] Bozukluhan K, Atakisi E, Atakisi O. Nitric oxide levels, total antioxidant and oxidant capacity in cattle with foot-and-mouth-disease. Kafkas Üniversitesi Veteriner Fakültesi Dergisi. 2013;**19**:179–181. DOI: 10.9775/kvfd.2012.7244

[17] Atakisi E, Kirmizigul AH, Atakisi O, Karadag Sari E, Ogun M, Marasli S, et al. Plasma nitric oxide (NO) and tumor necrosis factor-α (TNF-α) levels, adenosine deaminase (ADA), gamma glutamyl transferase (GGT) activities and to determine the rate of lymphocytes in the peripheral blood leukocytes alpha naphthyl acetate esterase (ANAE) in cattle with leptospirosis. Kafkas Üniversitesi Veteriner Fakültesi Dergisi. 2014;**20**:451–455. DOI: 10.9775/kvfd.2013.10427

[18] Ogun M, Ozcan A, Karaman M, Merhan O, Ozen H, Kukurt A, et al. Oleuropein ameliorates arsenic induced oxidative stress in mice. Journal of Trace Elements in Medicine and Biology. 2016;**36**:1–6

[19] Lundberg JO, Weitzberg E. The biological role of nitrate and nitrite: The times they are a-changin'. Nitric Oxide. 2010;**22**:61–63. DOI: 10.1016/j.niox.2009.11.004

[20] Gangolli SD, van den Brandt PA, Feron VJ, Janzowsky C, Koeman JH, Speijers GJ, et al. Nitrate, nitrite and N-nitroso compounds. European Journal Pharmacology. 1994;**292**:1-38

[21] Lundberg JO, Weitzberg E, Gladwin MT. The nitrate–nitrite–nitric oxide pathway in physiology and therapeutics. Nature Reviews Drug Discovery. 2008;**7**:156–167. DOI: 10.1038/nrd2466

[22] Briggs D. Environmental pollution and the global burden of disease. British Medical Bulletin. 2003;**68**:1–24. DOI: 10.1093/bmb/ldg019

[23] Tilman D, Cassman KG, Matson PA, Naylor R, Polasky S. Agricultural sustainability and intensive production practices. Nature. 2002;**418**:671–677. DOI: 10.1038/nature01014

[24] Gilchrist M, Shore AC, Benjamin N. Inorganic nitrate and nitrite and control of blood pressure. Cardiovascular Research. 2011;**89**:492–498. DOI: 10.1093/cvr/cvq309

[25] Gai X, Liu H, Zhai L, Tan G, Liu J, Ren T, et al. Vegetable yields and soil biochemical properties as influenced by fertilization in Southern China. Applied Soil Ecology. 2016;**107**:170–181

[26] Correia M, Barroso A, Barroso MF, Soares D, Oliveira MBPP, Delerue-Matos C. Contribution of different vegetable types to exogenous nitrate and nitrite exposure. Food Chemistry. 2010;**120**:960–966. DOI:10.1016/j.foodchem.2009.11.030

[27] Tiso M, Schechter AN. Nitrate reduction to nitrite, nitric oxide and ammonia by gut bacteria under physiological conditions. PLoS One. 2015;**10(3)**:e0119712. DOI: 10.1371/journal. pone.0119712

[28] Bedale W, Sindelar JJ, Milkowski AL. Dietary nitrate and nitrite: Benefits, risks, and evolving perceptions. Meat Science. 2016;**120**:85–92. DOI: 10.1016/j.meatsci.2016.03.009

[29] Rocha BS, Gago B, Barbosa RM, Cavaleiro C, Laranjinha J. Ethyl nitrite is produced in the human stomach from dietary nitrate and ethanol, releasing nitric oxide at physiological pH: Potential impact on gastric motility. Free Radical Biology and Medicine. 2015;**82**:160–166. DOI: 10.1016/j.freeradbiomed.2015.01.021

[30] Petersson J, Jadert C, Phillipson M, Borniquel S, Lundberg JO, Holm L. Physiological recycling of endogenous nitrate by oral bacteria regulates gastric mucus thickness. Free Radical Biology and Medicine. 2015;**89**:241–247. DOI: 10.1016/j.freeradbiomed. 2015.07.003

[31] Hohensinn B, Haselgrübler R, Müller U, Stadlbauer V, Lanzerstorfer P, Lirk G, et al. Sustaining elevated levels of nitrite in the oral cavity through consumption of nitrate-rich beetroot juice in young healthy adults reduces salivary pH. Nitric Oxide. 2016;**60**:10–15. DOI: 10.1016/j.niox.2016.08.006

[32] Weitzberg E, Lundberg JO. Novel aspects of dietary nitrate and human health. Annual Review of Nutrition. 2013;**33**:129–159. DOI: 10.1146/annurev-nutr-071812-161159

[33] Worley BV, Soto RJ, Kinsley PC, Schoenfisch MH. Active release of nitric oxide-releasing dendrimers from electrospun polyurethane fibers. ACS Biomaterials Science & Engineering. 2016;**2**:426–437. DOI: 10.1021/acsbiomaterials.6b00032

[34] Ascenzi P, Bocedi A, Gradoni L. The anti-parasitic effects of nitric oxide. IUBMB Life. 2003;**55**:573–578. DOI: 10.1080/15216540310001639265

[35] Schairer DO, Chouake JS, Nosanchuk JD, Friedman AJ. The potential of nitric oxide releasing therapies as antimicrobial agents. Virulence. 2012;**3**:271–279. DOI: 10.4161/viru.20328

[36] Kevil CG, Kolluru GK, Pattillo CB, Giordano T. Inorganic nitrite therapy: Historical perspective and future directions. Free Radical Biology and Medicine. 2011;**51**:576–593. DOI: 10.1016/j.freeradbiomed.2011.04.042

[37] Lanas A. Role of nitric oxide in the gastrointestinal tract. Arthritis Research and Therapy. 2008;**10**(Suppl 2):S4, 1–6. DOI: 10.1186/ar2465

[38] Grishin A, Bowling J, Bell B, Wang J, Ford HR. Roles of nitric oxide and intestinal microbiota in the pathogenesis of necrotizing enterocolitis. Journal of Pediatric Surgery. 2016;**51**:13–17. DOI: 10.1016/j.jpedsurg.2015.10.006

[39] Nadler EP, Stanford A, Zhang XR, Schall LC, Alber SM, Watkins SC, et al. Intestinal cytokine gene expression in infants with acute necrotizing enterocolitis: Interleukin-11 mRNA expression inversely correlates with extent of disease. Journal of Pediatric Surgery. 2001;**36**:1122–1129. DOI: 10.1053/jpsu.2001.25726

[40] Kapral M, Wawszczyk J, Sosnicki S, Weglarz L. Down-regulation of inducible nitric oxide synthase expression by inisitol hexaphosphatate in human colon cancer cells. Acta Poloniae Pharmaceutica. 2015;**72**:705–711

[41] Weitzberg E, Hezel M, Lundberg JO. Nitrate-nitrite-nitric oxide pathway: Implications for anesthesiology and intensive care. Anesthesiology. 2010;**113**:1460–1475. DOI: 10.1097/ALN. 0b013e3181fcf3cc

[42] Kapil V, Haydar SM, Pearl V, Lundberg JO, Weitzberg E, Ahluwalia A. Physiological role for nitrate-reducing oral bacteria in blood pressure control. Free Radical Biology and Medicine. 2013;**55**:93–100. DOI: 10.1016/j.freeradbiomed.2012.11.013

[43] Abdollahi M, Khaksar MR. Sodium nitrite. Encyclopedia of Toxicology. 2014;**4**:334–337. http://dx.doi.org/10.1016/B978-0-12-386454-3.01206-9

[44] Greenberg M, Birnkrant WB, Schiftner JJ. Outbreak of sodium nitrite poisoning. American Journal of Public Health and the Nation's Health. 1945;**35**:1217–1220

[45] Katabami K, Hayakawa M, Gando S. Severe methemoglobinemia due to sodium nitrite poisoning. Case Reports in Emergency Medicine. 2016;**9013816**. DOI: 10.1155/2016/9013816

[46] WHO. Nitrate and nitrite in drinking-water. Background document for development of WHO guidelines for drinking-water quality. Geneva: World Health Organization; 2011

[47] Jones JA, Hopper AO, Power GG, Blood AB. Dietary intake and bio-activation of nitrite and nitrate in newborn infants. Pediatric Research. 2015;**77**:173–181. DOI: 10.1038/pr.2014.168

[48] Jones RR, Weyer PJ, DellaValle CT, Inoue-Choi M, Anderson KE, Cantor KP, et al. Nitrate from drinking water and diet and bladder cancer among postmenopausal women in Iowa. Environmental Health Perspectives. 2016;**124**:1751–1758. DOI: 10.1289/EHP191

[49] Ansari FA, Mahmood R. Sodium nitrite enhances generation of reactive oxygen species that decrease antioxidant power and inhibit plasma membrane redox system of human erythrocytes. Cell Biology International. 2016;**40**:887–894. DOI: 10.1002/cbin.10628

[50] Jeddi S, Khalifi S, Ghanbari M, Bageripour F, Ghasemi A. Effects of nitrate intake on myocardial ischemia-reperfusion injury in diabetic rats. Arquivos brasileiros de cardiologia. 2016;**107**:339–347. DOI: 10.5935/abc.20160137

[51] Ling WC, Murugan DD, Lau YS, Vanhoutte PM, Mustafa MR. Sodium nitrite exerts an antihypertensive effect and improves endothelial function through activation of eNOS in the SHR. Scientific Reports. 2016;**6**:33048. DOI: 10.1038/srep33048

[52] El-Sheikh NM, Khalil FA. L-arginine and L-glutamine as immunonutrients and modulating agents for oxidative stress and toxicity induced by sodium nitrite in rats. Food and Chemical Toxicology. 2011;**49**:758–762. DOI: 10.1016/j.fct.2010.11.039

[53] Hassan HA, Yousef MI. Ameliorating effect of chiory (*Cichorium intybus* L.)-supplemented diet against nitrosamine precursors-induced liver injury and oxidative stress in male rats. Food and Chemical Toxicology. 2010;**48**:2163–2169. DOI: 10.1016/j.fct.2010.05.023

[54] Ozen H, Kamber U, Karaman M, Gul S, Atakisi E, Ozcan K, et al. Histopathologic, biochemical and genotoxic investigations on chronic sodium nitrite toxicity in mice. Experimental and Toxicologic Pathology. 2014;**66**:367–375. DOI: 10.1016/j.etp.2014.05.003

[55] Moncada S, Higgs A. The L-arginine-nitric oxide pathway. The New England Journal of Medicine. 1993;**329**:2002–2012. DOI: 10.1056/NEJM199312303292706

[56] Ischiropoulos H, Zhu L, Chen J, Tsai M, Martin JC, Smith CD, et al. Peroxynitrite-mediated tyrosine nitration catalyzed by superoxide dismutase. Archives Biochemistry and Biophysics. 1992;**298**:431–437

[57] Nordberg J, Arner ES. Reactive oxygen species, antioxidants, and the mammalian thioredoxin system. Free Radical Biology and Medicine. 2001;**31**:1287–1312

[58] Parnaud G, Pignatelli B, Peiffer G, Tache S, Corpet DE. Endogenous N-nitroso compounds, and their precursors, present in bacon, do not initiate or promote aberrant crypt foci in the colon of rats. Nutrition and Cancer. 2000;**38**:74–80. DOI: 10.1207/S15327914NC381_11

[59] Haorah J, Zhou L, Wang X, Xu G, Sidney S, Mirvish SS. Determination of total N-nitroso compounds and their precursors in frankfurters, fresh meat, dried salted fish, sauces,

tobacco, and tobacco smoke particulates. Journal of Agricultural and Food Chemistry. 2001;**49**:6068–6078

[60] Rostkowska K, Zwierz K, Rozanski A, Moniuszko-Jakoniuk J, Roszczenko A. Formation and metabolism of N-nitrosamines. Polish Journal of Environmental Studies. 1998; **7**:321–325

[61] Qajarbeygi P, Ahmadi M, Hoseini AH, Poorasl AM, Mahmoudi R, Ataee M. Evaluation of N-Nitrosamine formation in routine potato cooking. Biotech Health Science. 2015;**2**: e29887. DOI: 10.17795/bhs-29887

[62] Park J, Seo J, Lee J, Kwon H. Distribution of seven N-nitrosamines in food. Toxicological Research. 2015;**31**:279–288

[63] Wong HL, Murphy SE, Wang M, Hecht SS. Comparative metabolism of *N*-nitrosopiperidine and *N*-nitrosopyrrolidine by rat liver and esophageal microsomes and cytochrome P450 2A3. Carcinogenesis. 2003;**24**:291–300. DOI: https://doi.org/10.1093/carcin/24.2.291

[64] Veena S, Rashmi S. A review on mechanism of nitrosamine formation, metabolism and toxicity in in vivo. International Journal of Toxicological and Pharmacological Research. 2014;**6**:86–96

[65] Tricker AR, Preussmann R. Carcinogenic N-nitrosamines in the diet: Occurrence, formation, mechanisms and carcinogenic potential. Mutation Research. 1991;**259**:277–289

[66] Walker R. Nitrates, nitrites and N-nitrosocompounds: A review of the occurrence in food and diet and the toxicological implications. Food Additives and Contaminants. 1990;**7**:717–768

[67] Hecht SS. Approaches to cancer prevention based on an understanding of N-nitrosamine carcinogenesis. Proceedings of the Society for Experimental Biology and Medicine. 1997;**216**:181–191

[68] Mitacek EJ, Brunnemann KD, Suttajit M, Martin N, Limsila T, Ohshima H, et al. Exposure to N-nitroso compounds in a population of high liver cancer regions in Thailand: Volatile nitrosamine (VNA) levels in Thai Food. Food and Chemical Toxicology. 1999;**37**:297–305

[69] Shalaby AR. Significance of biogenic amines to food safety and human health. Food Research International. 1996;**29**:675–690

[70] Labroo RB, Paine MF, Thummel KE, Kharasch ED. Fentanyl metabolism by human hepatic and intestinal cytochrome p450 3a4: Implications for interindividual variability in disposition, efficacy, and drug interactions. Drug Metabolism and Disposition. 1997;**25**:1072–1080

[71] Scanlan RA. Formation and occurrence of nitrosamines in food. Cancer Research. 1983;**43**(5 Suppl):2435–2440

[72] Sebranek JG, Bacus JN. Cured meat products without direct addition of nitrate or nitrite: What are the issues? Meat Science. 2007;**77**:136–147

[73] Sindelar JJ, Milkowski AL. Human safety controversies surrounding nitrate and nitrite in the diet. Nitric Oxide. 2012;**26**:259–266

[74] Kim JH, Ahn HJ, Jo C, Park HJ, Chung YJ, Byun MW. Radiolysis of biogenic amines in model system by gamma irradiation. Food Control. 2004;**15**:405–408. DOI: 10.1016/S0956-7135(03)00102-6

[75] Linares DM, Martin MC, Ladero V, Alvarez MA, Fernandez M. Biogenic amines in dairy products. Critical Reviews in Food Science Nutrition. 2011;**51**:691–703. DOI: 10.1080/10408398.2011.582813

[76] Linares DM, Del Rio B, Ladero V, Martinez N, Fernandez M, Martin MC, et al. Factors influencing biogenic amines accumulation in dairy products. Frontiers in Microbiology. Microbiology. 2012;**3**:1–10. DOI: 10.3389/fmicb.2012.00180

[77] Chung KT. The etiology of bladder cancer and its prevention. Journal of Cancer Science and Therapy. 2013;**5**:346–361

[78] Oliveira CP, Gloria BMA, Barbour JF, Scanlan RA. Nitrate, nitrite, and volatile nitrosamines in whey-containing food products. Journal of Agricultural and Food Chemistry. 1995;**43**:967–969

[79] Song P, Wu L, Guan W. Dietary nitrates, nitrites, and nitrosamines intake and the risk of gastric cancer: A meta-analysis. Nutrients. 2015;**7**:9872–9895. DOI: 10.3390/nu7125505

[80] Krul CA, Zeilmaker MJ, Schothorst RC, Havenaar R. Intragastric formation and modulation of N-nitrosodimethylamine in a dynamic in vitro gastrointestinal model under human physiological conditions. Food and Chemical Toxicology. 2004;**42**:51–63

[81] Honikel KO. The use and control of nitrate and nitrite for the processing of meat products. Meat Science. 2008;**78**:68–76. DOI: 10.1016/j.meatsci.2007.05.030

[82] Vermeer IT, Engels LG, Pachen DM, Dallinga JW, Kleinjans JC, Maanen JM. Intragastric volatile N-nitrosamines, nitrite, pH, and helicobacter pylori during long-term treatment with omeprazole. Gastroenterology. 2001;**121**:517–525

[83] Lin K, Shen W, Shen Z, Wu Y, Lu S. Dietary exposure and urinary excretion of total N-nitroso compounds, nitrosamino acids and volatile nitrosamine in inhabitants of high- and low-risk areas for esophageal cancer in southern China. International Journal of Cancer. 2002;**102**:207–211

[84] Zeng T, Mitch WA. Oral intake of ranitidine increases urinary excretion of N-nitrosodimethylamine. Carcinogenesis. 2016;**37**:625–634. DOI: 10.1093/carcin/bgw034

[85] Vakkala M, Kahlos K, Lakari E, Paakko P, Kinnula V, Soini Y. Inducible nitric oxide synthase expression, apoptosis, and angiogenesis in in situ and invasive breast carcinomas. Clinical Cancer Research. 2000;**6**:2408–2416

[86] Xu W, Liu LZ, Loizidou M, Ahmed M, Charles IG. The role of nitric oxide in cancer. Cell Research. 2002;**12**:311–320. DOI: 10.1038/sj.cr.7290133

[87] Sheweita SA, El-Bendery HA, Mostafa MH. Novel study on N-nitrosamines as risk factors of cardiovascular diseases. BioMed Research International. 2014;Article ID817019, 10 pages. http://dx.doi.org/10.1155/2014/817019

[88] Yoshioka T, Uematsu T. Formation of N-hydroxy-N-arylacetamides from nitroso aromatic compounds by the mammalian pyruvate dehydrogenase complex. The Biochemical Journal. 1993;**290**:783–790

[89] Boldyrev AA, Bulygina ER, Kramarenko GG, Vanin AF. Effect of nitroso compounds on Na/K-ATPase. Biochimica et Biophysica Acta. 1997;**1321**:243–251

[90] Pichandi S, Pasupathi P, Rao YY, Farook J, Ponnusha BS, Ambika A, et al. The effect of smoking on cancer: A review. International Journal of Biological & Medical Research. 2011;**2**:593–602

[91] Cammack R, Joannou CL, Cui XY, Martinez CT, Maraj SR, Hughes MN. Nitrite and nitrosyl compounds in food preservation. Biochimica et Biophysica Acta 1999;**1411**:475–488

[92] Mirvish SS. Role of N-nitroso compounds (NOC) and N-nitrosation in etiology of gastric, esophageal, nasopharyngeal and bladder cancer and contribution to cancer of known exposures to NOC. Cancer Letters. 1995;**93**:17–48

[93] Pinto LF, Moraes E, Albano RM, Silva MC, Godoy W, Glisovic T, et al. Rat oesophageal cytochrome P450 (CYP) monooxygenase system: Comparison to the liver and relevance in N-nitrosodiethylamine carcinogenesis. Carcinogenesis. 2001;**22**:1877–1883

[94] Aiub CAF, Pinto LFR. N-Nitrosodiethylamine mutagenicity at low concentrations. Toxicology Letters. 2003;**145**:36–45. DOI: 10.1016/S0378-4274(03)00263-7

[95] Karaca EG, Baysu Sozbilir N. Investigation of the protective role of α-lipoic acid on rats given diethylnitrosamine. The Medical Journal of Kocatepe. 2007;**7**:11–17

[96] Atakisi E, Ozcan A. Investigation of the protective role of omega-3 fatty acids rich fish oil on rats given diethylnitrosamine. Turkish Journal of Biochemistry. 2005;**30**:279–284

[97] Sadik NAH, EL-Maraghy SA, Ismail MF. Diethylnitrosamine-induced hepatocarcinogenesis in rats: Possible chemoprevention by blueberries. African Journal of Biochemistry Research. 2008;**2**:81–87

[98] Merhan O, Ozcan A, Atakisi E, Ogun M, Kukurt A. The effect of β-carotene on acute phase response in diethylnitrosamine given rabbits. Kafkas Üniversitesi Veteriner Fakültesi Dergisi. 2016;**22**:533–537. DOI: 10.9775/kvfd.2016.14995

Section 5

Nitric Oxide Synthase Affecting Agents

Chapter 12

Nitric Oxide Synthase Inhibitors

Elizabeth Igne Ferreira and

Ricardo Augusto Massarico Serafim

Additional information is available at the end of the chapter

http://dx.doi.org/10.5772/67027

Abstract

Nitric oxide (NO) is an endogenic product from plants, bacteria, and animal cells that has many important effects in those organisms. It is produced by nitric oxide synthase (NOS), which is found in main three isoforms, namely endothelial NOS (eNOS), inducible NOS (iNOS), and neuronal NOS (nNOS). It has an important role in homeostasis in different physiological systems, such as micro- and macro-vascularization, inhibition of platelet aggregation, and neurotransmission regulation in the central nervous, gastrointestinal, respiratory, and genitourinary systems. However, its overproduction has been associated with diseases such as arthritis, asthma, cerebral ischemia, Parkinson's disease, neurodegeneration, and seizures. For this reason, and due to better understanding of the molecular mechanisms by which NO provokes those diseases, the interest on the design of NOS inhibitors with therapeutic purposes has highly increased. Based on the foregoing considerations, the proposal of this chapter is to show an overview about the design strategies, mechanism of action at the molecular level, and the main advances toward the search for selective NOS inhibitors available in the literature.

Keywords: nitric oxide synthase isoforms, structure-based drug design, enzymatic inhibition, selectivity, heterocyclic compounds

1. Introduction

Nitric oxide (NO) is a diatomic neutral molecule, produced by bacteria, plants, and animals. Having one unpaired electron, its effect in biological system is related to the stabilization of this electron. It acts as dissolved nonelectrolyte in the organisms, except for the lungs, where it is found in gaseous state [1–3].

NO has gain importance mainly in the 1990s, and from then on, it has been studied to obtain interesting pharmacological effects. A review by Serafim and collaborators describes the state of the art of this compound use in drug design [4].

The basal NO production has an important contribution to homeostasis in different physiological systems, such as micro- and macro-vascularization, inhibition of platelet aggregation, and neurotransmission regulation in central nervous, gastrointestinal, respiratory, and genitourinary systems. However, NO overproduction has been strongly associated with some diseases such as arthritis, asthma, cerebral ischemia, Parkinson's disease, neurodegeneration, and seizures [5–9].

Nitric oxide synthase (NOS) is the enzyme responsible for NO biosynthesis, and there are three main kinds of NOS isoforms [endothelial NOS (eNOS), inducible NOS (iNOS), and neuronal NOS (nNOS)]. They are tetramers, constituted of two NOS monomers associated with two calmodulin monomers (CaM), and contain relatively tightly bound cofactors, BH_4, FAD, FMN, and iron protoporphyrin IX (heme group). Their chemical function is to catalyze the reaction of L-arginine, NADPH, and oxygen to synthesize free radical NO, L-citrulline, and NADP (**Figure 1**) [10].

The substrate L-arginine establishes H-bond networks inside the catalytic site of NOS isoforms with the heme group, mainly due to the guanidine group, which is crucial to bind tightly using a salt-bridge interaction with the conserved carboxylate of Glu597 in human nNOS, Glu377 for iNOS and Glu361 for eNOS. In addition, L-arginine establishes H-bonds with the amide carbonyl from Trp592, in nNOS; with Trp372, in iNOS; and with Trp356, in eNOS. Moreover, α-amino group of L-arginine interacts through H-bond with the heme propionate side chain, and the guanidine *N*-terminal nitrogen of this amino acid coordinates with Fe^{II} (**Figure 2**). This ligand-receptor interaction profile is similar to all isoforms, which generates a challenge to selectivity [11].

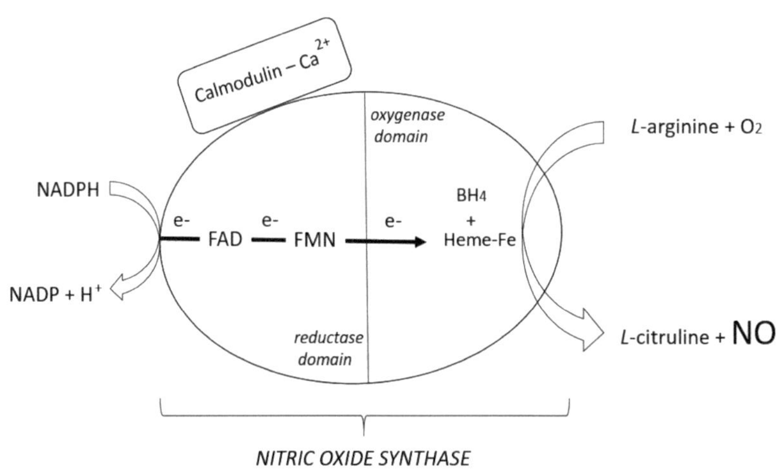

Figure 1. General reaction of NO formation by NOS. Adapted from [10].

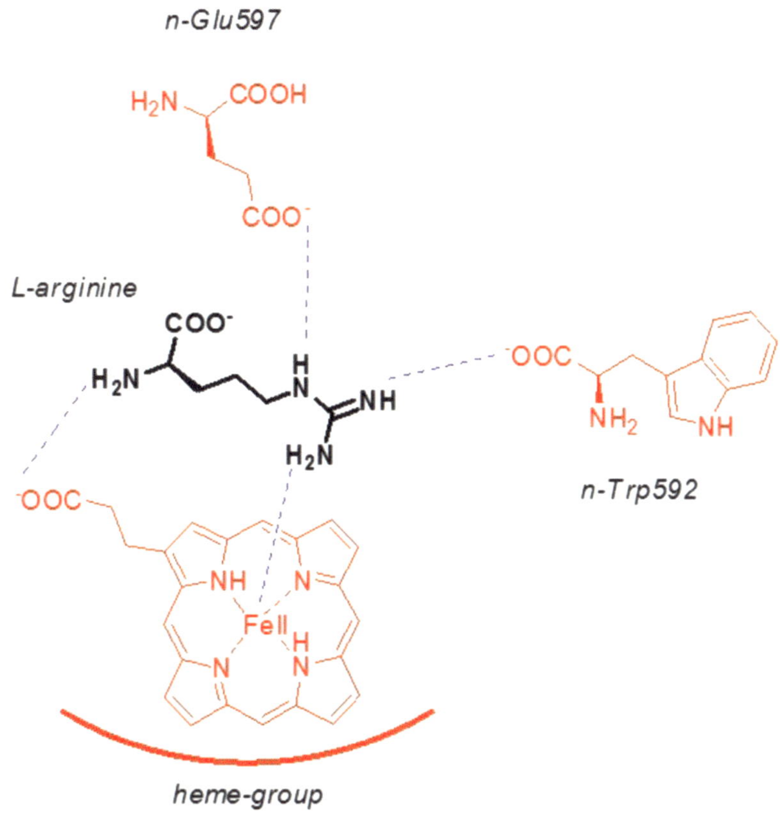

Figure 2. Scheme of nNOS-binding site.

Exacerbated induction of iNOS is associated with septic shock, inflammatory, and noninflammatory impairment processes in different tissues/organs, and, likewise, the nNOS is triggered in neurotoxicity, neurodegeneration process, and proliferation increase of some neoplastic cell lines. Depending on the clinical condition, decreasing NO levels is necessary, and excellent benefits might be achieved using NOS inhibitors. However, it is much important not to inhibit eNOS, because of its central role in smooth muscle relax, controlling vascular tone and blood pressure [12–14].

The first inhibitors designed (during the1980s and early 1990s) were based on L-arginine, the substrate of the enzyme, and this approach led to potent compounds but with poor selectivity level among the isoforms. In the late 1990s, the first crystal structure of iNOS and eNOS was unveiled, showing the high degree of similarity particularly in the active site of both isoforms. The nNOS crystal structure was reported in 2002, allowing the design of selective inhibitors [15, 16]. It is worth noting that changes in some amino acids of the isoforms lead to differences in electronic and steric effects on the binding site region, which can be interesting for designing selective inhibitors [11, 15]. The active collaboration between Richard Silverman and Thomas Poulos' groups has significantly contributed to this field, and some of their papers are discussed in this chapter.

NOS isoforms were validated as target for new drugs soon after their X-ray crystallography was available. From then on the design of effective and selective inhibitors has been an important approach in modern drug discovery involving NO biochemical pathways related to many dysfunctions of the human organism [12, 17–19].

2. Experimental studies

2.1. Inducible nitric oxide synthase (iNOS) inhibitors

Garvey and collaborators (1994) were the first to report highly selective iNOS inhibitors. The compounds were isothiourea derivatives (**Figure 3—1**) designed as L-arginine-competitive reversible inhibitors of human iNOS, with a K_i = 47 nM and a 190-fold selectivity over eNOS but only ~5-fold over nNOS [10, 20]. Further studies of the group led to the design of compound **1400W** (**Figure 3**), which is highly selective over eNOS and nNOS and able to penetrate into cells and tissues [10, 21].

The selective iNOS inhibition by aminoguanidine (**Figure 3—AG**) showed that NO can mediate the disruption of hematopoiesis during acute graft-versus-host disease (GVHD), also decreasing the endogenous bacterial infections in the spleen and liver in mice receiving the inhibitor [22]. The deleterious neuro-inflammation effects of iNOS/NO system stimulated by lipopolysaccharide (LPS) on learning and memory were evaluated in rats. Aminoguanidine decreases TNFα levels, oxidative stress indicators, and NO metabolites [23]. In addition, this compound seems to significantly relieve periapical inflammation in the canine teeth of cats and to reduce histological multiple organ damage in rats [24, 25].

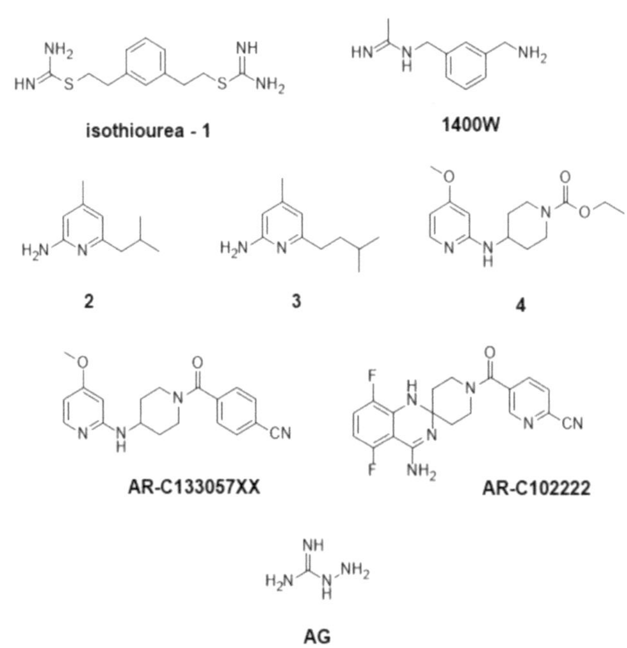

Figure 3. iNOS inhibitors (part 1).

http://dx.doi.org/10.5772/67027

In 2000, Hagmann and collaborators explored the structure-activity relationships of a series of substituted 2-aminopyridines. Compounds 4,6-dialkyl substituted (**Figure 3—2** and **3**) were found to be the most potent inhibitors of iNOS, presenting a significant degree of selectivity for this isoform [26]. Exploring the same scaffold, the synthesis of the derivatives *N*-4-piperidinyl-2-aminopyridine led to compound **4** (**Figure 3**), a 4-methoxy substituted derivative, 4-fold more potent in iNOS inhibition compared to the 4-methyl compound. Moreover, 4-cyanobenzamide derivative (**Figure 3—AR-C133057XX**) presented IC_{50} = 0.071 μM, being 1400-fold and around 100-fold selective for eNOS and nNOS, respectively. X-ray crystallography of **AR-C133057XX** showed that pyridine moiety binds deeply to the heme pocket, while the exocyclic ring interacts with another binding pocket. This difference in interaction could explain the good selectivity of this molecule [27].

On the other hand, the 1,2-dihydro-4-quinazolinamine compound **AR-C102222** (**Figure 3**) showed a dose-dependent inhibition on NO production induced by lipopolysaccharide (LPS). At the highest dose tested (100 μmol/kg) in rats, this compound completely abolished the chronic inflammatory arthritis development all over the 20-day experiment, thus confirming the in vivo efficacy of the class [28].

Structural-based approach using crystal structure and mutagenesis have identified specific induced-fit binding mode, which can generate some conformational changes toward a new specific cavity. Garcin and coworkers showed the *cis*-amidine moiety of quinazoline and aminopyridine chemotypes in the compounds **AR-C133057XX** and **AR-C102222**, respectively, promoted interactions of hydrogen with Glu (Glu371 and Glu377, mouse and human, respectively) at the binding site and with the heme group (mimicking the L-arginine substrate) (**Figure 4**). Those interactions increased the affinity of the inhibitor-containing bulky groups for iNOS. This occurs when Gln rotates on its own axis, accommodating the rigid bulky moieties of the inhibitors and exposing a new specific pocket to interact. The Gln-open conformation can create a cascade of conformational changes, leading to the generation of this new interaction site and directing the selectivity to the aminopyridine and quinazoline scaffolds [29]. Quantitative structure-activity relationships (QSAR) of quinazoline derivatives have been performed to evaluate the structural features required to interact with the active site of iNOS, allowing the design of more effective inhibitors [30].

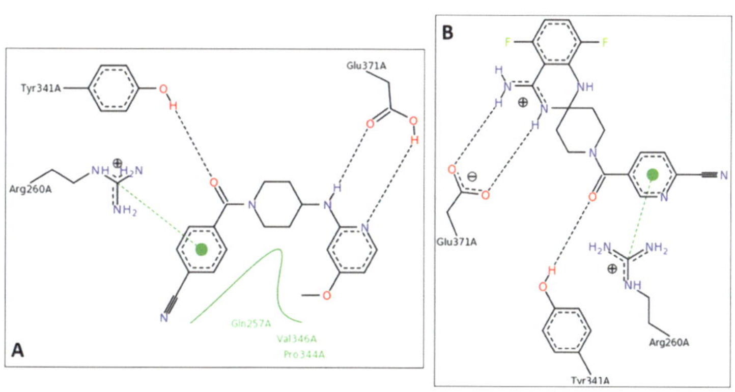

Figure 4. (A) iNOS-binding profile of **AR-C133057XX**, *PDB code*: 3EAI; (B) iNOS-binding profile of **AR-C102222**, *PDB code*: 3E7T.

Other compounds, as acetamidine derivatives (**Figure 5**—**5** and **6**), designed to inhibit iNOS, showed submicromolar activities (IC_{50} = 0.428 and 0.165 μM, respectively) and excellent selectivity over eNOS (>2300 and 550-fold more selective, respectively). In silico findings revealed that the activity drastically changes when ending amino groups are located instead of carboxylic function in the acceptor H-bond region, which is adjacent to the lipophilic region. This occurs because the charged amino group and its alkyl chain are not able to be stabilized inside the pocket interaction region of the enzyme, decreasing the binding efficiency [31].

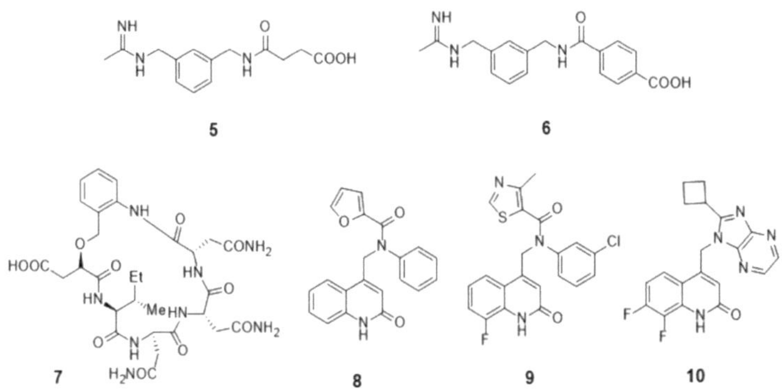

Figure 5. iNOS inhibitors (part 2).

A protein called SPSB2 plays an important role in modulating the activity of iNOS through its proteasomal degradation in defense cells. Since this complex is blocked, the NO production from iNOS is prolonged, increasing the killing activity against pathogen microorganisms, making it an interesting anti-infective target [32]. Some cyclic peptidomimetic compounds were designed using this strategy, and the most potent compound **7** (**Figure 5**) showed strong inhibition of SPSB2-iNOS complex in macrophage cell lysates and potent affinity value (K_D = 29 nM) [33].

High-throughput screening (HTS) strategy has been used to identify new iNOS inhibitors hits such as the compound **8** (**Figure 5**). After structural optimization process, the quinolone amide derivative **9** (**Figure 5**) was generated. This derivative has 2000-fold selectivity over human eNOS, besides being very potent (iNOS IC_{50} = 11 nM) and presenting oral bioavailability in mouse although its clearance (Clp > 100 mL/min/kg) has shown to be uninteresting [34]. The efforts to overcome this effect was to improve the pharmacokinetic properties, leading to the compound **10** (**Figure 5**), a 4,7-imidazopyrazine derivative. This is a dual iNOS/nNOS inhibitor, showing high potency in human iNOS (IC_{50} = 0.091 μM) and activity over nNOS (0.30 μM) while maintaining the desired selectivity over eNOS (180-fold). Its mechanism of action at molecular level is based on the inhibition of the iNOS dimerization process. Furthermore, compound **10** was effective using in vivo models of neuropathic pain, presenting desired clearance value (4–9 mL/min/kg), good oral bioavailability, and no tolerance after repeated doses [35].

Phenylpyrroles, pyrazoles, urea kynurenamines, ethynylcyanodienones, and amidine derivatives (**Figure 6**—**11**, **12**, **13**, **14**, and **15**) have also been interesting scaffolds to generate iNOS inhibitors [36–40]. The first derivative reduced significantly the iNOS activity to control values in MPTP-Parkinson`s disease model, showing a potential to act in central nervous system (CNS) disorders [36].

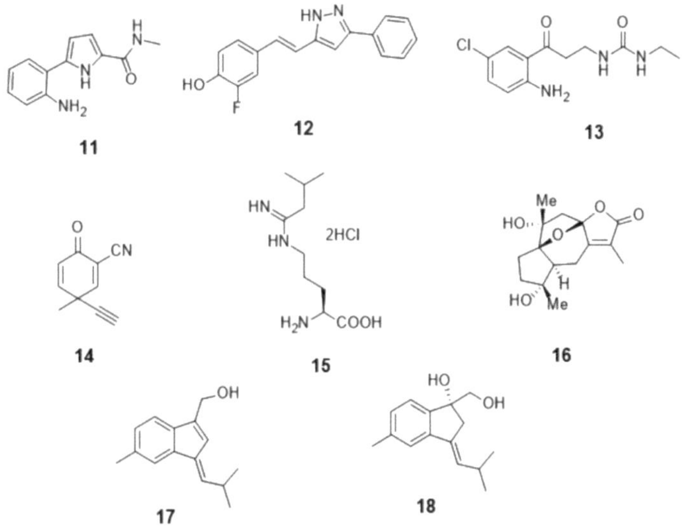

Figure 6. iNOS inhibitors (part 3).

Natural products have been a rich source of new bioactive molecules. Some examples in the NOS inhibition are a sesquiterpenoid, isolated from *Curcuma wenyujin* (**Figure 6**—**16**) and its isomer. They were strong inhibitors of NO production by LPS in iNOS (IC_{50} = 7.6 and 8.5 μM, respectively). Anmindenols A and B (**Figure 6**—**17** and **18**), from marine-derived bacterium *Streptomyces* sp., also demonstrated a relevant inhibitory activity in macrophage cells NO production (IC_{50} = 23 and 19 μM, respectively) [41, 42].

2.2. Neuronal nitric oxide synthase (nNOS) inhibitors

In the beginning of the 1990s, efforts to design selective nNOS inhibitor compounds were addressed, using the substrate L-arginine as the prototype molecule. Series of analogs was synthesized to evaluate which molecular change could interfere in the ligand activity and selectivity over other isoforms. The first selective compound over nNOS was L-nitroarginine (**Figure 7**), producing hypertension in animals due to the lack of selectivity over eNOS. In addition, many peptide analogs were synthesized trying to obtain more promising compounds. After the X-ray crystal complex elucidation, structure-activity relationship findings of several scaffolds have been explored to identify the molecular basis of improving the selectivity toward neuronal isoform [15, 19].

The non-arginine-based compound 7-nitroindazole (**Figure 7**—**NI**) showed little nNOS in vitro selectivity but high in vivo selectivity. Its mechanism of action involves competitive inhibition of H_4B cofactor, and series of related structures have been designed [11]. This compound has suppressed open-field behavior expressed as distance moved, exploratory rearing and grooming, suggesting that this compound can increase cortical excitability and interfere with some physiological and behavioral parameters [43]. Nevertheless, its anticonvulsant activity should be better understood, since studies in rodents reveal a beneficial activity although proconvulsant effect can be found in kainite-, nicotine- and soman-induced convulsions [44].

Entrena and collaborators, by using kynurenamine scaffold (**Figure 7—19**) as a template, carried out the synthesis of a series of new candidates to neuroprotective compounds, showing a pharmacophore model to interact with nNOS catalytic site [45].

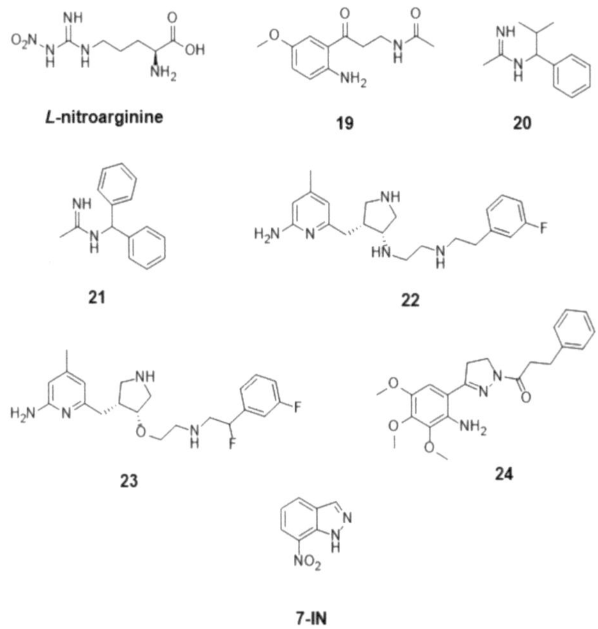

Figure 7. nNOS inhibitors (part 1).

N-Substituted acetamidines (**Figure 7—20** and **21**) showed nNOS inhibition activity (IC_{50} = 0.2 and 0.3 μM) with good selectivity index (500 and 1166-fold selectivity over eNOS, respectively, and 50 and 100-fold, over iNOS, respectively). In silico studies were useful to understand the fit of these scaffolds inside the catalytic site. nNOS contains a bigger heme group cavity compared with other isoforms, which could explain why bulky groups better accommodate in the neuronal isoform [46].

Aminopyridine is an attractive pharmacophoric group to bind in different regions of nNOS through H-bond. Using this moiety, compound **22** (**Figure 7**) and its optical isomer showed to be selective against nNOS, although unable to penetrate the blood-brain-barrier (BBB) in in vivo studies [15, 47, 48]. Trying to increase the CNS permeability, prodrug design approach was used in primary and secondary amines. However, this strategy did not increase the BBB penetration, even masking the charge by carbamate and azide functions [49]. On the other hand, using these compounds containing basic nitrogen, Xue and coworkers attached electron-withdrawing groups (**Figure 7—23**) close to these amine functions, decreasing their p*K*a values and improving the membrane permeability in cell-based assays [50].

Concerning 4,5-dihydro-1-*H*-pyrazole derivatives, they were confirmed as selective nNOS inhibitors. Compound **24** (**Figure 7**), the most active of the derivatives (82% of inhibition),

http://dx.doi.org/10.5772/67027

showed that its methoxy electron-donating group is important to improve potency and selectivity. By molecular modeling, it was possible to identify that the phenyl moiety can fit below the heme group, establishing π-π interaction. The methoxy groups adopt a conformation that allows them to interact with Arg481 by H-bonds. Moreover, the amine group interacts by H-bond with one of the carboxylate moieties of the heme group. On the contrary, electron-withdrawing groups are better to generate inhibitors for iNOS [51].

Other interesting structures such as 2-aminoquinolines are effective scaffold to be included in the structure of nNOS inhibitors. Crystallography studies showed that those compounds act as competitive arginine mimics. This scaffold makes important H-bonds with the active-site Glu residue, and the non-coordinating aryl rings are stabilized in a hydrophobic pocket in the extremity of the substrate access channel. Moreover, this structural class showed good pharmacokinetic properties (**Figure 8—25**), such as brain penetration and oral bioavailability according to the permeability results in Caco-2 cell assay [52]. Trying to optimize this class of compounds, chlorine was added on the phenyl ether central aryl ring (**Figure 8—26**). This substitution was found to be selective and highly potent in the design of nNOS inhibitors while retaining CNS penetration and showing a diminished off-target interaction. In complex with human nNOS, this compound showed a phenyl ring orientation where the alkyl amine makes an H-bond with the H_4B site [53].

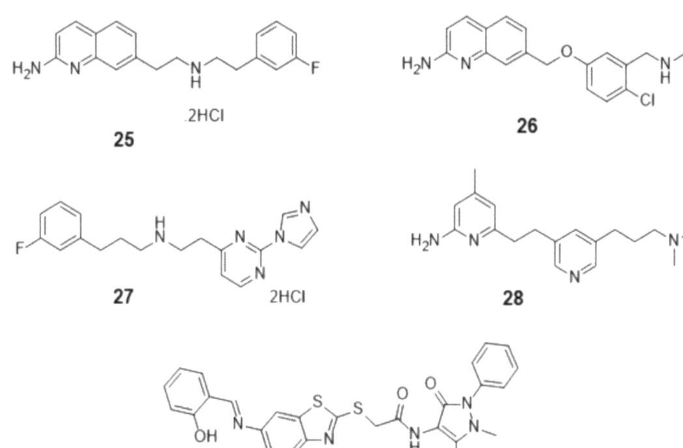

Figure 8. nNOS inhibitors (part 2).

Exploring the heme-coordinating potential of imidazole group, a series of 2,4-disubtituted pyrimidine compounds (**Figure 8—27**) was designed. They presented a nanomolar affinity to both rat and human nNOS (>200-fold and >100-fold selectivity over eNOS and iNOS, respectively), exhibiting a minimal off-target binding to 50 CNS receptors. Crystal structures of the complex (nNOS-**27**) indicate that heme Fe coordinates by the 2-imidazolyl group, and the non-coordinating aryl rings are stabilized in a hydrophobic pocket at the far end of the substrate access channel. The fluorine atom in compound **27** also interacts into the hydrophobic pocket, and the secondary amine of this derivative establishes dual ionic interaction with both heme propionates (**Figure 9**). This molecular interaction profile is important to obtain potency and selectivity. In addition, nitrogen from pyrimidine ring performs an H-bond with the His342 side chain. The imidazole ring of the most active compound acts as a weak CYP3A4 inhibitor, suggesting that modulating hydrophobicity and bulkiness can be useful to attenuate the effects in CYP isoforms [54].

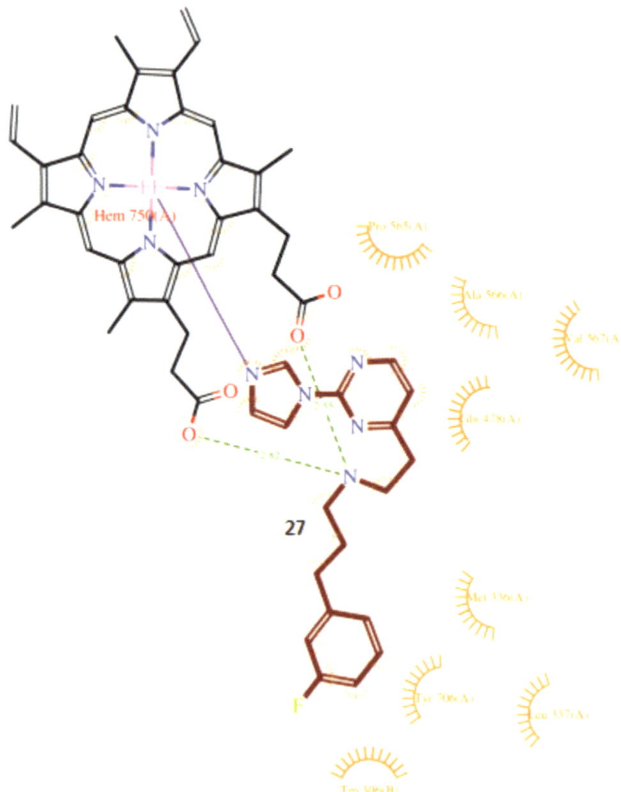

Figure 9. Rat nNOS-binding profile of **27**, *PDB code*: 4V3X.

Studies using aminopyridine-based scaffold with pyridine linker (**Figure 8—28**) showed that difference in the position of an amino acid, Asp597 of nNOS versus Asn368 of eNOS, controls the affinity and binding mode of this class of nNOS inhibitors. While the central pyridine is at least partially protonated to points up toward Tyr562 for optimal electrostatic interactions, the Asp597 provides additional and important electrostatic stabilization to the other part of the inhibitor [55–57].

Rational strategy for identifying new nNOS inhibitors using a combination of virtual screening approach based on 3D pharmacophore model and molecular docking was able to identify a hit compound structurally different from the available inhibitors (**Figure 8—29**). This strategy can be useful to design novel optimized analogs [58].

http://dx.doi.org/10.5772/67027

2.2.1. Double-headed nNOS inhibitors

Double-headed compounds have been explored by researches with the aim of obtaining high affinity binding in nNOS. Attaching a double-headed aminopyridine moiety in a compound led to a very potent (K_i = 25 nM) and selective (107-fold selective over nNOS and eNOS) compound (**Figure 10**—**30**). Both aminopyridine moieties interact in different positions with the enzyme—Glu592 in the active site and the heme group. According to the X-ray crystals, there is a second **30** molecule binding also in the H_4B site, specifically, and the pyridine moiety coordinates with the Zn atom. This interaction does not take place in eNOS. Cellular permeability studies confirmed compound **30** as an interesting lead [59, 60].

Other symmetric double-headed aminopyridine series without charge groups were designed to contain a tail on the central aromatic ring. The objective is to achieve an interaction in the electronegative region in the catalytic site, since only the neuronal isoform has Asp597 in this region. Derivative **31** (**Figure 10**) was very potent (K_i = 56 nM) and highly selective over other isoforms (472-fold selective for eNOS and 239-fold for iNOS). This occurs because an electropositive functional group (Ciano group) is preferred near Asp597 in nNOS, explaining the selectivity of this compound [61].

Furthermore, double-headed inhibitors containing chiral linkers derived from natural amino acids were designed and synthesized. The best compound (**Figure 10**—**32**) showed high potency (K_i = 32 nM) against nNOS and a good selectivity profile (475-fold selective and 244-fold over eNOS and iNOS, respectively). The aminomethyl moiety was crucial in this compound, allowing it to bind to the heme propionates in nNOS and leading to a high selectivity level [62].

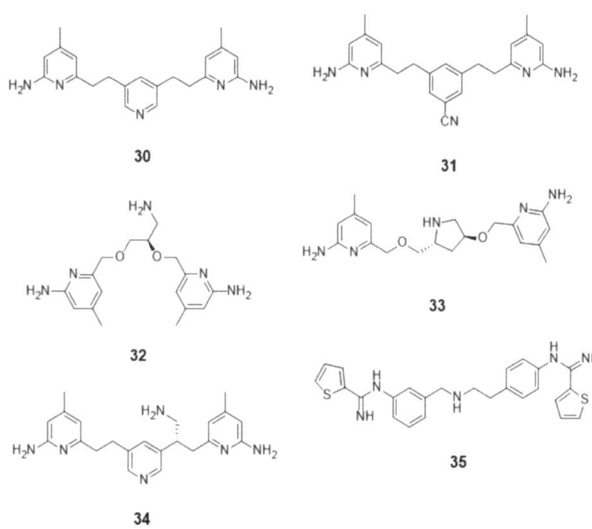

Figure 10. Double-headed nNOS inhibitors.

2-Amino-4-methylpyridine groups with a chiral linker derived from proline were designed as selective nNOS inhibitors. They showed to be interesting as they can interact in a unique orientation, what led to selectivity toward neuronal isoform. The aminopyridine groups interact with a Glu592 residue and the heme propionate in nNOS active site. In addition, the nitrogen from pyrrolidine linker is important to contribute to additional hydrogen bonds to the heme propionate, resulting in the most potent compound (K_i = 9.7 nM) (**Figure 10—33**). The finding that the isomer activities are different also reinforces the importance of chirality control of this kind of inhibitors and shows the dynamism of the target [63].

In addition, using chiral double-headed inhibitors, the α-amino-functionalized aminopyridine derivative **34** (**Figure 10**) was more potent than other chiral compounds (K_i value of 24 nM for nNOS, with 273-fold and 2822-fold selectivity against iNOS and eNOS, respectively). Structure-activity relationship studies reveal that the α-amino group close to the center phenyl ring is crucial to stabilize the double-headed binding. Those studies also showed that by changing to aminomethyl group the potency is improved. The inhibitor is able to make H-bonds with both the H_4B binding site and the propionate of the heme A-ring, which is essential to obtain selectivity over other isoforms. It is also important to note that the distinct electrostatic environments in different isoforms resulted in lower binding free energy in nNOS, which also can explain the potency difference [64].

The non-chiral double-headed thiophene-2-carboximidamide compound (**Figure 10—35**) exhibited an excellent inhibitory potency and selectivity (K_i = 5 nM; 540-fold and 340-fold selective over eNOS and iNOS, respectively). This compound also showed to be active in metastatic melanoma A375 cells, exhibiting EC_{50} values of 1.3 μM, better that of the drug cisplatin (EC_{50} = 4.2 μM) [65].

2.3. Bacterial nitric oxide synthase (bNOS) inhibitors

Bacterial nitric oxide synthase (bNOS) is present in many Gram-positive microorganisms and has been described as part of their defense system against other species and the oxidative stress provoked by antibiotics through NO releasing. Therefore, bNOS inhibition can increase the antibiotic potential and be harmful to bacterial cell [66].

A screening showed that some known nNOS inhibitors can decrease significantly the percent survival of *Bacillus subtilis* WT treated with the antimicrobial acriflavine. This potential is consistent with NO production inhibition, as it decreases the bacterial resistance against that compound [67].

Exploring the potential of bNOS as a drug target, high selectivity levels are necessary to its inhibitors. In this context, the design of compounds that target the active and pterin-binding site has been considered an important strategy (**Figure 11—36**). These compounds are able to carry out an unexpected rotameric position of residue Arg247 in the active site, besides interacting with the important residue of Glu243 from the same site [68].

In addition, with the goal to identify the differences among bNOS and other isoforms, crystallography studies were performed using different inhibitor chemotypes. Researchers observed that Tyr706 from nNOS is conserved in bNOS (Tyr 357) and both have the same rotameric behavior, which is very different, compared with eNOS. This molecular feature can be useful to design new selective bNOS over eNOS inhibitor. Since the pharmacokinetic properties are very different between bNOS and nNOS, selectivity over the latter is not a trouble. However,

http://dx.doi.org/10.5772/67027

due to steric hindrance in the tail end of thiophenecarboximidamide analogs, this scaffold can bind differently to bNOS comparatively to nNOS [69].

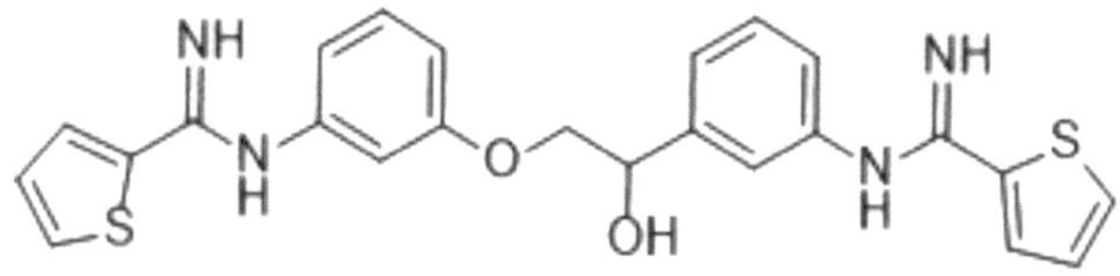

36

Figure 11. bNOS inhibitor.

3. Clinical studies

A nonselective compound L-NMMA (**Figure 12**), also known as tilarginine, was evaluated clinically in Translational Research Investigating Underlying Disparities in Acute Myocardial Infarction Patients' Health Status (TRIUMPH) study in North America and Europe with planned enrollment of 658 patients at 130 centers. The period of study was between January 2005 and August 2006 (the study was terminated early). Using 1 mg/kg bolus and 5 h infusion did not decrease the mortality rates in patients with refractory cardiogenic shock complicating myocardial infarction despite an open infarct artery. Although the good results showed in phase II, it has failed in phase III [70, 71]. In another study L-NMMA resulted in no differences in mean arterial pressure (MAP) after 2 h compared with placebo group [72].

Evaluating another inhibitor, N(G)-nitro-L-arginine methyl ester (**Figure 12**—L-NAME), in the treatment of refractory cardiogenic shock, the death at 1 month was 27% in the L-NAME group versus 67% in the control group [73]. Additional studies have been performed to further examination, concluding that TRIUMPH strongly indicated that nonselective NOS inhibitors are not clinically interesting [74].

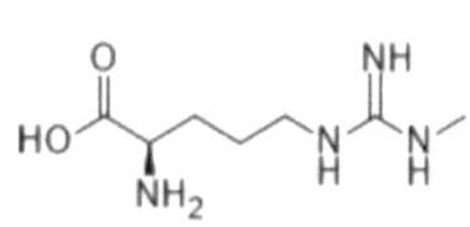

***L*-NMMA**

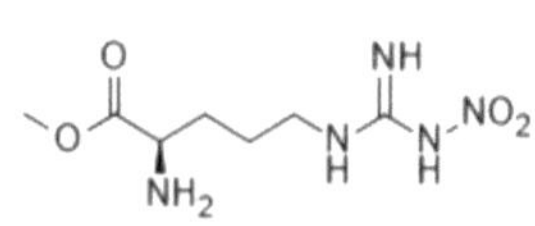

***L*-NAME**

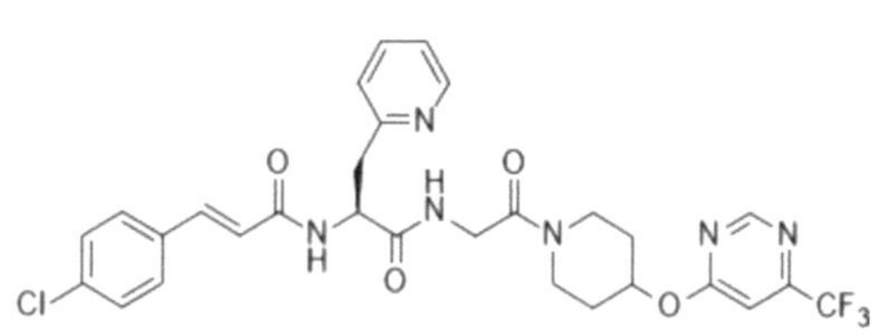

ASP9853

Figure 12. Clinically evaluated compounds.

Recent phase I study in advanced solid tumors with the iNOS inhibitor **ASP9853** (**Figure 12**) showed that the efficacy dose predicted in preclinical studies was not achieved due to overall toxicity limitations. In summary all these clinical information showed that the manipulation of the NOS pathway, with or without chemotherapy, appears to be more challenging than expected [75]. While designing new selective NOS inhibitors which should be highlighted, deeply studies to evaluate clinical benefits are also required.

4. Conclusions

Many scaffolds have been found to inhibit nitric oxide synthases. Some of them were presented in this chapter as promising for important therapeutic activity. It must be emphasized that the research about nitric oxide synthase inhibitors has expressively advanced thanks to the X-ray crystallographic studies of this enzyme. This helps the structure-based design approach toward the search for selective inhibitors of this enzyme and the comprehension of their mechanism of action. Notwithstanding, efforts have been made for imparting them with a drug-like profile.

Ackowledgements

The authors thank CAPES, for RAM Serafim scholarship, and CNPq, for EI Ferreira fellowship.

Author details

Elizabeth Igne Ferreira* and Ricardo Augusto Massarico Serafim

*Address all correspondence to: elizabeth.igne@gmail.com

LAPEN, Department of Pharmacy, Faculty of Pharmaceutical Sciences, University of Sao Paulo – FCF/USP, Sao Paulo, Brazil

References

[1] Kerwin JF, Lancaster JR, Feldman PL. Nitric oxide: a new paradigm for second messengers. Journal of Medicinal Chemistry. 1995;**38**:4343–4362. DOI: 10.1021/jm00022a001

[2] Al-sadoni HH, Ferro A. S-nitrosothiols: a class of nitric oxide-donor drugs. Clinical Science. 2000;**98**:507–520. DOI: 10.1042/cs0980507

[3] Daff S. NO synthase: structures and mechanisms. Nitric Oxide. 2010;**23**:1–11. DOI: 10.1016/j.niox.2010.03.001

[4] Serafim RAM, Primi MC, Trossini GHG, Ferreira EI. Nitric oxide: state of the art in drug design. Current Medicinal Chemistry. 2012;**19**:386–405. DOI: 10.2174/092986712803414321

[5] Hobbs AJ, Higgs A, Moncada S. Inhibition of nitric oxide synthase as a potential therapeutic target. Annual Review of Pharmacology and Toxicology. 1999;**39**:191–220. DOI: 10.1146/annurev.pharmtox.39.1.191

[6] Rehni AK, Singh TG, Kalra R, Singh N. Pharmacological inhibition of inducible nitric oxide synthase attenuates the development of seizure in mice. Nitric Oxide. 2009;**21**: 120–125. DOI: 10.1016/j.niox.2009.06.001

[7] Kavya R, Saluja R, Singh S, Dikshit M. Nitric oxide synthase regulation and diversity: implications in Parkinson's disease. Nitric Oxide. 2006;**15**:280–294. DOI: 10.1016/j.niox.2006.07.003

[8] Zheng B, Zheng T, Wang L, Chen X, Shi C, Zhao S. Aminoguanidine inhibition of iNOS activity ameliorates cerebral vasospasm after subarachnoid hemorrhage in rabbits via restoration of dysfunctional endothelial cells. Journal of the Neurological Sciences. 2010;**295**:97–103. DOI: 10.1016/j.jns.2010.04.012

[9] Ricciardolo FLM, Nijkamp FP, Folkerts G. Nitric oxide synthase (NOS) as therapeutic target for asthma and chronic obstructive pulmonary disease. Current Drug Targets. 2006;**7**:721–735. DOI: 10.2174/138945006777435290

[10] Alderton WK, Cooper CE, Knowles RG. Nitric oxide synthases: structure, function and inhibition. Biochemical Journal. 2001;**357**:593–615. DOI: 10.1042/bj3570593

[11] Mukherjee P, Cinelli MA, Kang S, Silverman RB. Development of nitric oxide synthase inhibitors for neurodegeneration and neuropathic pain. Chemical Society Reviews. 2014;**43**:6814–6838. DOI: 10.1039/c3cs60467e

[12] Vallance P, Leiper J. Blocking NO synthesis: how, where and why? Nature Review Drug Discovery. 2002;**1**:939–950. DOI: 10.1038/nrd960

[13] Vallance, P. Nitric oxide: therapeutic opportunities. Fundamental and Clinical Pharmacology. 2003;**17**:1–10. DOI: 10.1046/j.1472-8206.2003.00124.x

[14] Yang Z, Misner B, Ji H, Poulos TL, Silverman RB, Meyskens FL, Yang S. Targeting nitric oxide signaling with nNOS inhibitors as a novel strategy for the therapy and prevention of human melanoma. Antioxidants and Redox Signaling. 2013;**19**:433–447. DOI: 10.1089/ars.2012.4563

[15] Silverman RB. Design of selective neuronal nitric oxide synthase inhibitors for the prevention and treatment of neurodegenerative diseases. Accounts of Chemical Research. 2009;**42**:439–451. DOI: 10.1021/ar800201v

[16] Annedi SC. Cell-permeable inhibitors of neuronal nitric oxide synthase open new prospects for the treatment of neurological disorders. Journal of Medicinal Chemistry. 2015;**58**:1064–1066. DOI: 10.1021/acs.jmedchem.5b00057

[17] Paige J, Jaffrey SR. Pharmacologic manipulation of nitric oxide signaling: targeting NOS dimerization and protein-protein interactions. Current Topics in Medicinal Chemistry. 2007;**7**:97–114. DOI: 10.2174/156802607779318253

[18] Groves JT, Wang CCY. Nitric oxide synthase: models and mechanism. Current Opinion in Chemical Biology. 2000;**4**:687–695. DOI: 10.1016/S1367-5931(00)00146-0

[19] Poulos TL, Li H. Structural basis for isoform-selective inhibition in nitric oxide synthase. Accounts of Chemical Research. 2013;**46**:390–398. DOI: 10.1021/ar300175n

[20] Garvey EP, Oplinger JA, Tanoury GJ, Sherman PA, Fowler M, Marshall S, Harmon MF, Paith JE, Furfine ES. Potent and selective inhibition of human nitric oxide synthases. Inhibition by non-amino acid isothioureas. The Journal of Biological Chemistry. 1994;**269**:26669–26676.

[21] Garvey EP, Oplinger JA, Furfine ES, Kiff RJ, Laszlo F, Whittle BJ, Knowles RG. 1400W is a slow, tight binding, and highly selective inhibitor of inducible nitric oxide synthase in vitro and in vivo. The Journal of Biological Chemistry. 1997;**272**:4959–4963.

[22] Hoffman RA, Nussler NC, Gleixner SL, Zhang G, Ford HR, Langrehr JM, Demetris AJ, Simmons RL. Attenuation of lethal graft-versus-host disease by inhibition of nitric oxide synthase. Transplantation. 1997;**63**:94–100.

[23] Anaeigoudari A, Soukhtanloo M, Reisi P, Beheshti F, Hosseini M. Inducible nitric oxide inhibitor aminoguanidine, ameliorates deleterious effects of lipopolysaccharide on memory and long term potentiation in rat. Life Sciences. 2016;**158**:22–30. DOI: 10.1016/j.lfs.2016.06.019

[24] Farhad AR, Razavi S, Jahadi S, Saatchi M. Use of aminoguanidine, a selective inducible nitric oxide synthase inhibitor, to evaluate the role of nitric oxide in periapical inflammation. Journal of Oral Science. 2011;**53**:225–230.

[25] Soliman MM. Effects of aminoguanidine, a potent nitric oxide synthase inhibitor, on myocardial and organ structure in a rat model of hemorrhagic shock. Journal of Emergencies, Trauma and Shock. 2014;7:**190**–195. DOI: 10.4103/0974–2700.136864

[26] Hagmann WK, Caldwell CG, Chen PL, Durette PL, Esser CK, Lanza TJ, Kopka IE, Guthikonda R, Shah SK, Maccoss M, Chabin RM, Fletcher D, Grant SK, Green BG, Humes JL, Kelly TM, Luell S, Meurer R, Moore V, Pacholok SG, Pavia T, Williams HR, Wong KK. Substituted 2-aminopyridines as inhibitors of nitric oxide synthases. Bioorganic and Medicinal Chemistry. 2000;**10**:1975–1978. DOI: 10.1016/S0960-894X(00)00389-9

[27] Connolly S, Aberg A, Arvai A, Beaton HG, Cheshire DR, Cook AR, Cooper S, Cox D, Hamley P, Mallinder P, Millichip I, Nicholls DJ, Rosenfeld RJ, St-Gallay SA, Tainer J, Tinker AC, Wallace AV. 2-aminopyridines as highly selective inducible nitric oxide synthase inhibitors. Differential binding modes dependent on nitrogen substitution. Journal of Medicinal Chemistry. 2004;**47**:3320–3323. DOI: 10.1021/jm031035n

[28] Tinker AC, Beaton HG, Boughton-Smith N, Cook TR, Cooper SL, Fraser-Rae L, Hallam K, Hamley P, Mcinally T, Nicholls DJ, Pimm AD, Wallace AV. 1,2-Dihydro-4-quinazolinamines: potent, highly selective inhibitors of inducible nitric oxide synthase which show antiinflammatory activity in vivo. Journal of Medicinal Chemistry. 2003;**46**:913–916. DOI: 10.1021/jm0255926

[29] Garcin ED, Arvai AS, Rosenfeld RJ, Kroeger MD, Crane BR, Andersson G, Andrews G, Hamley PJ, Mallinder PR, Nicholls DJ, St-Gallay SA, Tinker AC, Gensmantel NP, Mete A, Cheshire DR, Connolly S, Stuehr DJ, Aberg A, Wallace AV, Tainer JA, Getzoff ED. Anchored plasticity opens doors for selective inhibitor design in nitric oxide synthase. Nature Chemical Biology. 2008;**4**:700–707. DOI: 10.1038/nchembio.115

[30] Sharma MC, Sharma S. Investigation on quantitative structure activity relationships of a series of inducible nitric oxide. Interdisciplinary Sciences. DOI: 10.1007/s12539-016-0176-5

[31] Maccallini C, Montagnani M, Paciotti R, Ammazzalorso A, De Filippis B, Di Matteo M, Di Silvestre S, Fantacuzzi M, Giampietro L, Potenza MA, Re N, Pandolfi A, Amoroso R. Selective acetamidine-based nitric oxide synthase inhibitors: synthesis, docking, and biological studies. ACS Medicinal Chemistry Letters. 2015;**6**:635–640. DOI: 10.1021/acsmedchemlett.5b00149

[32] Yap BK, Harjani JR, Leung EW, Nicholson SE, Scanlon MJ, Chalmers DK, Thompson PE, Baell JB, Norton RS. Redox-stable cyclic peptide inhibitors of the SPSB2-iNOS interactions. FEBS Letters. 2016;**590**:696–704. DOI: 10.1002/1873-3468.12115

[33] Harjani JR, Yap BK, Leung EWW, Lucke A, Nicholson SE, Scanlon MJ, Chalmers DK, Thompson PE, Norton RS, Baell JB. Design, synthesis, and characterization of cyclic peptidomimetics of the inducible nitric oxide synthase binding epitope that disrupt the protein–protein interaction involving SPRY domain-containing suppressor of cytokine signaling box protein (SPSB) 2 and inducible nitric oxide synthase. Journal of Medicinal Chemistry. 2016;**59**: 5799–5809. DOI: 10.1021/acs.jmedchem.6b00386

[34] Bonnefous C, Payne JE, Roppe J, Zhuang H, Chen X, Symons KT, Nguyen PM, Sablad M, Rozenkrants N, Zhang Y, Wang L, Severance D, Walsh JP, Yazdani N, Shiau AK, Noble SA, Rix P, Rao TS, Hassig CA, Smith ND. Discovery of inducible nitric oxide synthase (iNOS) inhibitor development candidate KD7332, part 1: identification of a novel, potent, and selective series of quinolinone iNOS dimerization inhibitors that are orally active in rodent pain models. Journal of Medicinal Chemistry. 2009;**52**:3047–3062. DOI: 10.1021/jm900173b

[35] Payne JE, Bonnefous C, Symons KT, Nguyen PM, Sablad M, Rozenkrants N, Zhang Y, Wang L, Yazdani N, Shiau AK, Noble SA, Rix P, Rao TS, Hassig CA, Smith ND. Discovery of dual inducible/neuronal nitric oxide synthase (iNOS/nNOS) inhibitor development candidate 4-((2-cyclobutyl-1H-imidazo[4,5-b]pyrazin-1yl)methyl)-7,8-difluoroquinolin-2(1H)-one (KD7332) part 2: identification of a novel, potent, and selective series of benzimidazole-quinolinone iNOS/nNOS dimerization inhibitors that are orally active in pain models. Journal of Medicinal Chemistry. 2010;**53**:7739–7755. DOI: 10.1021/jm100828n

[36] Cara LCL, Camacho ME, Carrión MD, Tapias V, Gallo MA, Escames G, Acuña-Castroviejo D, Espinosa A, Entrena A. Phenylpyrrole derivatives as neuronal and inducible nitric oxide synthase (nNOS and iNOS) inhibitors. European Journal of Medicinal Chemistry. 2009;**44**:2655–2666. DOI: 10.1016/j.ejmech.2008.11.013

[37] Nieto CI, Cabildo MP, Cornago MP, Sanz D, Claramunt RM, Torralba MC, Torres MR, Elguero J, García JA, López A, Acuña-Castroviejo D. Fluorination effects on NOS inhibitory activity of pyrazoles related to curcumin. Molecules. 2015;**20**:15643–15665. DOI: 10.3390/molecules200915643

[38] Chayah M, Carrión MD, Gallo MA, Jiménez R, Duarte J, Camacho ME. Development of urea and thiourea kynurenamine derivatives: synthesis, molecular modeling, and biological evaluation as nitric oxide synthase inhibitors. ChemMedChem. 2015;**10**:874–882. DOI: 10.1002/cmdc.201500007

[39] Li W, Zheng S, Higgins M, Morra Jr RP, Mendis AT, Chien CW, Ojima I, Mierke DF, Dinkova-Kostova AT, Honda T. New monocyclic, bicyclic, and tricyclic ethynylcyanodienones as activators of the Keap1/Nrf2/ARE pathway and inhibitors of inducible nitric oxide synthase. Journal of Medicinal Chemistry. 2015;**58**:4738–4748. DOI: 10.1021/acs.jmedchem.5b00393

[40] Tang W, Li H, Poulos TL, Silverman RB. Mechanistic studies of inactivation of inducible nitric oxide synthase by amidines. Biochemistry. 2015;**54**:2530–2538. DOI: 10.1021/acs.biochem.5b00135

[41] Yin GP, Li LC, Zhang QZ, An YW, Zhu JJ, Wang ZM, Chou GX, Wang ZT. iNOS inhibitory activity of sesquiterpenoids and a monoterpenoid from the rhizomes of *Curcuma wenyujin*. Journal of Natural Products. 2014;**77**:2161–2169. DOI: 10.1021/np400984c

[42] Lee J, Kim H, Lee TG, Yang I, Won DH, Choi H, Nam SJ, Kang H. Anmindenols A and B, inducible nitric oxide synthase inhibitors from a marine-derived *Streptomyces* sp. Journal of Natural Products. 2014;**77**:1528–1531. DOI: 10.1021/np500285a

[43] Brozíckova C, Mikulecka A, Otáhal J. Effect of 7-nitroindazole, a neuronal nitric oxide synthase inhibitor, on behavior and physiological parameters. Physiological Research. 2014;**63**:637–648.

[44] Banach M, Piskorska B, Czuczwar SJ, Borowicz KK, Nitric oxide, epileptic seizures, and action of antiepileptic drugs. CNS Neurological Disorders Drug Discovery. 2011;**10**:808–819. DOI: 10.2174/187152711798072347.

[45] Entrena A, Camacho E, Carrión D, López-Cara LC, Velasco G, León J, Escames G, Acuña-Castroviejo D, Tapias V, Gallo MA, Vivo A, Espinosa A. Kynurenamines as neural nitric oxide synthase inhibitors. Journal of Medicinal Chemistry. 2005;**48**:8174–8181. DOI: 10.1021/jm050740o

[46] Maccallini C, Patruno A, Lannutti F, Ammazzalorso A, Filippis BD, Fantacuzzi M, Franceschelli S, Giampietro L, Masella S, Felaco M, Re N, Amoroso R. *N*-substituted acetamidine and 2-methylimidazole derivates as selective inhibitors of neuronal nitric oxide synthase. Bioorganic and Medicinal Chemistry Letters. 2010;**20**:6495–6499. DOI: 10.1016/j.bmcl.2010.09.059

[47] Delker SL, Ji H, Li H, Jamal J, Fang J, Xue F, Silverman RB, Poulos TL. Unexpected binding modes of nitric oxide synthase inhibitors effective in the prevention of cerebral palsy phenotype in an animal model. Journal of the American Chemical Society. 2010;**132**:5437–5442. DOI: 10.1021/ja910228a

[48] Lawton GR, Ranaivo HR, Wing LK, Ji H, Xue F, Martesek P, Roman LJ, Watterson DM, Silverman RB. Analogues of 2-aminopyridine-based selective inhibitors of neuronal nitric oxide synthase with increased bioavailability. Bioorganic and Medicinal Chemistry. 2009;**17**:2371–2380. DOI: 10.1016/j.bmc.2009.02.017

[49] Silverman RB, Lawton GR, Ranaivo HR, Chico LK, Seo J, Watterson DM. Effect of potential amine prodrugs of selective neuronal nitric oxide synthase inhibitors on blood-brain barrier penetration. Bioorganic and Medicinal Chemistry. 2009;**17**:7593–7605. DOI: 10.1016/j.bmc.2009.08.065

[50] Xue F, Fang J, Lewis WW, Martásek P, Roman LJ, Silverman RB. Potent and selective neuronal nitric oxide synthase inhibitors with improved cellular permeability. Bioorganic and Medicinal Chemistry Letters. 2010;**20**:554–557. DOI: 10.1016/j.bmcl.2009.11.086

[51] Carrión D, Chayah M, Entrena A, López A, Gallo MA, Acuña-Castroviejo D, Camacho ME. Synthesis and biological evaluation of 4,5-dihydro-1H-pyrazole derivatives as potential nNOS/iNOS selective inhibitors. Part 2: influence of diverse substituents in both the phenyl moiety and the acyl group. Bioorganic and Medicinal Chemistry. 2013;**21**:4132–4142. DOI: 10.1016/j.bmc.2013.05.016

[52] Cinelli MA, Li H, Chreifi G, Martásek P, Roman LJ, Poulos TL, Silverman RB. Simplified 2-aminoquinoline-based scaffold for potent and selective neuronal nitric oxide synthase inhibition. Journal of Medicinal Chemistry. 2014;**57**:1513–1530. DOI: 10.1021/jm401838x

[53] Cinelli MA, Li H, Pensa AV, Kang S, Roman LJ, Martásek P, Poulos TL, Silverman RB. Phenyl ether- and aniline-containing 2-aminoquinolines as potent and selective inhibitors of neuronal nitric oxide synthase. Journal of Medicinal Chemistry. 2015;**58**:8694–8712. DOI: 10.1021/acs.jmedchem.5b01330

[54] Mukherjee P, Li H, Sevrioukova I, Chreifi G, Martásek P, Roman LJ, Poulos TL, Silverman RB. Novel 2,4-disubstituted pyrimidines as potent, selective, and cell-permeable inhibitors of neuronal nitric oxide synthase. Journal of Medicinal Chemistry. 2015;**58**:1067-1088. DOI: 10.1021/jm501719e

[55] Li H, Wang HY, Kang S, Silverman RB, Poulos TL. Electrostatic control of isoform selective inhibitor binding in nitric oxide synthase. Biochemistry. 2016;**55**:3702–3707. DOI: 10.1021/acs.biochem.6b00261

[56] Kang S, Li H, Tang W, Martasek P, Roman LJ, Poulos TL, Silverman RB. 2-aminopyridines with a truncated side chain to improve human neuronal nitric oxide synthase inhibitory potency and selectivity. Journal of Medicinal Chemistry. 2015;**58**:5548–5560. DOI: 10.1021/acs.jmedchem.5b00573

[57] Wang HY, Qin Y, Li H, Roman LJ, Martasek P, Poulos TL, Silverman RB. Potent and selective human neuronal nitric oxide synthase inhibition by optimization of the 2-aminopyridine-based scaffold with a pyridine linker. Journal of Medicinal Chemistry. 2016;**59**:4913–4925. DOI: 10.1021/acs.jmedchem.6b00273

[58] Xu G, Chen Y, Shen K, Wang X, Li F, He Y. The discovery of potentially selective human neuronal nitric oxide synthase (nNOS) inhibitors: a combination of pharmacophore modelling, CoMFA, virtual screening and molecular docking studies. International Journal of Molecular Sciences. 2014;**15**:8553–8569. DOI: 10.3390/ijms15058553

[59] Xue F, Fang J, Delker SL, Li H, Martasek P, Roman LJ, Poulos TL, Silverman RB. Symmetric double-headed aminopyridines, a novel strategy for potent and membrane-permeable inhibitors of neuronal nitric oxide synthase. Journal of Medicinal Chemistry. 2011;**54**:2039–2048. DOI: 10.1021/jm101071n

[60] Delker SL, Xue F, Li H, Jamal J, Silverman RB, Poulos TL. Role of zinc in isoform-selective inhibitor binding to neuronal nitric oxide synthase. Biochemistry. 2010;**49**:10803–10810. DOI: 10.1021/bi1013479

[61] Huang H, Li H, Martásek P, Roman LJ, Poulos TL, Silverman RB. Structure-guided design of selective inhibitors of neuronal nitric oxide synthase. Journal of Medicinal Chemistry. 2013;**56**:3024–3032. DOI: 10.1021/jm4000984

[62] Jing Q, Li H, Chreifi G, Roman LJ, Martásek P, Poulos TL, Silverman RB. Chiral linkers to improve selectivity of double-headed neuronal nitric oxide synthase inhibitors. Bioorganic and Medicinal Chemistry Letters. 2013;**23**:5674–5679. DOI: 10.1016/j.bmcl.2013.08.034

[63] Jing Q, Li H, Roman LJ, Martásek P, Poulos TL, Silverman RB. Accessible chiral linker to enhance potency and selectivity of neuronal nitric oxide synthase inhibitors. ACS Medicinal Chemistry Letters. 2014;**5**:56–60. DOI: 10.1021/ml400381s

[64] Kang S, Tang W, Li H, Chreifi G, Martásek P, Roman LJ, Poulos TL, Silverman RB. Nitric oxide synthase inhibitors that interact with both heme propionate and tetrahydrobiopterin show high isoform selectivity. Journal of Medicinal Chemistry. 2014;**57**:4382–4396. DOI: 10.1021/jm5004182

[65] Huang H, Li H, Yang S, Chreifi G, Martásek P, Roman LJ, Meyskens FL, Poulos TL, Silverman RB. Potent and selective double-headed thiophene-2-carboximidamide inhibitors of neuronal nitric oxide synthase for the treatment of melanoma. Journal of Medicinal Chemistry. 2014;**57**;686–700. DOI: 10.1021/jm401252e

[66] Gusarov I, Shatalin K, Starodubtseva M, Nudler E. Endogenous nitric oxide protects bacteria against a wide spectrum of antibiotics. Science. 2009;**325**:1380–1384. DOI: 10.1126/science.1175439.

[67] Holden JK, Li H, Jing Q, Kang S, Richo J, Silverman RB, Poulos TL. Structural and biological studies on bacterial nitric oxide synthase inhibitors. Proceedings of the National Academy of Sciences of the United States of America. 2013;**110**:18127–12131. DOI: 10.1073/pnas.1314080110

[68] Holden JK, Kang S, Hollingsworth SA, Li H, Lim N, Chen S, Huang H, Xue F, Tang W, Silverman RB, Poulos TL. Structure-based design of bacterial nitric oxide synthase inhibitors. Journal of Medicinal Chemistry. 2015;**58**: 994–1004. DOI: 10.1021/jm501723p

[69] Holden JK, Dejam D, Lewis MC, Huang H, Kang S, Jing Q, Xue F, Silverman RB, Poulos TL. Inhibitor bound crystal structures of bacterial nitric oxide synthase. Biochemistry. 2015;**54**:4075–4082. DOI: 10.1021/acs.biochem.5b00431.

[70] Alexander JH, Reynolds HR, Stebbins AL, Dzavik V, Harrington RA, Van der Werf F, Hechman JS.Effect of tilarginine acetate in patients with acute myocardial infarction and cardiogenic shock: the TRIUMPH randomized controlled trial. Journal of American Medical Association. 2007;**297**:1657–1666.

[71] Wong VW, Lerner E. Nitric oxide inhibition strategies. Future Sciences OA. 2015;**1**:pii-FSO35. DOI: 10.4155/FSO.15.35

[72] Dzavik V, Cotter G, Reynolds HR, Alexander JH, Ramanathan K, Stebbis AL, Hathaway D, Farkouh ME, Ohman EM, Baran DA, Prondzinsky R, Panza JA, Cantor WJ, Vered Z, Buller CE, Kleiman NS, Webb JG, Holmes DR, Parrillo JE, Hazen SL, Gross SS, Harrington RA, Hachman JS. Effect of nitric oxide synthase inhibition on haemodynamics and outcome of patients with persistent cardiogenic shock complicating acute myocardial infarction: a phase II dose-ranging study. European Heart Journal. 2007;**28**:1109–1116. DOI: 10.1093/eurheartj/ehm075.

[73] Cottera G, Kaluskia E, Miloa O, Blatta A, Salaha A, Hendlera A, Krakovera R, Golickb A, Vereda Z. LINCS: *L*-NAME (a NO synthase inhibitor) in the treatment of refractory cardiogenic shock: a prospective randomized study. European Heart Journal. 2003;**24**:1287–1295. DOI: 10.1016/S0195-668X(03)00193-3.

[74] Salem R, Mebazaa A. Nitric oxide inhibition rapidly increases blood pressure with no change in outcome in cardiogenic shock: the TRIUMPH trial. Critical Care. 2007;**11**;1–2. DOI: 10.1186/cc5925.

[75] Luke JJ, LoRusso P, Shapiro GI, Krivoshik A, Schuster R, Yamazaki T, Arai Y, Fakhoury A, Dmuchowski C, Infante JR. ASP9853, an inhibitor of inducible nitric oxide synthase dimerization, in combination with docetaxel: preclinical investigation and a Phase I study in advanced solid tumors. Cancer Chemotherapy Pharmacology. 2016;**77**:549–558. DOI: 10.1007/s00280-016-2967-0.